Periparturient Diseases of Dairy Cows

Burim N. Ametaj
Editor

Periparturient Diseases of Dairy Cows

A Systems Biology Approach

Springer

Editor
Burim N. Ametaj
Department of Agriculture
Food and Nutritional Science
University of Alberta
Edmonton, Alberta, Canada

ISBN 978-3-319-43031-7 ISBN 978-3-319-43033-1 (eBook)
DOI 10.1007/978-3-319-43033-1

Library of Congress Control Number: 2017959645

Printed on acid-free paper

This Springer imprint is published by Springer Nature
The registered company is Springer International Publishing AG
The registered company address is: Gewerbestrasse 11, 6330 Cham, Switzerland

Preface

There are several reasons for undertaking the editing and writing of this book. The first reason is that there is a need to address the definition of metabolic diseases, metabolic disorders, production diseases, or transition cow diseases. Which one to choose and which one is more accurate? In fact, the definitions of those diseases were developed during the last half century and given that there are new developments and contributions in this particular area of science it is imperative that we discuss the drawbacks of those definitions. Based on new research reports of multiple labs around the world, we could say that the concepts of metabolic disease or production disease are not accurately defined and we need to address several other issues that influence our approach to the causality and pathogenesis of those diseases. Defining them as simply metabolic or simply production diseases has affected our efforts to identify the real causal agents and the development of new prevention strategies. Second, the incidence rates of periparturient diseases have been increasing steadily during the last two decades and the culling rate has reached more than 50%, shortening the productive life of dairy cows to less than 2 years. If we choose to do nothing, then the health status of dairy cows will continue to decline and the culling rates will further increase causing big losses to the dairy industry. The increasing rates of culling indicate that there is something missing in our understanding of the cause(s) of these diseases and that the methodology and the philosophy that we have been using to approach the etio-pathobiology deserve to be revisited and redimensionalized. The title of this book has been intentionally selected as "periparturient diseases of dairy cows." The reason for this choice is to include in the book not only those diseases that traditionally have been defined as metabolic in nature like milk fever or ketosis but also those that have been defined as bacterial in nature like metritis and mastitis. It should be noted that metabolic alterations and bacterial involvement are present in all periparturient diseases whether they have been defined as metabolic or infective in nature. The reason for this is that they seem to be interrelated to each other and one cow might be affected by more than one of these traditional diseases. The title of the book also has included a "systems biology approach" to periparturient diseases. Most readers are aware that a new group of sciences under the name of systems biology or systems veterinary approach has emerged during the last decade. These new sciences include genomics, proteomics, transcriptomics, and metabolomics. Although contribution of omics sciences have been at the pioneering level and not very well-organized contribution that they are giving to the understanding of periparturient diseases of dairy cows is tremendous. In this book we bring forward the contribution of omics sciences to the understanding of periparturient diseases of dairy cows. We hope that this book will serve as a stimulus to further research in clarifying the pathomechanisms and causality of periparturient diseases of dairy cows and in developing new preventive strategies for lowering the incidence rate of those diseases in the future. This book has been written with the intention to reach every undergraduate and graduate students as well as instructors in animal science departments and faculties of veterinary medicine that study or teach periparturient diseases of dairy cows.

Structure of the Book

The book is structured in such a way that the reader is first introduced into the concept of omics sciences. Since this book is about the application of systems biology in better understanding the etio-pathobiology of periparturient diseases of dairy cows, the reader is introduced first to those concepts. The second chapter discusses myths established during the last half century with regard to the definition of metabolic and production diseases and presents the benefits of the new philosophy of disease approach, the systems biology or veterinary approach as well as advantages of the predictive, preventive, and individualized medicine versus traditional approach of reactive medicine. Immunity around parturition is a very important topic that has been indicated as key to multiple diseases. In the third chapter, new concepts of potential effects of external factors on immunity around parturition are discussed. The fourth chapter deals with ruminal acidosis. Rumen is very important not only for digestion of feed material but also for generation of multiple bacterial products that can harm the health of cows. The authors describe ruminal acidosis from the systems biology approach perspective. The authors of the fifth chapter discuss microbiota of the rumen and intestines and their contribution to diseases. Also, all the new knowledge about utilization of omics sciences in approaching microbiota in the GI tract is discussed. The sixth chapter deals with the number one health problem, i.e., dairy cows' infertility. This is the main reason for culling cows in a dairy herd. The authors discuss both male and female fertility and the contribution of omics sciences in better understanding the reasons of infertility. Chapter 7 deals with retained placenta. Although retained placenta is not the second most important disease of dairy cows, it is listed immediately after infertility because retained placenta cows are affected significantly by infertility. Chapter 8 deals with mastitis, the second most important disease of dairy cows from the culling perspective. It is a very difficult disease and there is much activity from various labs to better understand the pathomechanism of the disease. Laminitis is discussed in Chap. 9. The authors discuss the most recent knowledge about the application of systems biology approach to laminitis in dairy cows. Ketosis is discussed in Chap. 10. Ketosis is a silent disease that affects more than 40% of dairy cows in a subclinical way. Understanding ketosis from the omics perspective is the subject of that chapter. Fatty liver also is a silent disease and requires a liver biopsy to be diagnosed. Almost 50% of cows are affected by fatty liver. It has multiple implications and is associated with several other periparturient diseases, especially metritis and mastitis. The authors bring the most up-to-date information about the application of omics sciences in this area of research. Finally, the book concludes with one of the most studied and most controversial diseases in the area of cow health, milk fever, or periparturient hypocalcemia. Is hypocalcemia a deficiency or part of immune response of the host during endotoxemias? The book discusses some of the most known hypotheses on milk fever and brings forward the omics research work in this field of study. We hope that the reader will find this book interesting and up-to-date and will use the knowledge in their research and teaching to the new generation.

Edmonton, AB, Canada Burim N. Ametaj

Contents

Editor and Contributors

About the Editor

Burim N. Ametaj, D.V.M., Ph.D., Ph.D. is a Professor of Animal Physiology and Bovine Periparturient Diseases and Immunology in the Department of Agricultural, Food and Nutritional Science at the University of Alberta, Canada. He received his veterinary degree (D.V.M.) at the Faculty of Veterinary Medicine, University of Agriculture, Tirana, Albania. He worked there as a Professor of Animal Physiology and conducted a Ph.D. on metabolic diseases of dairy cows. He served as Vice-Dean for Research and Graduate studies between 1997 and 2002. In 2004 Dr. Ametaj moved to Iowa State University and National Animal Disease Center, IA, USA, where he did a 1 year Fulbright Scholarship. In 1995 he started a second Ph.D. on immunobiology of dairy cows and graduated in 1999. Dr. Ametaj served three post-doctoral fellow positions at Iowa State University, Ames, Iowa, Purdue University, Lafayette, Indiana, and Cornel University, Ithaca, New York, USA. In 2004, Dr. Ametaj moved as Assistant Professor to the University of Alberta, Edmonton, Alberta, Canada. His research activities involve studying the pathomechanism of periparturient diseases and development of new technologies for their prevention. Dr. Ametaj's lab has developed a new grain processing technology that prevents ruminal acidosis, two oro-parenteral vaccines against bacterial endotoxins, one intravaginal probiotic technology that lowers the incidence of uterine infections, and has applied metabolomics and microbiome studies to develop biomarkers of six periparturient diseases of dairy cows. His lab also has been involved in understanding the role of bacterial endotoxins in the pathobiology of prion disease.

Contributors

Burim N. Ametaj, D.V.M., Ph.D., Ph.D. Department of Agricultural, Food and Nutritional Science, University of Alberta, Edmonton, AB, Canada

The Faculty of Veterinary Medicine, University of Agriculture, Tirana, Albania

Fabrizio Ceciliani, D.V.M., Ph.D. Department of Veterinary Medicine, Università degli Studi di Milano, Milano, Italy

Esterina De Carlo, Ph.D. Istituto Zooprofilattico Sperimentale del Mezzogiorno—Centro di Referenza Nazionale sull'Igiene e le Tecnologie dell'Allevamento e delle Produzioni Bufaline—Sezione di Salerno, Salerno, Italy

Elda Dervishi, Ph.D. Department of Agricultural, Food and Nutritional Science, University of Alberta, Edmonton, AB, Canada

André Luiz Garcia Dias, Ph.D. Department of Agricultural, Food and Nutritional Science, University of Alberta, Edmonton, AB, Canada

Emily F. Eckel, Ph.D. Department of Agricultural, Food and Nutritional Science, University of Alberta, Edmonton, AB, Canada

Peter David Eckersall, Ph.D. College of Medical, Veterinary and Life Science, Institute of Biodiversity Animal Health and Comparative Medicine, University of Glasgow, Glasgow, UK

Ashley F. Egyedy, M.Sc. Department of Agricultural, Food and Nutritional Science, University of Alberta, Edmonton, AB, Canada

Morteza H. Ghaffari Department of Agricultural, Food and Nutritional Science, University of Alberta, Edmonton, AB, Canada

Ehsan Khafipour, Ph.D. Departments of Animal Science and Medical Microbiology, University of Manitoba, Winnipeg, MB, Canada

Cristina Lecchi, D.V.M., Ph.D. Department of Veterinary Medicine, Università degli Studi di Milano, Milano, Italy

Juan J. Loor, Ph.D. Department of Animal Sciences, University of Illinois, Urbana-Champaign, IL, USA

Alessandra Martucciello, Ph.D. Istituto Zooprofilattico Sperimentale del Mezzogiorno—Centro di Referenza Nazionale sull'Igiene e le Tecnologie dell'Allevamento e delle Produzioni Bufaline—Sezione di Salerno, Salerno, Italy

Manikhandan Mudaliar, Ph.D. Glasgow Molecular Pathology Node, College of Medical, Veterinary and Life Science, University of Glasgow, Glasgow, UK

Valentino Palombo, Ph.D. Department of Animal Sciences, University of Illinois, Urbana-Champaign, IL, USA

Michael A. Steele, Ph.D. Department of Agricultural, Food and Nutritional Science, University of Alberta, Edmonton, AB, Canada

Funmilola Clara Thomas, Ph.D. Biochemistry Unit, Department of Veterinary Physiology and Pharmacology, College of Veterinary Medicine, Federal University of Agriculture, Abeokuta, Nigeria

Richard R.E. Uwiera, D.V.M., Ph.D. Department of Agricultural, Food and Nutritional Science, University of Alberta, Edmonton, AB, Canada

Mario Vailati-Riboni, Ph.D. Department of Animal Sciences, University of Illinois, Urbana-Champaign, IL, USA

Domenico Vecchio, Ph.D. Istituto Zooprofilattico Sperimentale del Mezzogiorno—Centro di Referenza Nazionale sull'Igiene e le Tecnologie dell'Allevamento e delle Produzioni Bufaline—Sezione di Salerno, Salerno, Italy

Guanshi Zhang, D.V.M., Ph.D. Institute of Metabolomic Medicine, Center for Renal Translational Medicine, School of Medicine, University of California San Diego, La Jolla, CA, USA

What Are Omics Sciences?

1

Mario Vailati-Riboni, Valentino Palombo, and Juan J. Loor

Abstract

The word omics refers to a field of study in biological sciences that ends with *-omics*, such as genomics, transcriptomics, proteomics, or metabolomics. The ending -ome is used to address the objects of study of such fields, such as the genome, proteome, transcriptome, or metabolome, respectively. More specifically genomics is the science that studies the structure, function, evolution, and mapping of genomes and aims at characterization and quantification of genes, which direct the production of proteins with the assistance of enzymes and messenger molecules. Transcriptome is the set of all messenger RNA molecules in one cell, tissue, or organism. It includes the amount or concentration of each RNA molecule in addition to the molecular identities. The term proteome refers to the sum of all the proteins in a cell, tissue, or organism. Proteomics is the science that studies those proteins as related to their biochemical properties and functional roles, and how their quantities, modifications, and structures change during growth and in response to internal and external stimuli. The metabolome represents the collection of all metabolites in a biological cell, tissue, organ, or organism, which are the end products of cellular processes. Metabolomics is the science that studies all chemical processes involving metabolites. More specifically, metabolomics is the study of chemical fingerprints that specific cellular processes establish during their activity; it is the study of all small-molecule metabolite profiles. Overall, the objective of omics sciences is to identify, characterize, and quantify all biological molecules that are involved in the structure, function, and dynamics of a cell, tissue, or organism.

M. Vailati-Riboni, Ph.D. • V. Palombo, Ph.D. • J.J. Loor, Ph.D. (✉)
Division of Nutritional Sciences, Department of Animal Sciences, University of Illinois at Urbana-Champaign, Urbana, IL 61801, USA
e-mail: vailati2@illinois.edu; valentino.palombo@gmail.com; jloor@illinois.edu

B.N. Ametaj (ed.), *Periparturient Diseases of Dairy Cows*,
DOI 10.1007/978-3-319-43033-1_1

1.1 Introduction

The trademark characteristic of omics technologies is their holistic capability in the context of the cell, tissue, or organism. They are aimed primarily at the universal detection of genes (genomics), mRNA (transcriptomics), proteins (proteomics), and metabolites (metabolomics) in a specific biologic sample in a non-targeted and non-biased manner. The basic aspect of these approaches is that a complex system can be understood more thoroughly if considered as a whole. The omics approach is suitable for hypothesis-generating experiments, as holistic approaches acquire and analyze all available data to define a hypothesis, which can be further tested, in situations when no hypothesis is known or prescribed due to lack of data. When applied to well-studied scenarios, omics are still applicable to test and prove the connections and interrelationships among the many faces of a complex physiologic state, and to discover missing pieces in the current knowledge.

The first omics technologies were the automated DNA sequencer and the ink-jet DNA synthesizer developed in the early 1990s by Leroy Hood and colleagues as a tool for global gene expression analysis (e.g., transcriptomics) (Hood 2002). Around the same time, Hood's group also introduced the protein sequencer and the protein synthesizer to study protein expression at the cellular level, a process known as "proteomics." Furthermore, the concomitant emergence of metabolomics studies started by Frank Baganz and his group (Oliver et al. 1998) completed the physiologic flow of biological information processing and synthesis, from gene expression, to protein synthesis, and metabolites changes.

1.2 Genomics

Genomics pertains to the study of the complete set of DNA in an organism, including all of its genes, i.e., the "genome." With the advent of next-generation sequencing (NGS) technology the acquisition of genome-scale data has never been easier, expanding our ability to analyze and understand whole genomes and decreasing the existing gap between genotype and phenotype. Genetics and genomics sound alike but they have specific distinctions. Genetics is the study of heredity, or how the characteristics of living organisms are transmitted from one generation to the next via DNA. It involves studies focusing on specific and limited numbers of genes, or part of genes with known function, to understand how these influence particular traits of interest. At present, high-throughput technology and advances in computational biology have changed this paradigm enabling the study of organisms in terms of genome structure, addressing biological questions at a genome-wide scale, i.e., genetics is being progressively "contaminated" with genomics.

The advent of genomics turned genome-wide association studies (GWAS) into the gold standard method to identify candidate regions associated with complex traits of interest (quantitative trait loci—QTL), both in humans and other species (Gondro et al. 2013). Probe-based chips developed by various commercial companies and with a large number of single nucleotide polymorphism (SNP) markers spread across the genome are currently being used to help uncover associations between genes and traits of interest. Species-specific arrays encompassing 10,000 up to 800,000 SNP are currently available. Such coverage ensures that any QTL will be closely linked with at least one marker. For this reason, GWAS became powerful enough to map causal genes with modest effects, i.e., disease-related quantitative traits. With the large number of genes studied simultaneously, genomic studies, in fact, can overcome the limitations of traditional genetic association approaches, enhancing our understanding of peripartal diseases (Loor 2010).

The interpretation of GWAS results still represents an important challenge. For instance, if a robust association between a phenotype and a list of genes is uncovered, one can have more confidence about the possibility for discovery of novel candidate genes. Despite the power of GWAS for discovery, studies to confirm the role of genes associated with the trait of interest should be performed to confirm functional relationships. To address this issue, gene-based software offers an effective solution in post-GWAS analysis (Capomaccio et al. 2015). Several SNP array data management tools have been developed in recent years and among these PLINK (Purcell et al. 2007), due to its speed and stability, is standard for data management. Currently, the entire GWAS pipeline can be easily executable by *ad hoc* computer programs, which, in the majority of cases, are open-source multiplatform software packages often developed in the R environment (Nicolazzi et al. 2015).

In the context of animal breeding for a given trait or traits, "genomic selection" deserves special mention (Meuwissen et al. 2001). This approach is a form of "marker-assisted selection" in which a large number of genetic markers, covering the whole genome, are used to estimate animal breeding values (EBV), i.e., the genetic value of young animals based on their genotype. This can lower the generation interval and increase the rate of genetic progress in different animal populations, traditionally based on progeny testing (Goddard and Hayes 2007). The continued progress in DNA sequencing efficiency in the near future will allow for sequencing complete genomes of individual animals, hence allowing the selection of animals with favorable QTL's alleles. Clearly, we are at the beginning of an era where individual genome sequencing will allow not only the study of domestication and selection of breeds, but also the understanding of quantitative differences associated with environmental factors, all of which will help guide experimental design for more effective animal disease control (Bai et al. 2012).

1.3 Transcriptomics

The transcriptome is the total RNA (i.e., mRNA, noncoding RNA, rRNA, and tRNA) expressed by a cell or tissue, thus representing a snapshot of cellular metabolism. The transcriptome era started when Schena et al. (1995) developed the "microarray" technology using the ink-jet DNA synthesizer, allowing for the analysis of a predetermined set (from hundreds to thousands) of cellular mRNA on a large scale. However, the recent introduction of high-throughput next-generation DNA sequencing (NGS) technology has revolutionized transcriptomics by allowing RNA analysis through cDNA sequencing on a massive scale (RNAseq) (Voelkerding et al. 2009). This technology eliminated several challenges posed by microarray technologies, including the limited dynamic range of detection, while providing further knowledge of the qualitative, and not only quantitative, aspects of transcriptome: (1) transcription initiation sites, (2) sense and antisense transcripts, (3) alternative splicing events, and (4) gene fusion.

As it also provides detailed information on the noncoding RNA portion of the total RNA, RNAseq has enabled the understanding of complex regulatory mechanisms (e.g., epigenetics). Since the early twenty-first century, among the various epigenetic mechanisms, microRNA (miRNA), a class of small noncoding RNA (18-25 nucleotides), have received the greatest notoriety. Such attention is well founded because miRNA play a major role in controlling posttranscriptional regulation by preventing translation of mRNA (Romao et al. 2011). Furthermore, miRNA are not only part of the epigenetic machinery, but also are involved in its regulation, underscoring their pivotal role as epigenetic mediators (Poddar et al. 2017). In addition, through RNAseq or miRNA-designed microarrays, the miRNome (the total mRNA expressed by a cell at a given time) also can be analyzed.

1.4 Proteomics

The term "proteome" was defined as the characterization and quantification of all sets of proteins in a cell, organ, or organism at a specific time and was coined by Wasinger et al. (1995). Thus, a proteomic analysis provides the protein inventory of a cell or tissue at a defined time point, facilitating discovery of novel biomarkers, identification and localization of post-translational modifications, and study of protein–protein interactions (Chandramouli and Qian 2009). Powerful techniques have been established to identify and differentially quantify protein species of complex biological samples, and proteomic is being adopted by livestock researchers (Lippolis and Reinhardt 2008; Sauerwein et al. 2014).

The core of modern proteomics is mass spectrometry (MS) (Aebersold and Mann 2003), a technique in which all chemical compounds in a sample are ionized and the resulting charged molecules (ions) are analyzed according to their mass-to-charge (m/z) ratios. For a simple pre-separation of complex protein mixtures before MS analysis one- or two-dimensional polyacrylamide gel electrophoresis (1D-PAGE, 2D-PAGE) is often used. But to further enhance automation in the process and create a streamed pipeline analyses, different types of liquid chromatography (LC or HPLC) are used to complement or substitute gel-based separation techniques.

Identification of the proteins among treatments or conditions is performed by comparison against a database of proteins "digested *in silico*," meaning that the raw data are directly compared with theoretically generated data from protein databases. Reliable quantification of the identified protein also is possible with several MS-based quantification methods including chemical, metabolic, enzymatic labeling, and label-free (May et al. 2011). Proteomic advances made absolute quantification of proteins possible through the AQUA (Absolute quantification of proteins), QConCat (artificial proteins comprised of concatenated peptides), and protein standard for absolute quantification (PSAQ) approaches (Rivers et al. 2007; Brun et al. 2007).

1.5 Metabolomics

The metabolome consists of the global profiling of metabolites in a biological sample. A metabolomics analysis may be conducted on a variety of biological fluids and tissue types and may utilize a number of different technology platforms. Metabolomics typically uses high-resolution analysis together with statistical tools such as principal component analysis (PCA) and partial least squares (PLS) to derive an integrated picture of the metabolome (Zhang et al. 2012). As one of the most-common spectroscopic analytical techniques, nuclear magnetic resonance (NMR) can uniquely identify and simultaneously quantify a wide range of organic compounds in the micro-molar range, providing unbiased information about metabolite profiles. The wide spectrum of molecules detectable by this approach includes peptides, amino acids, nucleic acids, carbohydrates, organic acids, vitamins, polyphenols, alkaloids, and inorganic species. Application of MS is gaining increased interest in high-throughput metabolomics, often coupled with other techniques such as chromatography (GE-MS, LC-MS, UPLS-MS) or electrophoretic techniques (CE-MS). Due to its high sensitivity and wide range of covered metabolites, MS has become the technique of choice in many metabolomics studies (Zhang et al. 2012).

1.6 Perspectives

Omics technologies have contributed widely to the understanding of the delicate physiologic equilibrium that allows for a successful transition into lactation (Vailati-Riboni et al. 2016). Their application to the study of peripartal disease pathophysiology is spreading across research groups worldwide. Despite this, a reductionist approach focusing on parts and sections of the physiology (rather than considering it as a whole) is still the main approach used by scientists when handling this holistic output. We are still considering single organs as the "system" to study, subsequently inferring the connection with the rest of the organism based on the existing literature. The physiologic and metabolic complexity of these diseases unavoidably requires a systems biology approach, i.e., a way to systematically study the complex interactions in the cow using a method of integration instead of reduction. Only in this way researchers will be able to uncover the underlying links (pathways, regulatory networks, and structural organization) within and between tissues (e.g., adipose and liver; skeletal muscle and adipose; gut microorganisms and epithelia), and to detect new emergent properties that may arise from examining the interactions between all components of a system.

The systems approach in its purest connotation has not yet been applied to the field of dairy science. This is largely due to the fact that when integrating multiple datasets, one tends to generate bare numerical relationships rather than meaningful biological connections among organs. Therefore, as a future frontier, the dairy science community must address the need for "useful" approaches (e.g., modeling, bioinformatics) to integrate knowledge derived from multiple "omics" analyses within and between tissues, focusing both on the classical flow of genetic information (transcriptome, proteome, metabolome), and what lies above it (epigenetic).

References

Aebersold R, Mann M (2003) Mass spectrometry-based proteomics. Nature 422(6928):198–207. doi:10.1038/nature01511

Bai YS, Sartor M, Cavalcoli J (2012) Current status and future perspectives for sequencing livestock genomes. J Anim Sci Biotechnol 3:8. doi:10.1186/2049-1891-3-8

Brun V, Dupuis A, Adrait A, Marcellin M, Thomas D, Court M, Vandenesch F, Garin J (2007) Isotope-labeled protein standards: toward absolute quantitative proteomics. Mol Cell Proteomics 6(12):2139–2149. doi:10.1074/mcp.M700163-MCP200

Capomaccio S, Milanesi M, Bomba L, Vajana E, Ajmone-Marsan P (2015) MUGBAS: a species free gene-based programme suite for post-GWAS analysis. Bioinformatics 31(14):2380–2381. doi:10.1093/bioinformatics/btv144

Chandramouli K, Qian PY (2009) Proteomics: challenges, techniques and possibilities to overcome biological sample complexity. Hum Genomics Proteomics 2009. doi:10.4061/2009/239204

Goddard ME, Hayes BJ (2007) Genomic selection. J Anim Breed Genet 124(6):323–330. doi:10.1111/j.1439-0388.2007.00702.x

Gondro C, JVD W, Hayes B (2013) Genome-wide association studies and genomic prediction, Methods of molecular biology, vol 1019. Humana Press, New York

Hood L (2002) A personal view of molecular technology and how it has changed biology. J Proteome Res 1(5):399–409. doi:10.1021/pr020299f

Lippolis JD, Reinhardt TA (2008) Centennial paper: Proteomics in animal science. J Anim Sci 86(9):2430–2441. doi:10.2527/jas.2008-0921

Loor JJ (2010) Genomics of metabolic adaptations in the peripartal cow. Animal 4(7):1110–1139. doi:10.1017/S1751731110000960

May C, Brosseron F, Chartowski P, Schumbrutzki C, Schoenebeck B, Marcus K (2011) Instruments and methods in proteomics. Methods Mol Biol 696:3–26. doi:10.1007/978-1-60761-987-1

Meuwissen THE, Hayes BJ, Goddard ME (2001) Prediction of total genetic value using genome-wide dense marker maps. Genetics 157(4):1819–1829

Nicolazzi EL, Biffani S, Biscarini F, Orozco Ter Wengel P, Caprera A, Nazzicari N, Stella A (2015) Software solutions for the livestock genomics SNP array revolution. Anim Genet 46(4):343–353. doi:10.1111/age.12295

Oliver SG, Winson MK, Kell DB, Baganz F (1998) Systematic functional analysis of the yeast genome. Trends Biotechnol 16(9):373–378

Poddar S, Kesharwani D, Datta M (2017) Interplay between the miRNome and the epigenetic machinery: implications in health and disease. J Cell Physiol. doi:10.1002/jcp.25819

Purcell S, Neale B, Todd-Brown K, Thomas L, Ferreira MA, Bender D, Maller J, Sklar P, de Bakker PI, Daly MJ, Sham PC (2007) PLINK: a tool set for whole-genome association and population-based linkage analyses. Am J Hum Genet 81(3):559–575. doi:10.1086/519795

Rivers J, Simpson DM, Robertson DH, Gaskell SJ, Beynon RJ (2007) Absolute multiplexed quantitative analysis of protein expression during muscle development using QconCAT. Mol Cell Proteomics 6(8):1416–1427. doi:10.1074/mcp.M600456-MCP200

Romao JM, Jin W, Dodson MV, Hausman GJ, Moore SS, Guan LL (2011) MicroRNA regulation in mammalian adipogenesis. Exp Biol Med 236(9):997–1004. doi:10.1258/ebm.2011.011101

Sauerwein H, Bendixen E, Restelli L, Ceciliani F (2014) The adipose tissue in farm animals: a proteomic approach. Curr Protein Pept Sci 15(2):146–155

Schena M, Shalon D, Davis RW, Brown PO (1995) Quantitative monitoring of gene expression patterns with a complementary DNA microarray. Science 270(5235):467–470

Vailati-Riboni M, Elolimy A, Loor JJ (2016) Nutritional systems biology to elucidate adaptations in lactation physiology of dairy cows. In: Kadarmideen HN (ed) Systems biology in animal production and health, vol 2. Springer International Publishing, Cham, pp 97–125. doi:10.1007/978-3-319-43332-5

Voelkerding KV, Dames SA, Durtschi JD (2009) Next-generation sequencing: from basic research to diagnostics. Clin Chem 55(4):641–658. doi:10.1373/clinchem.2008.112789

Wasinger VC, Cordwell SJ, Cerpa-Poljak A, Yan JX, Gooley AA, Wilkins MR, Duncan MW, Harris R, Williams KL, Humphery-Smith I (1995) Progress with gene-product mapping of the Mollicutes: Mycoplasma genitalium. Electrophoresis 16(7):1090–1094

Zhang A, Sun H, Wang P, Han Y, Wang X (2012) Modern analytical techniques in metabolomics analysis. Analyst 137(2):293–300. doi:10.1039/c1an15605e

Demystifying the Myths: Switching Paradigms from Reductionism to Systems Veterinary in Approaching Transition Dairy Cow Diseases

2

Burim N. Ametaj

Abstract

Animal scientists have made tremendous progress and put their best efforts, during the last 50 years, in regard to selecting dairy cows for high milk yield and in designing the best rations for high milk production. Moreover, dairy cow health scientists have persistently studied health issues and offered the best solutions to numerous transition cow diseases. Indeed, one dairy cow today produces an amount of milk that is equal to six cows 50 years ago. This is an accomplishment of both geneticists and animal nutritionist that deserve to be credited. However, there is still one grey area that both animal and health scientists have not been able to solve, the reason for high incidence of periparturient diseases and the high cull rates. Indeed, despite much research work and progress made with regard to cow health there are still various health concerns that continue to affect dairy cows, which cost dairy industry billions of dollars in economic losses on a yearly basis. The number of cull cows is increasing and so has the incidence of several diseases including uterine infections and infertility, mastitis, laminitis, ketosis, and retained placenta. This has raised questions whether the approach, the philosophy, or the scientific methodology that we have been using to address diagnosis, treatment, and prevention of transition cow diseases is appropriate. There is broad discussion among scientists in biological and medical sciences that suggests that the reductionist philosophy that has led biological sciences and medicine for centuries has failed to find optimal solutions to the many unresolved health issues of humans and animals. A new philosophy, known as systems biology approach, is emerging that has been embraced by various leading groups in the world that seems very promising and that proposes to look at the animal as a whole and the disease as a complex interaction among genotype,

B.N. Ametaj, D.V.M., Ph.D., Ph.D.
Department of Agricultural, Food and Nutritional Science, 4-10G, Agriculture/Forestry Centre, University of Alberta, Edmonton, Alberta T6G 2P5, Canada
e-mail: burim.ametaj@ualberta.ca

B.N. Ametaj (ed.), *Periparturient Diseases of Dairy Cows*,
DOI 10.1007/978-3-319-43033-1_2

phenotype, and environment. This discussion, recently, has involved animal and veterinary scientists and we are bringing those debates to the readers of this book. In this chapter, we examine some of the concepts that have dominated the animal and veterinary sciences during the last century and have become myths and suggest that it is time to revisit and redefine those concepts in order to improve cows' health, their welfare and well-being as well as the profitability of dairy industry. In the last section of the chapter a comparison between the reductionist approach and the systems biology or veterinary approach is made and a discussion of the advantages and the drawbacks are presented so that the readers better understand the philosophy of veterinary medicine and adopt the best approach in their research activities.

2.1 Introduction

According to CanWest DHI Canada (2015) the incidence rate of periparturient diseases of dairy cows has been increasing during the last 14 years (2001–2014). Based on their records and other surveys conducted by various research labs worldwide the three most predominant periparturient diseases of transition dairy cows are metritis (inflammation of the uterus), mastitis (inflammation of the udder), and laminitis (inflammation of the hoof laminae). Other major periparturient diseases reported include ketosis, milk fever, retained placenta, fatty liver, displaced abomasum, and ruminal acidosis.

The culling rates (for different reasons) also have been increasing from around 24% during 2011 to almost 51% during 2014. This has shortened the productive life of dairy cows to less than 2 years. Increasing trends of disease incidence and culling rates have been reported in the US, European countries, and beyond (USDA 2007a, b; Maher et al. 2008; Rushen 2013). The increasing number of cull cows is a major setback for the dairy industry and if this trend will continue during the next decade, then, the productive life of a cow might even shorten to 1 year. Occurrence of disease is associated with lower milk production, lower pregnancy rate and infertility, and lower profitability for dairy producers. This state of cows' health influences greatly the profitability and the future of dairy industry. Moreover, sickness is associated with poor welfare and well-being of dairy cattle and certainly poor products.

It should be noted that almost 30–50% of cows in a herd are affected by one or more periparturient disease at the same time (Ametaj et al. 2010; LeBlanc 2010; CanWest DHI Canada 2015). The unsolved question is to why this high incidence of disease and what is/are the cause(s) or the etiological factors of the disease state? This question has been raised and addressed by cow health researchers during the last 100 years and although the knowledge about pathobiology of diseases in transition dairy cows is increased tremendously we still do not have a complete understanding of the causes and etiopathogenesis of transition cow diseases. Lack of full knowledge of disease process and causative agents has affected the appropriate treatment of disease and development of prevention strategies. Some of the reasons for this state of knowledge are discussed below.

An important reason for the present state of cow's health status is the philosophy that the scientists have used to diagnose disease, treat the sick cows, or develop prevention interventions. The dominant philosophy of science in the last few centuries has been the "Reductionist" approach. The basis of reductionist philosophy of disease has been to theoretically split the whole organism of the animal into smaller and smaller parts and analyze and study the details of those parts in order to better understand the components of the whole. This approach has generated a tremendous amount of knowledge about the structure and functions of various body units including various organs, tissues as well as cells, enzymes, proteins, carbohydrates, lipids, minerals, vitamins, or other components of the body. However, it is obvious that the reductionist approach has not been successful in addressing complex health issues of dairy cows and the issues of the organism as a whole.

Other important reasons for this state of health status in dairy cows are the feeding (i.e., environment) systems used and the genotype of the cows. During the last century there was not very much knowledge about how to properly feed dairy cows to keep them healthy and productive. Currently the science of nutrition is advancing; however, it is not yet at the level of individual feeding and feeding for both health and productivity. Developing a diet that will support both productivity and health would be the future challenge of animal nutritionists.

Also, not very much knowledge exists with regard to selection of dairy cows for health traits (i.e., healthy genotype). The only focus of cow geneticists in the past few decades has been selecting cows for high milk yield. Presently there is a great interest from scientists and dairy industry to select dairy cows for health traits.

Another area that deserves more efforts from research scientists is the etiopathology of disease process around parturition. There are opportunities in the area of early diagnosis of disease, better prognosis of disease outcome as well as development of new preventive interventions. However, these efforts are at their pioneering stage.

It should be noted that a new philosophy has emerged in relation to how we study periparturient diseases of dairy cows: the philosophy of systems biology or veterinary approach. This new philosophy proposes a new methodology of how we diagnose disease, how we treat sick animals, and how we prevent disease from occurring. Indeed, presently we are able to use multiple sophisticated instruments that the scientists of the past did not have. These instruments are helping us to better approach sick animals and better understand the disease processes and causality. Instruments like NMR (nuclear magnetic resonance), DI-MS (Desorption Ionization Mass Spectrometry), GC-MS (Gas Chromatography Mass Spectrometry), ICP-MS (Inductively Coupled Plasma Mass Spectrometry), and many other modern instruments related to RNA, DNA, proteomics, and microbiome research make the basis for a higher level of approach to disease state.

Another advantage of systems biology approach is the current evolvement of computers, bioinformatics, and the means of communication. Scientists of the past had to use mostly hand writing, book writing, letter writing, telephoning (if that was available), library loans, and borrowing of information from one another or from libraries of other countries. Presently, scientists can obtain information and learn about research conducted in any part of the globe in a matter of seconds or minutes and not days or months as during the last century. They can process large amounts of data in a matter of minutes with more advanced bioinformatics tools. Indeed, this is the century of electronics and vast and fast information and communication. This facilitates interaction among scientists and expedites the process of innovation.

In this chapter, issues related to drawbacks of reductionist approach will be discussed in more detail later. Moreover, the novel philosophy of "systems veterinary approach" will be presented. Although the original terminology for this new philosophy is "systems biology approach," it would be more appropriate to use the term "systems veterinary approach" for veterinary medicine. The reason for that is that the whole science of veterinary as related to monitoring of animal health, diagnosis of disease, treatment of disease as well as prevention strategies and prediction of risk for diseases needs to embrace the systems approach in order to improve the health status of animals in general and more specifically the health of dairy cows.

Multiple hypotheses have been proposed during the years for each periparturient disease of dairy cows with milk fever championing all other diseases with more than 30 different hypotheses during the last century. Although hypotheses put forward about the etiology of various diseases have their merits, they still have not been able to solve the multiple issues related to cow's health. Cows still continue to be sick in large numbers and culled in high rates and unfortunately we are at a standstill situation that requires solutions. Because many scientists of the last century have based their approaches to disease diagnosis, treatment, and prevention based on the methodology of reductionism or reactive veterinary medicine, it would be of great interest to discuss some of the myths created during the last century and the need for their demystification in order to open the way to new approaches and new solutions.

2.2 Myth Number 1: A Metabolic Disease Is a Disturbance of the Internal Homeostasis of a Metabolite or Several Metabolic Processes

According to Jack Payne (1977) "*a metabolic disease is an abnormal change in the internal homeostasis of a metabolite or several metabolic processes*."

Payne made this statement almost half a century ago when there was not very much knowledge about metabolic diseases. The scientific study of metabolic diseases was at its early stages. The definition given by Payne on metabolic diseases was correct, based on the knowledge of that time. However, since that time there has been major developments in the understanding of metabolic and periparturient diseases of dairy cows. For example, for several decades it was thought that lower iron in the blood (i.e., hypoferremia) was an indication of iron deficiency. However, during the last decade scientific research has proven that during bacterial infections iron is withdrawn from blood systemic circulation, also known as anemia of inflammation, in order to make it difficult for pathogenic bacteria to freely access iron, which is essential for their growth and proliferation (Ganz 2009; Wander et al. 2009). Mounting evidence also suggests that the same might be true in regard to calcium (Ca) concentration in the systemic circulation or other body fluids or cells. Several recent review articles suggest that lowering of calcium in the blood, during bacterial infections or endotoxemia, might be part of the immune response for safely removing bacterial endotoxins from blood circulation (Waldron et al. 2003; Ametaj et al. 2010; Eckel and Ametaj 2016). Moreover, sensing an increase in the concentration of Ca inside the pathogenic bacterial cells of the lungs may play a role in the activation of virulence genes, particularly for expression of pili and neuraminidase. Pili are involved in infection of the lungs, and elevated intracellular Ca is one of the most important signals for activating expression of pili, a condition that could begin as bacteria transition into the particularly high Ca environment of the lung (Rosch et al. 2008). Therefore, high extracellular (i.e., blood) Ca might help pathogenic bacteria to activate their virulence genes and withdrawal of Ca might be a protective response of the host against bacterial infection.

During the last decade there has been a boom in application of omics technologies in the study of transition cow diseases. Exciting results have been reported by multiple labs around the globe. An important lesson learned from metabolomics studies is identification of multiple metabolite alterations in the body fluids (blood, urine, or milk) during various periparturient diseases of transition dairy cows. This has raised the question, whether we can define or associate one metabolic disease with perturbation of one single metabolite or even a few metabolites in the body fluids of the host? Recently our group has been able to identify and measure hundreds of metabolites and multiple metabolic pathways that are altered before, during, and after clinical appearance of six important periparturient diseases of dairy cows including uterine infections, mastitis, laminitis, ketosis, milk fever, and retained placenta (Dervishi et al. 2016a, b; Zhang et al. 2017). These data suggest that each periparturient disease whether typically infectious (metritis or mastitis) or metabolic in nature (ketosis or milk fever) is characterized by multiple changes in the concentrations of multiple metabolites and mineral elements. Therefore, identifying one metabolic disease (e.g., ketosis, fatty liver, or milk fever) with one perturbed metabolite (BHBA, NEFA, or Ca, respectively) doesn't seem to be any longer acceptable scientifically. By the same token, it would not be scientifically accurate to identify a few perturbed metabolites with one metabolic disease because in most of the major periparturient diseases there are tens of metabolites and multiple metabolic pathways that are perturbed.

Another interesting development in life sciences is that in most human modern metabolic diseases including type 2 diabetes, obesity, and metabolic syndrome there is always present a chronic inflammatory state (Monteiro and Azevedo 2010). Recently, data from our laboratory also show that all six major periparturient diseases of dairy cows, including metritis, mastitis, laminitis, ketosis, milk fever, and retained placenta (Dervishi et al. 2015, Dervishi et al. 2016a, b; Zhang et al. 2015, 2016), are preceded and associated by chronic inflammation. These data suggest that there are no simply metabolic or inflammatory states but more combined host metabolic and immune responses to the pathological state. This also suggests that the etiology, pathobiology, and definition of a metabolic disease should be revisited and redefined.

Since the time Payne presented his definition of a metabolic disease various research groups have demonstrated that numerous inflammatory intermediates stimulate or inhibit various metabolic responses. For example, tumor necrosis factor (TNF) and interleukin (IL)-1β have been shown to have lipolytic activities and inhibitory effects on gluconeogenesis, whereas IL-6 influences hepatic protein synthesis (Raj et al. 2008). Additionally, TNF was demonstrated to lower insulin sensitivity of the liver, adipose tissue, and skeletal muscles (Hotamisligil et al. 1993; Weisberg et al. 2003). Furthermore, TNF and IL-1β suppress expression of GLUT2 (glucose receptor) and glucokinase in pancreatic β-cells, thus making them less sensitive to the blood glucose level (Park et al. 1999). TNF also was demonstrated to cause fatty liver in dairy cows (Bradford et al. 2009) and administration of endotoxin lowered plasma Ca (Waldron et al. 2003). These examples indicate that there is no clear separation between metabolic and inflammatory responses and that diseases are far more complex than previously thought and cannot be classified any longer as solely metabolic or merely inflammatory.

Another issue that deserves more discussion from the dairy cow health community is related to definition of what is and what is not a metabolic disease. As a starting point for a discussion is the recent reported findings of the number of metabolites present in the body fluids of humans. For example, Psychogios et al. (2011) reported a total of 4229 various metabolites in the serum of human subjects. It is anticipated that the same number of metabolites will be identified and quantified in the blood of different livestock animals, including cattle. There are several challenging questions for the human and animal health scientists that need to be addressed with regard to definition of a metabolic disease. For example, does the host have >4000 homeostatic control mechanisms to regulate concentrations of all metabolites present in the body fluids or some or most of them are just passing by through the host because they are part of the food products consumed? Are there essential metabolites that are strictly regulated, and others that are not so important? Moreover, if homeostasis of one or a few or all of those metabolites are disturbed (increased or decreased) at a certain time point, then, should we define a total of 4229 metabolic diseases in humans or animals (based on reductionist concept of one perturbed metabolite one disease)?

Another issue worth to be discussed is what is called "*an abnormal change in the internal homeostasis*" as suggested by Payne? Can we call fluctuations (i.e., increases or decreases) of blood metabolites in the body fluids "*abnormal changes of the internal homeostasis*"? It should be emphasized that most metabolites that circulate in the blood or other body fluids during disease process flow to various organs, tissues, and cells to participate in numerous activities like synthesis of larger molecules such as glycogen, phospholipids, triglycerides, enzymes, proteins, immunoglobulins, acute phase proteins, antimicrobial compounds, and many others. Some of the metabolites participate in energy production, others become part of cell membranes, some others serve as building blocks for enzymes or fuel for the cells, several metabolites participate as components of the immune response and many other functions. Moreover, metabolites are increased or decreased in the body fluids based on the needs of specific sites, tissues, or organs for these compounds at certain time points or activities. They are absorbed from the GI tract or released from body storage sites into the systemic circulation and are transported from one part of the body to another organ, tissue, or cell. Therefore, the challenging question is how do we distinguish the normal flow of metabolites from an abnormal one? It might be that metabolites are flowing to the mammary gland to provide nutrients for milk synthesis, or they might be flowing to the rumen to support development of an immune response. So, what is an abnormal and what is a physiological flow of metabolites and nutrients?

Another important question is what is considered a normal concentration of a blood metabolite? This is a very complex question and very difficult to be answered. There are multiple variables that need to be controlled in order to determine the normal concentration of a blood metabolite. Here are some of them: (1) The animal has to be absolutely free of any disease process. The animal should be free of parasites, infectious diseases, viral diseases, clinical metabolic diseases, subclinical diseases including subclinical ketosis, subclinical hypocalcemia, subclinical mastitis, or any other ailment, which is very difficult to be achieved; (2) The animal should have absolutely no deficiencies or sufficiency of any nutrient(s) in the diet. Feeding lower amounts of nutrients than the NRC recommended amounts or more than the recommended amounts would affect their blood concentrations and therefore the definition of normality; (3) Environmental conditions should be such that they will not influence in any way the health or metabolic processes of the host. Temperature, humidity, barn and stall space, and season of the year; (4) The animals should be free of stress (i.e., optimal well-being and welfare) like stress of being tied in stalls, stress of management, stress of hierarchy, stress of accessing foodstuff or water and other potential influential variables; (5) The source of water and mineral mix also is important because they vary from one dairy farm to the other and from one location to the other and contain different amounts of minerals or other compounds; (6) Physiological stage and age of the animals also are major factors that influence concentration of various metabolites in the body fluids. For instance, cows during the dry off have a different diet, a different physiological stage compared to cows in early lactation or mid- or late-lactation and have completely different nutrient requirements. Moreover, young animals as well as heifers versus cows have different physiologies and certainly different concentrations of blood variables; (7) Feedstuff used for the animals should be absolutely free of mycotoxins, endotoxins, heavy metals, or other toxic compounds. It is known that various fungi and bacteria grow under humid and optimal temperatures in the feedstuff used for dairy cattle or other livestock animals. Metabolic and immunity effects of some of those toxic compounds have been reported in the scientific journals. Presence of those compounds certainly will affect the normality or abnormality of metabolites in the body fluids of cattle. Overall, defining what is normal and what is abnormal with regard to the concentration of a metabolite or multiple metabolites in various body fluids of a dairy cow or other animals is a very challenging task.

The methodology that the scientists have used to determine whether an abnormal change has occurred in the internal homeostasis of a metabolite in a cow or a group of cows has been mainly by comparing clinically healthy cows versus those displaying clinical or subclinical signs of a disease and under comparable parities, physiological stages as well as similar feeding and management systems. However, generally, no specific evaluations of control animals are conducted to rule out that they did not have subclinical health issues. It might take some time until studies that will control all the variables described above could be conducted and advanced instrumentation will be used to determine normal concentrations and fluctuations of thousands of metabolites present in the body fluids of dairy cows.

It should be noted that the concept of a metabolic disease in dairy cows is quite different from that of humans, which defines metabolic disease as "any of the diseases or disorders that disrupt normal metabolism, the process of converting food to energy on a cellular level (Enns 2016)." In human medicine metabolic diseases are defined mainly as hereditary or inborn errors of metabolism (Enns 2016). It is interesting also to point out that except for diabetes, a disease related to alteration of glucose and insulin metabolism, and metabolic syndrome there are no other defined metabolic diseases of humans. In contrast in veterinary medicine there are several diseases that have been defined as metabolic disorders including milk fever, ketosis, fatty liver, and several others less important diseases like hypophosphatemia, grass tetany or hypomagnesemic tetany, and pregnancy toxemia.

Finally, given the most recent developments in systems biology, metabolic diseases should be viewed more as perturbations of multidimensional and integrated cellular and organ level genomics, transcriptomics, proteomics, and metabolomics networks in interaction with the environmental factors. These interactions should be viewed as multilayered and different in individual animals with the same disease. This complexity requires a new methodology of handling the challenges associated with the disease process. It also requires development of a new philosophy on how to prevent occurrence of disease in livestock animals.

Overall the concept of metabolic diseases or disorders for livestock animals in general and dairy cattle in particular needs to be revisited and redefined. This is a necessity deriving from new developments in the area of dairy cow health. This is also an obligation of the current research community to the dairy industry, the future generations of students and researchers, and to the dairy cattle well-being and welfare.

2.3 Myth Number 2: The Concept of Production Diseases

The term production disease has been coined and widely accepted by the scientific community since its first introduction by Jack Payne in the early 1970s. Sixteen international conferences have been organized so far starting from 1972 until the most recent one on 2016, under the name of "International Conference on Production Diseases of Farm Animals." However, since the introduction of this concept there has been many new developments in the science of biology and veterinary medicine and the philosophy of approaching a disease state. Therefore, it is time to revisit this concept and adjust the concept of production disease to the new era of systems biology approach.

What was the original definition of production diseases? The concept of production diseases was introduced by Jack Payne (1977). He indicated that: "*as intensification and production levels increase so too do the problems of metabolic disorders.*" He further continues to elaborate the concept indicating that: "*The modern view is that metabolic diseases of farm ruminants are not primarily due to inherent defects in the animal's biochemistry. Rather they result from a breakdown in the animal's ability to cope with the metabolic demands of high production, coupled with the strains of modern intensive husbandry and feeding. In other words, metabolic disease is a failure to compensate for imposed and man-made demands on farm livestock. This idea has led to the introduction of a new collective name for metabolic disorders of farm ruminants—**production disease**.*"

Payne proposed that metabolic diseases are related to the inability of the cows to cope with the high nutrient demands of milk production. However, dairy cows that produce high amounts of milk have inherited the trait of high milk production from their parents. Therefore, they have or own the capacity to produce high amounts of milk. If we offer cows all the necessary nutrients required for the level of milk production that they are capable off, then, the cows must reach that potential. This is supported by the fact that around half of high producing dairy cows in a herd are affected by metabolic or periparturient diseases during the transition period. The other half remains healthy and productive. They can reach their production capacity with no health issues. This is an important indication that the environment is not the only or the main actor in the development of most metabolic or periparturient diseases of dairy cows, as indicated by Payne (1977). Moreover, if half of the cows in a herd consume the same amount of feed and similar ingredients and are not affected by metabolic diseases, this means that the genotype is also a very significant factor that should be taken into consideration when studying metabolic diseases. However, Payne in his book (Payne 1977) indicates that "*metabolic disease is not an inherent defect in the animal's biochemistry.*" Therefore, Payne excludes the possibility that metabolic diseases in dairy cows are related to inborn errors of metabolism or genotype of the cow.

Let's take another example to better understand the role of genotype in the pathobiology of metabolic diseases. Feeding high amounts of grain immediately after calving, to provide cows the necessary energy and nutrients required for milk production, is associated with low rumen fluid pH. This low pH affects microbiome composition and the bacterial toxic compounds released in the rumen. For example, the amount of endotoxin in the rumen fluid of cows fed high amounts of grain is increased almost 14-fold (Emmanuel et al. 2008). Endotoxin is able to translocate through rumen and colon tissues and to enter into systemic circulation triggering a whole variety of metabolic and immune responses, even diseases like fatty liver, laminitis or retained placenta (Emmanuel et al. 2007). There is increasing evidence that mammals and cows respond differently to endotoxin (Jacobsen et al. 2007, 2008). Some individuals are more susceptible to endotoxin and succumb to sickness or even death. Other individuals are more resistant to the same dose of endotoxin and overcome the challenge of endotoxin with a slight fever. There is also mounting evidence that endotoxins are involved in multiple periparturient diseases of dairy cows including fatty liver, milk fever, ketosis, retained placenta, displaced abomasum, laminitis, and several others diseases (Ametaj et al. 2010, 2015). In this case it is obvious that the susceptibility or resistance of some of the cows to endotoxin-related diseases is strongly related to the genotype of the animal.

Payne (1977) and other scientists also believed that "*metabolic disease is a failure to compensate for imposed and man-made demands on farm livestock.*" Based on this definition it seems like men have created metabolic diseases by demanding or selecting cows for greater milk production. However, there is no solid evidence to indicate that selection of cows for high milk production is associated with greater incidence of periparturient diseases (including those called metabolic diseases). One clear example that increased milk production is not associated with increased incidence of periparturient diseases is milk fever. Milk fever was reported for the first time during the eighteenth century. It's been almost one century that we study milk fever, however, the incidence of this disease in North America, Europe, or elsewhere has fluctuated around 5–10%. This clearly shows that milk fever is not a man-made disease, not related to selection of cows for high milk production. On the other hand, some breeds of cows like Jerseys are more prone to milk fever than Holsteins. This also supports the idea that genotype of the cows is very important in disease development.

Payne (1977) further explains definition of a production disease: "*Firstly, it implies that the disease is likely to occur when the demand for production exceeds the animal's metabolic capacity. Secondly, it draws attention to the fact that, in the interests of high production, animals are exposed to metabolic hazard because they are not always fed or managed appropriately for their specialized physiology and metabolic needs.*"

If Payne's statement would be true, then 100% of the cows in a herd should succumb to metabolic or periparturient disease around calving. However, this does not happen although cows are fed a similar ration. In fact, only up to 30–50% of the cows are affected by one or multiple periparturient diseases concurrently. This again suggests that genotype is very important in the susceptibility of the cows to periparturient disease.

Another aspect of Payne statement (Payne 1977) that deserves to be discussed is that he is defining metabolic disease in a mechanical way. He states: "*All production systems have three basic components. All have inputs of raw materials, a central processing system and an output of the finished product. This is the basic pattern of production not only for the dairy cow but also for manufacturing systems of all kinds. All are prone to similar 'diseases'.*" It is obvious that Payne is considering the animal a machine ignoring that biology is quite different from a human or electrically or electronically controlled machine. A cow is far more complex being than a human made machine. Another aspect that was ignored by Payne's definition is that there are tenfold or more bacteria in the GI tract of a cow than her own body cells. Those bacteria and their products have been reported to be involved in various pathologies in both humans and animals; however, this information was not yet available to that generation of scientists.

Payne also proposed the concept of imbalance between input and output. He calls metabolic diseases as put-put diseases similar with the functioning of a car factory. [Of note Payne's definition about production diseases has been made at a time when there was not very much knowledge about the role of bacterial endotoxins in transition cow diseases.] Presently, there is mounting evidence that feeding high amounts of grain or concentrate immediately postpartum is associated with multiple abnormal alterations in the rumen fluid, rumen wall as well as enhanced release of endotoxins in the rumen fluid. There is increasing evidence that endotoxins released in the rumen, in the infected mammary gland, or in the infected uterus play a significant role in the pathobiology of multiple periparturient diseases. This example clearly indicates that "production diseases" are not simply put-put diseases or related to the imbalance between input and output. There are other important factors that play significant roles in health and disease that cannot be ignored when discussing about transition cow diseases. Therefore, the mechanical model of Payne cannot be applied to a dairy cow which is a far more complex organism than a man-made car manufacturing factory.

Additionally, there are several authors that have studied and reported that there is a relationship between milk production and the greater risk of metabolic or production diseases in dairy cows; however, most of them have reported that there is no such risk. Additionally, the most comprehensive review regarding association of high milk production with greater incidence of disease was conducted by Ingvartsen et al. (2003). The authors conducted a meta-analytical investigation of 11 epidemiological and 14 genetic studies related to an association between high milk yield and increased risk of dystocia, parturient paresis (milk fever), ketosis, displaced abomasum, retained placenta, ovarian cyst, metritis, mastitis, and lameness. They concluded that "there is no clear evidence that high-yielding cows have an increased risk of production diseases, apart from mastitis."

Some studies have reported an effect of high milk yield on the risk of some diseases (Curtis et al. 1984, 1985; Gröhn et al. 1989; Gröhn et al. 1990a, b; Bigras-Poulin et al. 1990; Lyons et al. 1991). The results of those investigations are correct in relation with the hypotheses of the authors. However, it should be kept in mind that other factors interfere with the disease incidence that have not been considered by those authors. For example, feeding cows greater amounts of grain in the diet is associated with greater incidence of periparturient diseases. However, feeding increasing amounts of barley grain at 0, 15, 30, and 45% (i.e., DM-basis) was associated with greater release of endotoxin in the rumen fluid and activation of innate immunity in the systemic circulation when they were fed 30% or 45% of the diet DM in the form of barley grain. Feeding between 30 and 45% grain in the diet was associated with a 14-fold increase in the amount of endotoxin in the rumen fluid (Emmanuel et al. 2008). Also, it has been reported that endotoxin is able to pass through rumen and colon tissues of cattle, especially when they are inflamed, which commonly happens during high grain feeding (Emmanuel et al. 2007). Another important observation is that cows that produce more milk are fed more DMI compared to those that produce less milk. This is a common practice in a modern dairy farm where the producers follow milk production of a cow and feed them based on their current milk production. The more milk a cow produces the more feed is fed to that cow. Consuming more feed (i.e., TMR) means also ingesting greater amounts of grain or concentrate in the TMR compared with lower producing cows. Feeding greater amounts of grain will affect rumen microbiome and certainly the health-disease status of a cow. In conclusion, there is no strong evidence that increasing milk yield will increase the risk of periparturient diseases. Therefore, instead of calling those diseases put-put, man-made, or production diseases it would be more appropriate to call them simply periparturient diseases of transition dairy cows.

2.4 Myth Number 3: Cow Sacrifices Herself for Milk Production

Payne in his book on metabolic diseases of dairy cows (Payne 1977) gives another concept that became dominant and continues to dominate the animal and veterinary sciences. For example, he indicates that "*Dairy cow has an inherent defect biologically which puts her at a serious disadvantage. This disadvantage is peculiar to the farm animals and is not shared by man-made industrial processes. Quite simply the problem is that* ***the dairy cow's output tends to be obligatory and even in times of input shortage, the production receives priority even though the cow dies in the process***." It seems very idealistic and convincing to read that a sublime mother dies or sacrifices herself for the survival of the baby! Is the creation so weak!? If the mother dies in the process of milk production, then, the newborn also will die because we know that in nature no other cow will nurse the calf of another cow. Additionally, Payne is indicating that cows are a little bit different from car producing factories because cows can die in the process of milk production, a dominant feature indispensable for survival of the offspring, whereas factories do not (they are not perished in the process of a car production). Unfortunately, if the mother dies the offspring also will die, as explained above. Factories simply stop producing when they lack the required materials. Do sick cows really die to produce the milk necessary for the growth of the offspring? Or do they sacrifice themselves in the process? Is there scientific support for Payne's statement? Recent data from our lab and many other investigators show clearly that during six most prevalent periparturient diseases of dairy cows including metritis, mastitis, laminitis, ketosis, milk fever, and retained placenta there is a significant drop (almost 25%) in the milk production, but no death of the cows during the disease process (Dervishi et al. 2015, Dervishi et al. 2016a, b; Zhang et al. 2015, 2016). This is a clear evidence that the cow does not sacrifice herself in the process of milk production for the offspring, although she is clinically sick and lowers feed intake. Secondly, the amount of milk produced by the cow per day (even when she is sick) overcomes the need of the calf for growth. A calf cannot drink more than 10 kg of milk per day, while today's cows, in developed countries, produce from 40–50 kg of milk per day. Dropping the milk yield by 25% means a drop of 10–12 kg/day. There is still 30–40 kg milk per day available to feed the calf. Certainly the udder will continue to produce milk because the postpartum physiology (hormones that support and induce milk production like prolactin) of the cow immediately after calving is supportive of milk production; however, during disease the amount of milk produced is lower than the cow's potential but more than enough to feed the calf. Therefore, it does not make too much sense to indicate that "*production receives priority for the cow's even though she can die in the process*."

2.5 Myth Number 4: One Metabolite One Disease Concept

The dominant reductionist principle for establishing a metabolic disease has been "one altered metabolite, one metabolic disease." Indeed, all known metabolic diseases of dairy cows have been identified with one altered metabolite. For example, milk fever has been identified as periparturient hypocalcemia; ketosis as increased concentration of β-hydroxybutyrate (BHB) or ketone bodies; fatty liver as increased blood levels of non-esterified fatty acids (NEFA) postpartum and their storage in the liver as triacylglycerides. Based on this logic other metabolic diseases established over the years have been hypophosphatemia or chronic phosphorous deficiency as well as hypomagnesemic tetany or deficiency of magnesium.

In fact, recent data generated by various research labs through proteomics and metabolomics sciences have demonstrated that during major diseases of transition dairy cows including milk fever, ketosis, and fatty liver, as will be illustrated in the following chapters in this book, there are multiple alterations in the concentrations of various amino acids, lipids, phospholipids, sphingomyelines, acylcarnitines, and metals in the blood, urine, and milk of sick dairy cows (Klein et al. 2012; Hailemariam et al. 2014; Imhasly et al. 2014; Sun et al. 2014). Moreover, multiple metabolic pathways are perturbed during most prevalent infectious diseases of transition dairy cows including metritis, mastitis, and inflammatory-related lameness. This has prompted us to suggest that metabolic and inflammatory responses of dairy cows around calving have been misinterpreted as metabolic diseases or deficiencies (e.g., hypocalcemia) or overproduction (e.g., high blood NEFA) of certain metabolites (Ametaj 2015). The reason for this assumption is that researchers during the last century have been focused on the concept that "one disturbed metabolite is the main cause of one metabolic disease." Today's systems biology data have shown that the number of metabolites or proteins altered significantly during major periparturient diseases of dairy cows are in the tens or even hundreds. Based on the logic of one perturbed metabolite one disease, presently, we can easily invent hundreds of new metabolic diseases of transition dairy cows'. These new findings suggest that the scientific community should revisit and redefine the concept of metabolic diseases in dairy cows.

As explained previously there are no pure metabolic diseases in dairy cows because for each of the main periparturient diseases of dairy cows like metritis, mastitis, laminitis, ketosis, milk fever, and retained placenta there is an activation of innate immunity that precedes the disease process and is present during clinical disease, besides multiple metabolite and metabolic pathway perturbations (Dervishi et al. 2015, 2016a, b; Zhang et al. 2015, 2016). This suggests that the pathobiologies of periparturient diseases of dairy cows are far more complex than previously thought. It seems like metabolic fluctuations around calving are part of the host response to any periparturient disease given the fact that before, during, and after disease occurrence a large number of innate immunity and metabolites related to amino acid, carbohydrate, lipid, phospholipid, sphingomyeline, acylcarnitine, and mineral metabolism are required for mounting a response to the disease process and for healing.

In conclusion, in defining metabolic diseases we need to focus on the perturbed pathways and networks instead of a single gene, a single metabolite, or a single protein approach so that we can understand what is occurring in the whole organism more than in one single component of the body.

2.6 Reductionist Versus Systems Veterinary Approach

2.6.1 Predictive, Preventive, and Individualized Medicine

The present veterinary medicine has been mostly linked to diagnosis of subclinical or clinical disease at its latest stages. This means that each disease goes through its complete subclinical course unnoticed and then displays fully with its clinical symptoms. Even if the disease is in its subclinical form most diseases like milk fever, ketosis, or mastitis are diagnosed based on the "one metabolite, one disease" approach. Milk fever as lower concentration of Ca in the blood circulation, ketosis as greater concentration of BHBA in the blood, and mastitis as higher than a certain number of somatic cell count (SCC) in the milk. The problem with diagnosis of the subclinical disease at a late stage or of the clinical disease at the stage when symptoms are evident is that it is too late to reverse its progression. Moreover, treatment at this stage is more expensive and might result in death of the cow. As a consequence, a large number of cows do not respond well to the medical treatments. It is a known fact that up to 50% of dairy cows in a herd become non-economical and are culled each year in developed countries. Therefore, a means to early detect the disease in its very starting point and an early intervention to prevent the progression of the disease process would be more advantageous than treatment.

During the last century conventional veterinary medicine has been mostly a reactive medicine. This means that veterinary practitioners have been treating individual cows once they are diagnosed with a disease or when they display clear signs of a disease. Indeed, conventional cure has been and is focusing more on treating a symptom or a consequence of a causative agent related to a certain pathology in the body and elimination of that symptom is considered a cure. However, the cow might continue to be sick in a subclinical form. Indeed, recent metabolomics data from a large study conducted from our lab (Dervishi et al. 2016a, b; Zhang et al. 2017) indicate that even if the cows seems clinically normal there are major perturbations in multiple metabolites, pathways, and networks which later developed into metritis, mastitis, lameness, ketosis, milk fever, and retained placenta. The same study also showed that major metabolic pathway alterations and innate immunity reactants are present more than 2 months before cows show clinical signs of disease. Unfortunately, in most occasions the reactive medicine is unable to work with cows during the early stages of subclinical disease and frequently has not been able to recover dairy cows to the point that they might be functional and economical. This has resulted in almost half of the cows in a herd ending up culled each year.

Unlike reactive veterinary medicine which waits for the disease to begin its course and then rushes in to eliminate the symptoms, proactive or preventive medicine provides the opportunity for prevention of the disease and aims to treat the root cause of the disease. Recent developments of systems veterinary approach have made possible to start thinking and introducing a new philosophy in dealing with diseases. Systems biology sciences of genomics, transcriptomics, proteomics, and metabolomics have made it possible to develop new ways to screen for risk of disease and give veterinarians a possibility to prevent development of a specific disease and also approach the disease in quite a different way. However, prevention of a disease is a science in itself and needs a little bit more discussion so that we better understand what is meant by prevention of disease. There are two types of preventive interventions that should be taken into consideration with regard to animal health as discussed below.

The systems veterinary approach is suggesting a different approach to the disease process. It proposes to detect disease at its earliest stage when it is possible to reverse the disease course. It proposes to do this by identification and utilization of multiple screening, monitoring or predictive biomarkers. It also proposes to achieve that through screening of body fluids with the least invasive intervention by using easily accessible body fluids like urine or milk. A significant advantage of screening or monitoring biomarkers is that they can be applied directly by the producers themselves, at their own time and place. These new cow-side technologies, which are in the process of development, will be able to lower the cost of medication by reversing the disease process at its earliest stages and by minimum medical intervention. Predictive veterinary medicine also proposes to use green technologies like application of probiotics or new vaccines to prevent occurrence of disease (Ametaj et al. 2010, 2012a, b; Deng et al. 2015; Iqbal et al. 2014). Other technologies include processing of grains fed to cows so that they become healthier and friendlier to the physiology of the gastrointestinal tract of the cows (Iqbal et al. 2012). It should be noted that once biomarkers will be introduced and embraced by veterinary practitioners and dairy industry then a new era of "Individualized Medicine" might start to be implemented. There are three different approaches to preventive veterinary medicine: primary, secondary, and tertiary.

The primary veterinary prevention approach is related to preventing the disease from occurring, in the first place. The main purpose of this type of intervention is to prevent, or in the worst case scenario, lower the disease incidence. This type of approach aims at taking preventive measures before the appearance of disease. Examples include using biomarkers to predict the risk of developing, for example, uterine infections and then stratifying cows to be treated with probiotics to prevent occurrence of uterine infections, as reported recently by our lab (Ametaj et al. 2010; Deng et al. 2015, 2016). It can also include stratification of cows based on potential risk to develop a periparturient disease and then vaccinating them during the dry off period to prevent occurrence of endotoxin-related diseases or mastitis (Ametaj 2015). This type of preventive intervention has been developed and might start being applied soon.

The secondary veterinary prevention approach is another approach to lower occurrence of disease that aims at detecting the disease at its earliest stage or subclinical stage, before the signs of disease appear clinically in dairy cows, and either to stop further progression of the disease or to slow down its progression. This type of preventive intervention is again based on biomarkers for disease diagnosis at subclinical stages of development. It is always better to intervene when the problem is small rather than when the issues become aggravated and more difficult to tackle or reverse. This type of preventive intervention aims at using cow-side screening tests to detect biomarkers for the risk of the disease at its earliest possible time. An example of this type of medical intervention is development of screening biomarkers of disease risk through proteomics or metabolomics technologies. Our team is in the process of developing such cow-side tests for early diagnosis of metritis, mastitis, lameness, ketosis, milk fever, and retained placenta.

The third veterinary prevention strategy is related to economics of a treatment. Differently from human medicine, veterinary medicine can humanely cull an animal including a cow, if it can be predicted that it would be economically not feasible to treat a cow for a certain disease, if prognostic biomarkers would suggest that the treatment will fail. This might save billions of dollars of veterinary bills to dairy producers worldwide if such a preventive strategy will be introduced in the future.

Defining a healthy and a disease phenotype and developing the process of individual prevention, treatment, and feeding strategies to animals to prevent disease would be the challenge of this new millennium. Using a biomarker-based medicine would be able to stratify dairy cows into healthy phenotypes or those that are susceptible to disease. This will make possible to follow the efficiency of treatment based again on biomarker utilization for prognosis of curability. This whole concept of predicting the risk of disease and efficiency of treatment is part of the "Individualized Medicine." Application of these new concepts at the farm level by producers themselves and veterinary consultants will increase profitability and efficiency of dairy farming.

In addition to biomarkers for prediction of disease risk it would be interesting to follow the concept of individual feeding. Dairy cows do not respond similarly to the total mixed ration that they are offered at each dairy farm. There are indications that some cows are more susceptible to disease than others or that respond to dietary ingredients and amounts fed in a different way (García et al. 2007). Individualized feeding might address these issues in the future. This could be done through the science of nutrigenomics. The science of nutrigenomics contains the three omics disciplines genomics, proteomics, and metabolite profiling, as applied to the field of nutrition and health. Furthermore, nutrigenomics forms the scientific basis for developing nutrition adapted to the specific needs of specific cows, be they healthy, at risk, or sick. Again biomarkers, whether proteins or metabolites in various body fluids, that stratify cows in efficient and healthy phenotypes will be very important in the future individual feeding.

2.6.2 Switching Paradigms: From Reductionism to Systems Veterinary Approach

There is a general discussion among biologists whether the philosophy used to approach disease understanding, diagnosis, and treatment is still valid or should it be replaced by a new philosophy?

During many centuries, the dominant thinking of life sciences has been that in order to understand how the whole (i.e., the organism) works you can study the structure and functions of the parts and deduce from them how the whole works. This type of thinking was developed by two philosophers Descartes and Laplace and consisted in the mechanistic view of the world. The world was described as a machine, being conceived as one big mechanical device operating according to natural laws. Therefore, understanding the functioning of this machine was related to the understanding of how the parts of this machine were working. This philosophy of thinking was known as "Reductionism." This type of philosophy was embraced by life sciences for centuries, including animal and veterinary sciences.

Reductionist approach, during the centuries, evolved into three related but different sub-philosophies: ontological, methodological, and epistemic (Esfeld 2013). In medical sciences, ontological reductionism corresponds to the idea that each particular biological system (e.g., an organism) is constituted by molecules and their interactions and that biological properties are related to physical properties of the components. Therefore, understanding at the molecular level is sufficient to explain any biological phenomena.

On the other hand, methodological reductionism is the idea that biological systems are most efficiently investigated at the lowest possible level, and that the objective of the scientific investigation should be to detect molecular and biochemical causes. A common example of this type of strategy is the breakdown of a complex system into parts; a biologist might study the cellular parts of an organism in order to comprehend its behavior, or examine the biochemical components of a cell to understand its features (Peacocke 1985).

Finally, epistemic reductionism indicates that given the presupposition that living beings are made up of the same kind of atoms and molecules as the physics and chemistry knows them, and nothing else, then, having a comprehensive knowledge of the particular atoms and molecules in a given organism and their conformation, together with the knowledge of the laws of physics and chemistry, would in principle, be sufficient to redefine all the properties of this organism in atomic and molecular terms and to derive its behavior and all the laws that it obeys (Kaiser 2011).

Although there are small differences among the three reductionist sub-philosophies, they all have the same Descartian mechanical thinking in common, that the functioning of the whole organism is none but the sum of functions of all the parts. Based on the reductionist approach almost all research investigations in life sciences have "zoomed in" on a specific trait, cell, protein, gene, metabolite, or on a very narrow question. Taking ketosis as an example, focusing on ketone bodies to explain the whole disease of ketosis has been the basic approach to explaining the whole pathobiology of that disease during the last century.

During the last decade the philosophy of reductionism has been challenged by life scientists given the fact that it has been impossible to understand complex organisms by simple knowledge of the functions of the parts in isolation (Boogerd et al. 2007). Given that the old philosophy of reductionism fails to explain the functioning of the whole, life scientists have developed a new philosophy that is known as systems biology approach. The term "system" implies the whole object or organism, which can be divided into components whose essential properties cannot be fully explained merely from the knowledge of the parts. A system also implies certain number of different and interacting components. Objects that do not interact do not form a system (Maly 2009). In simple words, the systems biology believes that "the sum is more than the parts." An additional difference of systems approach is that it looks at components of a system as units that interact in a nonlinear way with each other, which means that new properties might arise as a consequence of those interactions. This also implies that studying components of an organism in isolation cannot revel presence of those new properties. It should be noted that interactions among the components with each other are limitless given the large number of cells, tissues, organs, genes, proteins, and metabolites present in an organism. Moreover, the systems approach believes that it is the whole organism that determines how its components behave and that the behavior of the parts within the system is qualitatively different from their behavior in isolation. The systems veterinary is none but application of systems biology approach to veterinary medical sciences.

The systems veterinary uses a group of interrelated sciences including genomics, transcriptomics, proteomics, metabolomics, computational biology, informatics, biostatistics, mathematics, and high-throughput technologies to integrate complex data about interactions in biological systems.

In the following chapters of this book many examples of how systems veterinary approach is contributing to better understanding of periparturient diseases of transition dairy cows will be given. It should be noted that the systems approach is just at its beginning and more is expected to come in the next decade or so. Hopefully better solutions to many cow health issues will be developed and our understanding of the pathobiologies of various diseases and how to prevent periparturient diseases will reach new heights.

Conclusions

Overall we have already started changing paradigms in the veterinary sciences in terms of the way how we are approaching the study of the pathobiology and prevention of periparturient diseases of dairy cows. Reductionist approach has contributed for centuries in studying various parts of the cow's organism, disease diagnosis, and treatment; however, this philosophy of science has not been capable to provide us full comprehension of the exact causes of disease and in developing effective strategies for prevention of periparturient diseases. Reductionism thinking has wrongly generated numerous myths in veterinary and animal sciences about definition of metabolic and production diseases which need to be revisited and redefined. A new methodology of life sciences, systems biology approach, has just started to be embraced by many labs throughout the world and is viewed as advantageous with respect to reductionism. This new approach to etiopathology of periparturient diseases is supported by multiple innovative instruments and a new methodology which looks at how all the parts interact during health or disease states. Moreover, utilization of this approach is expected to give us a better understanding of how the host responds to disease, determine what are the causes of disease, make possible early diagnosis of periparturient diseases, help in predicting the risk of occurrence of diseases, and designing and applying novel preventive strategies in the near future.

References

Ametaj BN (2015) A systems veterinary approach in understanding transition cow diseases: metabolomics. Proceedings of the 4th international symposium on dairy cow nutrition and milk quality, session 1, advances in fundamental research, May 8–10, Beijing, China, pp 78–85

Ametaj BN, Zebeli Q, Iqbal S (2010) Nutrition, microbiota, and endotoxin-related diseases in dairy cows. Rev Bras Zootec 39:433–444

Ametaj BN, Zebeli Q, Iqbal S, Dunn SM (2012a) Meeting the challenges of improving health in periparturient dairy cows. Adv Dairy Technol 24:287–317

Ametaj BN, Zebeli Q, Sivaraman S, Dunn SM (2012b) Repeated oral administration of lipopolysaccharide from *Escherichia coli* 0111:B4 modulated humoral immune responses in periparturient dairy cows. Innate Immun 18:638–647

Bigras-Poulin M, Meek AH, Martin SW (1990) Interrelationships among health problems and milk production from consecutive lactations in selected Ontario Holstein cows. Prev Vet Med 8:15–24

Boogerd FC, Bruggeman J, Hofmeyr JS, Westerhoff HV (2007) Towards philosophical foundations of systems biology: introduction, in systems biology: philosophical foundations. Elsevier, Amsterdam, pp 3–19

Bradford BJ, Mamedova LK, Minton JE, Drouillard JS, Johnson BJ (2009) Daily injection of tumor necrosis factor-α increases hepatic triglycerides and alters transcript abundance of metabolic genes in lactating dairy cattle. J Nutr 139:1451–1456

CanWest DHI and Valacta (2015) Culling and replacement rates in dairy herds in Canada. http://www.dairyinfo.gc.ca/pdf/genetics-cull_e.pdf

Curtis CR, Erb HN, Sniffen CJ, Smith RD (1984) Epidemiology of parturient paresis: predisposing factors with emphasis on dry cow feeding and management. J Dairy Sci 67:817–825

Curtis CR, Erb HN, Sniffen CJ, Smith RD, Kronfeld DS (1985) Path analysis of dry period nutrition, postpartum metabolic and reproductive disorders, and mastitis in Holstein cows. J Dairy Sci 68:2347–2360

Deng Q, Odhiambo JF, Farooq U, Lam T, Dunn SM, Wang Y, Gänzle M, Ametaj BN (2015) Intravaginal lactic acid bacteria modulated local and systemic immune responses and lowered the incidence of uterine infections in periparturient dairy cows. PLoS One 10(4):e0124167

Deng Q, Odhiambo JF, Farooq U, Lam T, Dunn SM, Wang Y, Gänzle M, Ametaj BN (2016) Intravaginal probiotics modulated metabolic status and improved productive performance of transition dairy cows. J Anim Sci 94:760–770

Dervishi E, Zhang G, Hailemariam D, Dunn SM, Ametaj BN (2015) Innate immunity and carbohydrate metabolism alterations precede occurrence of subclinical mastitis in transition dairy cows. J Anim Sci Technol 57:46–65

Dervishi E, Zhang G, Hailemariam D, Dunn SM, Ametaj BN (2016a) Occurrence of retained placenta is preceded by an inflammatory state and alterations of energy metabolism in transition dairy cows. J Anim Sci Biotechnol 7:26–39

Dervishi E, Zhang G, Hailemariam D, Goldansaz S, Deng Q, Dunn SM, Ametaj BN (2016b) Alterations in innate immunity reactants and carbohydrate and lipid metabolism precede occurrence of metritis in transition dairy cows. Res Vet Sci 104:30–39

Eckel EF, Ametaj BN (2016) Invited review: role of bacterial endotoxins in the etiopaathogenesis of periparturient diseases of transition dairy cows. J Dairy Sci 98:5967–5990

Emmanuel DGV, Madsen KL, Churchill TA, Dunn SM, Ametaj BN (2007) Acidosis and presence of lipopolysaccharide from *Escherichia coli* O55:B5 cause hyperpermeability of rumen and colon tissues. J Dairy Sci 90:5552–5557

Emmanuel DGV, Dunn SM, Ametaj BN (2008) Feeding high proportions of barley grain stimulates an inflammatory response in dairy cows. J Dairy Sci 91:606–614

Enns G (2016) Encyclopedia Britannica. Metabolic disease. https://www.britannica.com/science/metabolic-disease

Esfeld M (2013) Reductionism today. In: Galavotti MC, Nemeth E, Stadler F (eds) European philosophy of science—philosophy of science in Europe and the Viennesse heritage. Springer, Cham, pp 89–101

Ganz T (2009) Iron in innate immunity: starve the invaders. Curr Opin Immunol 21:63–67

García SC, Pedernera M, Fulkerson WJ, Horadagoda A, Nandra K (2007) Feeding concentrates based on individual cow requirements improves the yield of milk solids in dairy cows grazing restricted pasture. Aust J Exp Agric 47:502–508

Gröhn YT, Erb HN, McCulloch CE, Saloniemi HS (1989) Epidemiology of metabolic disorders in dairy cattle: association among host characteristics, disease, and production. J Dairy Sci 72:1876–1885

Gröhn YT, Erb HN, McCulloch CE, Saloniemi HS (1990a) Epidemiology of mammary gland disorders in multiparous Finnish Ayrshire cows. Prev Vet Med 8:241–252

Gröhn YT, Erb HN, McCulloch CE, Saloniemi HS (1990b) Epidemiology of reproductive disorders in dairy cattle: association among host characteristics, disease, and production. Prev Vet Med 8:25–39

Hailemariam D, Mandal R, Saleem F, Dunn SM, Wishart DS, Ametaj BN (2014) Metabolomics approach reveals altered plasma amino acid and sphingolipid profiles associated with pathological state in transition dairy cows. Curr Metabolomics 2:184–195

Hotamisligil GS, Shargill NS, Spiegelman BM (1993) Adipose expression of tumor necrosis factor-alpha: direct role in obesity-linked insulin resistance. Science 259:87–91

Imhasly S, Naegeli H, Baumann S, von Bergen M, Luch A, Jungnickel H, Potratz S, Gerspach C (2014) Metabolomics biomarkers correlating with hepatic lipidosis in dairy cows. BMC Vet Res 10:122–130

Ingvartsen KL, Dewhurst RJ, Friggens NC (2003) On the relationship between lactational performance and health: is it yield or metabolic imbalance that cause production diseases in dairy cattle? A position paper. Livestock Prod Sci 83(2–3):277–308

Iqbal S, Mazzolari A, Zebeli Q, Dunn SM, Ametaj BN (2012) Treating barley grain with lactic acid and heat prevented sub-acute ruminal acidosis and increased milk fat content in dairy cows. Anim Feed Sci Technol 172:141–149

Iqbal S, Zebeli Q, Mansmann DA, Dunn SM, Ametaj BN (2014) Oral administration of lipopolysaccharide and lipoteichoic acid prepartum modulated reactants of innate and humoral immunity in periparturient dairy cows. Innate Immun 20(4):390–400

Jacobsen S, Andersen PH, Aasted B (2007) The cytokine response of circulating peripheral blood mononuclear cells is changed after intravenous injection of lipopolysaccharide in cattle. Vet J 174(1):170–175

Kaiser M (2011) The limoits of reductionism in the life sciences. Hist Philos Life Sci 33:453–476

Klein MS, Buttchereit N, Miemczyk SP, Immervoll AK, Louis C, Wiedemann S, Junge W, Thaller G, Oefner PJ, Gronwald W (2012) NMR metabolomic analysis of dairy cows reveals milk glycerophosphocholine to phosphocholine ratio as prognostic biomarker for risk of ketosis. J Proteome Res 11:1373–1381

LeBlanc SJ (2010) Health in the transition period. Adv Dairy Technol 22:97–110

Lyons DT, Freeman AE, Kuck AL (1991) Genetics of health traits in Holstein cattle. J Dairy Sci 74:1092–1100

Maher P, Good M, More S (2008) Trends in cow numbers and culling rate in the Irish cattle population, 2003 to 2006. Ir Vet J 61:455–463

Maly IV (2009) Introduction: a practical guide to the systems approach in biology. In: Systems biology. Springer, Berlin, pp 1–11

Monteiro R, Azevedo I (2010) Chronic inflammation in obesity and the metabolic syndrome. Mediators Inflam. pii: 289645. doi:10.1155/2010/289645

Park C, Kim JR, Shim JK, Kang BS, Park YG, Nam KS, Lee YC, Kim CH (1999) Inhibitory effects of streptozotocin, tumor necrosis factor-alpha, and interleukin-1beta on glucokinase activity in pancreatic islets and gene expression of GLUT2 and glucokinase. Arch Biochem Biophys 362:217–224

Payne J (1977) Metabolic diseases in farm animals. William Heineman Medical Books Ltd., London, p 1

Peacocke A (1985) Reductionism in academic disciplines. Society for Research into Higher Education & NFER-Nelson, Guilford

Psychogios N, Hau DD, Peng J, Guo AC, Mandal R, Bouatra S, Sinelnikov I, Krishnamurthy R, Eisner R, Gautam B, Young N, Xia J, Knox C, Dong E, Huang P, Hollander Z, Pedersen TL, Smith SR, Bamforth F, Greiner R, McManus B, Newman JW, Goodfriend T, Wishart DS (2011) The human serum metabolome. PLoS One 6(2):e16957. doi:10.1371/journal.pone.0016957

Raj DS, Moseley P, Dominic EA, Onime A, Tzamaloukas AH, Boyd A, Shah VO, Glew R, Wolfe R, Ferrando A (2008) Interleukin-6 modulates hepatic and muscle protein synthesis during hemodialysis. Kidney Int 73:1054–1061

Rosch JW, Sublett J, Gao G, Wang YD, Tuomanen EI (2008) Calcium efflux is essential for bacterial survival in the eukaryotic host. Mol Microbiol 70(2):435–444

Rushen J (2013) The importance of improving cow longevity. Cow longevity conference. Hamra Farm, Sweden

Sun Y, Xu C, Li C, Xiaa C, Xu C, Wu L, Zhang H (2014) Characterization of metabolic profile of dairy cows with milk fever using 1H-NMR spectroscopy. Vet Q 34:159–163

USDA (2007a) Dairy 2007, Part 1: Reference of Dairy Cattle Health and Management Practices in the United States, 2007. USDA-APHIS-VS, CEAH, Fort Collins

USDA (2007b) Dairy 2007, Part II: Changes in the U.S. Dairy Cattle Industry, 1991–2007. USDA-APHIS-VS, CEAH, Fort Collins

Waldron MR, Nonnecke BJ, Nishida T, Horst RL, Overton TR (2003) Effect of lipopolysaccharide infusion on serum macromineral and vitamin D concentrations in dairy cows. J Dairy Sci 86:3440–3446

Wander K, Shell-Duncan B, McDade TW (2009) Evaluation of iron deficiency as a nutritional adaptation to infectious disease: an evolutionary medicine perspective. Am J Hum Biol 21:172–179

Weisberg SP, McCann D, Desai M, Rosenbaum M, Leibel RL, Ferrante AW Jr (2003) Obesity is associated with macrophage accumulation in adipose tissue. J Clin Invest 112:1796–1808

Zhang G, Hailemariam D, Dervishi E, Deng Q, Goldansaz S, Dunn SM, Ametaj BN (2015) Alterations of innate immunity reactants in transition dairy cows before clinical signs of lameness. Animals 5:717–747

Zhang G, Hailemariam D, Dervishi E, Deng Q, Goldansaz S, Dunn SM, Ametaj BN (2016) Dairy cows affected by ketosis show alterations in innate immunity and lipid and carbohydrate metabolism during the dry off period and postpartum. Res Vet Sci 107:246–256

Zhang G, Dervishi E, Dunn SM, Mandal R, Liu P, Han B, Wishart DS, Ametaj BN (2017) Metabotyping reveals distinct metabolic alterations in ketotic cows and identifies early predictive serum biomarkers for the risk of disease. Metabolomics 13:43–58

An Omics Approach to Transition Cow Immunity

3

Emily F. Eckel and Burim N. Ametaj

Abstract

Traditionally, research of transition cow immunity has focused on a reductionist approach trying to pinpoint a single factor that causes periparturient immunosuppression. Both previous and recent research has revealed that this phenomenon has a multifactorial etiology and that our current understanding remains insufficient to properly manage the high disease incidence at this time. In taking a systems biology approach through omics technologies we will be able to develop a fundamental understanding of the causal agents and the mechanisms underlying immunosuppression where preceding technologies have failed. Moreover, these new technologies have the potential to help us develop management techniques to restore the normality of immune response during transition period and lower the increased disease incidence that follows immunosuppression. It should be noted that currently application of omics approaches to transition cow immunity are at their pioneering level and further research is warranted to realize the importance of these sciences in the field of cow health and immunity.

3.1 Reductionist Approach to Understanding Immunity Around Calving

As with many other health issues of transition dairy cows, immunity around calving has mainly been approached by a reductionist view that focuses on finding one single factor that might trigger the well-documented immunosuppression around

E.F. Eckel, Ph.D. (✉) • B.N. Ametaj, D.V.M., Ph.D., Ph.D.
Department of Agricultural, Food and Nutritional Science, University of Alberta, Edmonton, AB, Canada, T6G 2P5
e-mail: eeckel@ualberta.ca; burim.ametaj@ualberta.ca

B.N. Ametaj (ed.), *Periparturient Diseases of Dairy Cows*,
DOI 10.1007/978-3-319-43033-1_3

calving. The bulk of this focus, until recently, has been on the link between nutrition and immunity and how nutritional management techniques can help reduce immune dysfunction. However, as many of these studies have determined, the bulk of immunosuppression cannot be pinned on a single factor. New and interesting findings also confirm the multifactorial nature of periparturient immunosuppression while conflicting classical views on its etiology. In this chapter we will discuss the contribution that has been made by the reductionist approach and outline the need for a more complex method to study immunosuppression. We will additionally summarize what the omics sciences have contributed so far to a better understanding of immunosuppression.

3.2 Basic Concepts of Immunology

In order to understand the physiological changes occurring around calving that contribute to immunosuppression it is important that we first understand how the immune system functions normally. Immunity is the body's natural defense against pathogens and immune responses can be generally separated into two categories, innate and adaptive, although these do not function independently of one another. Innate immunity refers to components that are present from birth which respond rapidly and show broad specificity in the sense that they are effective against a wide range of potentially infectious agents. Adaptive immunity, on the other hand, involves B and T cells which are capable of recognizing specific antigens. The onset of an adaptive response is slower than that of the innate response, while, unlike innate immunity, it provides long-term protection via immunological memory and leads to a heightened response upon second exposure to an antigen (Coico and Sunshine 2015; Williams 2012).

The innate immune system serves as a first line of defense, responding rapidly to pathogens and subsequently facilitating the development of adaptive immunity. There are many layers to innate immune mechanisms including barriers (e.g., skin, mucous membranes), humoral factors (e.g., acute-phase proteins, antimicrobial peptides, complement proteins), and cell-mediated responses (e.g., phagocytosis, cytotoxicity, antigen presentation), just to name a few. The three main goals of innate immunity are to: (1) distinguish self vs. non-self, (2) kill pathogens or infected cells and remove foreign particles, and (3) signal the induction of inflammation and adaptive immunity. Immune cell types that participate in the innate immune response include granulocytes, monocytes, macrophages, dendritic cells (DCs), and natural killer (NK) cells. Granulocytes include polymorphonuclear (PMN) leukocytes (consisting of neutrophils, basophils, and eosinophils) and mast cells and these are so named by the presence of cytoplasmic granules that contain proinflammatory mediators and/or antimicrobial factors which are released into the extracellular environment through the process of degranulation. The first step of an immune response is the recognition of a foreign pathogen which requires the expression of pattern recognition receptors (PRRs) such as toll-like receptors (TLRs) and NOD-like receptors (NLRs). These receptors respond to conserved microbial structures known as pathogen-associated molecular patterns (PAMPs). Pathogens present outside the cell (extracellular pathogens) are typically recognized by macrophages and mast cells which go on to produce the downstream effects leading to inflammation and the activation of adaptive immunity. Activated macrophages secrete proinflammatory cytokines which are soluble mediators that regulate the immune system by stimulating local inflammation, tissue repair, acute-phase protein (APP) synthesis, and recruit leukocytes such as neutrophils and monocytes to the site of infection. Mast cells, on the other hand, release histamine which causes vasodilation and increased vascular permeability. These effects increase blood flow to the site of infection and allows for the movement of proteins and leukocytes from the blood into the tissue which overall function to help resolve inflammation.

Induction of an adaptive response requires the help of professional antigen presenting cells (APCs), including DCs, macrophages, and B cells, which phagocytose pathogens and present antigens via major histocompatibility complex (MHC) II to $CD4^+$ T cells, also known as helper T cells. Presentation of an antigen to naive helper T cells causes their differentiation into one of the possible subsets, including T_H1, T_H2, T_H17, or Treg cells, depending on the cytokines present when this maturation is occurring. During the early innate response to intracellular pathogens the production of IL-12 (DCs), IFN-γ (NK cells), and TNF (macrophages) cause the differentiation of T_H1 helper cells that subsequently produce IFN-γ and IL-2. The T_H1 response stimulates cell-mediated immunity in order to eliminate the intracellular pathogen which initiated the response. Hallmarks of the T_H1 response include increased activity of $CD8^+$ cells, known as cytotoxic T cells (CTLs), and NK cell activity, and stimulates phagocytosis. The main function of CTLs and NK cells is to kill infected cells through stimulation of apoptosis. A small portion of activated helper T cells and CTLs survive past the resolution of inflammation and become memory T cells responsible for a more rapid and effective response upon second exposure to the same antigen. The response to an extracellular pathogen, on the other hand, stimulates the development of T_H2 cells via the presence of IL-4 and IL-4 with the absence of IL-12. The production of IL-4, IL-5, IL-6, IL-10, and IL-13 by T_H2 cells stimulates B cell proliferation, immunoglobulin class switching to IgE and activated eosinophils. This response is typically associated with the response to parasitic worms as well as allergens.

Antibody production by B cells is another important factor to adaptive immunity. Antibodies, also known as immunoglobulins, are responsible for the humoral adaptive response and are highly antigen specific. Activation by T cell dependent (TD) antigens occurs by helper T cell recognition of an antigen being presented by a B cell via MHC II leading to production of IL-4 and activation of B cells. Antigens that are able to stimulate B cell activation in the absence of helper T cells are known as T cell independent antigens (TI). The activation of B cells causes their proliferation, antibody class switching, and antibody production. Upon recognition of its specific antigen an antibody binds to the surface of the target cell, thus signaling for recognition by an effector cell leading to enhanced macrophage phagocytosis and NK cell antibody-dependent cell-mediated cytotoxicity (ADCC) (Coico and Sunshine 2015; Williams 2012).

3.3 What Is Immunosuppression?

The term immunosuppression refers to the temporary impairment of innate and adaptive immune responses around the time of calving, a phenomenon which has been documented in many studies over the last 40 years (Aleri et al. 2016; LeBlanc et al. 2006; Mallard et al. 1998). Our knowledge and understanding of the transition period as well as the mechanisms underlying immunosuppression at this time continues to grow and evolve as the transition period remains an important area of study (Aleri et al. 2016; Drackley 1999; LeBlanc et al. 2006). It has been well described in reviews by Drackley (1999) and LeBlanc et al. (2006) that research into the transition period and periparturient immunosuppression represents enormous potential for the improvement of disease prevention during transition as well as production, health, and performance throughout lactation.

Studies performed on dairy cattle nearly 30 years ago reported impaired neutrophil chemotaxis and antimicrobial functions as well as decreased lymphocyte proliferation during the periparturient period. These early investigations were performed to identify the cause of increased mastitis susceptibility during the periparturient period (Kehrli et al. 1989a, b). Although these results could only begin to implicate immune dysfunction in disease susceptibility around the time of calving (Kehrli et al. 1989a, b), several subsequent studies have confirmed the impairment of immune cell functions at this time (Kimura et al. 2002; Mallard et al. 1998; Nagahata et al. 1992; Wathes et al. 2009). Advances since these early studies have not only helped to confirm previous findings but have also revealed the interconnectivity of multiple systems in relation to immune dysfunction and suppression and in some cases have conflicted previous results.

For example, recent studies aimed at the identification of potential biomarkers to predict major diseases affecting dairy cattle around calving have discovered a general state of immune activation as early as 8 weeks before parturition and well before clinical signs of lameness (Zhang et al. 2015), metritis (Dervishi et al. 2016b), ketosis (Zhang et al. 2016), retained placenta (RP; Dervishi et al., 2016a), and subclinical mastitis (Dervishi et al. 2015). Results of these studies have indicated that innate immunity is activated, rather than suppressed, significantly before the onset of clinical disease signs. Elevated cytokines and APPs were observed in all cases as well as an elevation of serum lactate. While the classical view is that the immune response is suppressed around calving thus leading to an increase in disease incidence the new findings suggest that the onset of disease is associated with an early activation of innate immunity. These contradictory results begin to show that not all functions of innate immunity are suppressed at this time and that our understanding of immunosuppression can greatly benefit from expanding our view out from the reductionist approach to look at the larger picture.

3.4 Factors Contributing to Peripartal Immune Dysfunctions

Ongoing investigations continue to show the intricate connection between the extreme physiological changes that occur during transition and the development of immune dysfunction. One central concept that has emerged is that there is a delicate dynamic between nutrition, hormonal changes, and immunity (Aleri et al. 2016; Ingvartsen and Moyes 2015). A dramatic increase in nutritional requirements to support milk synthesis is the hallmark of the transition period with a threefold increase in demand for glucose, twofold for amino acids, and fivefold for fatty acids (Overton and Waldron 2004). Consequently, the dramatic increase in metabolic activity leads to oxidative stress via increased cellular respiration and β-oxidation of non-esterified fatty acids (NEFA) for energy at the peripheral tissues (Abuelo et al. 2015). Lastly, significant changes in the hormone profile of transition cows occur as specific shifts in reproductive hormones are required for healthy calving and the stressful nature of labor causes glucocorticoid release (Kindahl et al. 2002; Senger 2003). All of these major changes have been observed to alter factors of immunity in dairy cattle and will now be addressed.

3.4.1 Metabolic Stress

The dramatic increase in nutritional demands at the onset of lactation coupled with concurrent appetite suppression nearing calving leads to extreme metabolic stress. Excessive mobilization of adipose tissue occurs to meet the nutritional demands of lactation and results in elevated concentrations of NEFA and ketones, such as β-hydroxybutyric acid (BHBA), in the circulation. Extensive evidence has shown that elevated NEFA and BHBA can cause alterations in immune functions, contributing to immunosuppression (Ingvartsen and Moyes 2015; Sordillo and Raphael 2013). It has been observed that NEFA concentrations are elevated in the serum close to parturition with a peak shortly after calving. Serum NEFA is an indicator of the level of adipose mobilization following calving and is directly linked to an increase in BHBA when NEFA levels exceed what can be fully oxidized by the liver (Seifi et al. 2007). However, this response is variable and depends on several factors such as prepartum body condition score (Busato et al. 2002) and dry period diet (Dann et al. 2006). Although increased NEFA concentration is observed to be normal, the prepartum elevation of NEFA above 0.4 mmol/L close to calving (at roughly 1 week prepartum) is a significant risk factor for several diseases including displaced abomasum (DA) and RP as well as an increased likelihood of culling before 60 DIM (LeBlanc 2010; Mordak and Stewart 2015).

It has been outlined that although there is evidence that NEFA modulate the immune response, the effects remain poorly understood. The association between high prepartum NEFA concentration and an increase in common periparturient diseases as well as an increase in culling during early lactation (Ingvartsen and Moyes

2015; LeBlanc 2010) indicates NEFA is associated with impaired immune function. However, there are conflicting results on the role of NEFA in immunosuppression, possibly due to the variation in effect of individual fatty acids as, in general, unsaturated fatty acids impair the immune response while saturated fatty acids improve it (Ingvartsen and Moyes 2015; Sordillo and Raphael 2013). For example, Ster et al. (2012) observed that elevated NEFA concentrations impaired proliferation and function of peripheral blood mononuclear cells (PBMC) and reduced the oxidative burst of neutrophils. In contrast, a previous study by Scalia et al. (2006) found that phagocytosis-associated oxidative burst activities increased dramatically at high NEFA concentrations while cell viability was reduced. Interestingly, there was no difference in the NEFA mixture utilized by both Ster et al. (2012) and Scalia et al. (2006), suggesting there are other factors in the variable response.

Hyperketonemia, on the other hand, has been widely observed to negatively impact normal immune functions in dairy cattle. The presence of ketone bodies has been shown to significantly inhibit the proliferation of peripheral blood lymphocytes (Sato et al. 1995) as well as that of bovine bone marrow cells (Hoeben et al. 1999). Several studies have shown that ketotic or postpartum levels of BHBA can also inhibit several functions of bovine neutrophils, including chemotaxis (Hillreiner et al. 2016; Suriyasathaporn et al. 2000; Zarrin et al. 2014), phagocytosis (Suriyasathaporn et al. 2000), and antimicrobial functions such as oxidative burst or extracellular traps (Grinberg et al. 2008; Hoeben et al. 1999; Sordillo and Raphael 2013), which are critical components of innate immunity.

3.4.2 Oxidative Stress

Oxidative stress refers to the accumulation of reactive oxygen species (ROS) when their production exceeds the neutralizing capacity of antioxidant mechanisms and has been identified as a significant underlying factor of periparturient immunosuppression (Abuelo et al. 2015; Sordillo et al. 2009). A certain level of ROS production is essential to the immune response as they contribute, most notably, to the oxidative burst which kills pathogens that have been phagocytosed by neutrophils or macrophages, as well as other immune mechanisms. Unfortunately, ROS can also lead to damage of host cells if allowed to accumulate (Sordillo 2013). Oxidative stress at transition is fundamentally linked to excessive lipid mobilization, and thus to metabolic stress. When used as an energy source at the peripheral tissues NEFA increases the production of ROS during β-oxidation (Abuelo et al. 2015; Shi et al. 2015). Oxidative stress can contribute to immunosuppression indirectly by exacerbating the effects of increased NEFA and BHBA concentrations. Evidence suggests that ROS can activate NF-kB leading to release of TNF, a proinflammatory cytokine that increases mitochondrial ROS production and directly stimulates lipolysis while reducing DMI (Sordillo and Raphael 2013). These effects lead to a vicious cycle of ROS production (Abuelo et al. 2015) and the immunosuppressive effects of metabolic stress outlined above.

Further evidence that oxidative stress contributes to immune dysfunction is exhibited in studies of antioxidant supplementation (Sordillo and Aitken 2009; Spears and Weiss 2008) and the beneficial effects they have on transition cow immunity. Vitamin E and Se are both known for their antioxidant properties and their supplementation in dairy cattle has been shown to benefit immune functions including increased phagocytosis, bacterial killing, and oxidative metabolism of neutrophils compared to deficient cows. Additionally, vitamin E has been found to increase macrophage-derived IL-1 and MHC II expression, and mitogen-induced production of IgM by bovine PBMC while Se enhances chemotaxis of neutrophils to the mammary following *E. coli* challenge (Sordillo and Aitken 2009). These results identify oxidative stress as a possible underlying factor of immune dysfunction and indicate potential for the use of antioxidant supplementation to increase disease resistance.

3.4.3 Endocrine Factors

Significant alterations in sex steroid hormones during late gestation are required for healthy calving. Estradiol levels increase rapidly roughly 1 week prior to calving with a peak in the last 3 days and a subsequent drop-off following parturition. Progesterone that is maintained at high levels throughout pregnancy drops during late gestation with a dramatic decrease in the 2 days prior to calving. These extensive changes have been associated with compromised immunity at this time (Lamote et al. 2006; Senger 2003). Chacin et al. (1990) observed the association between progesterone treatment and inhibition of lymphocyte proliferation while a stimulatory effect on IgA secretion into the uterine lumen was observed. Estradiol was observed to impair the migration of neutrophils as well as impair their viability following migration (Lamote et al. 2006; Lamote et al. 2004). Additionally, low concentrations of estradiol inhibited proliferation of granulocyte progenitor cells in vitro (Van Merris et al. 2004). Lastly, a peak in pregnancy-associated glycoprotein, a marker for early pregnancy detection in cattle, has been observed to precede impaired PMN oxidative burst (Dosogne et al. 1999).

An increase in stress-related hormones is also common nearing parturition due to changes in environment around this time as well as the stressful nature of labor itself (Aleri et al. 2016; Kindahl et al. 2002). The association between stress, glucocorticoids, and immunosuppression in various species has been observed over many years (Griffin 1989; Mallard et al. 1998) with recent research only building on our knowledge of the mechanisms by which this occurs. Cortisol and dexamethasone (Dex) downregulate the adhesion molecules L-selectin and CD18 on the surface of bovine neutrophils, which impairs chemotaxis (Burton et al. 1995). Dexamethasone has also been found to impair IFN-γ and IgM production by PBML (Nonnecke et al. 1997), and deplete both T cell populations and NK cells (Maslanka 2014) while hydrocortisone was observed to reduce the growth of both granulocyte and monocyte colonies (Van Merris et al. 2004), further suggesting the role of glucocorticoids in immunosuppression. Epinephrine and norepinephrine are additional stress-related hormones and may contribute to immunosuppression through the stimulation of anti-inflammatory cytokines which suppress cellular immune responses (Aleri et al. 2016). However, given the short-term elevation of stress-hormones around parturition, it is unlikely these effects are the major contributor to periparturient immunosuppression (Ingvartsen and Moyes 2015).

3.4.4 Bacterial Toxins

The role of bacterial toxins in manipulating host defenses has been established in human medicine with little insight to date on the role this may play in immunosuppression of dairy cattle. It is an interesting prospect that there may be external factors that exacerbate immunosuppression. An early study of bacteria responsible for periodontal disease by Van Dyke et al. (1982) found that Gram-negative bacteria were able to inhibit neutrophil chemotaxis by releasing non-chemotactic peptides that bind the FMLP (N-formylmethionyl-leucyl-phenylalanine) receptor, exhibit antagonistic effects, and thus prevent detection of the chemokine gradient by neutrophils. The production of peptides which competed for FMLP receptor binding was observed in most inhibitory organisms evaluated in this study.

In humans, *Staphylococcus aureus* is a commensal bacterium of the nares and skin that is frequently the cause of soft tissue and bloodstream infections. These bacteria have been found to possess mechanisms to evade the innate immune system and prevent the development of adaptive immunity (Thammavongsa et al. 2015). As *S. aureus* is one of the main bacterial species associated with bovine mastitis (Tiwari et al. 2013), these mechanisms are possibly at play in cattle as well. In order to evade immune mechanisms that would normally eliminate bacterial infection, *S. aureus* secretes soluble factors capable of impairing neutrophils. For example, neutrophil chemotaxis is impaired by the secretion of staphylococcal SSL5 and extracellular adherence protein (Eap), thus blocking neutrophil interaction with adhesion molecules and inhibiting extravasation. Additional immune-evading mechanisms of soluble factors produced by *S. aureus* include binding of chemokine receptors with antagonistic proteins, impairment of the complement cascade through blocking or cleavage of proteins, inhibiting IgG opsonization and phagocytosis, degradation of neutrophil extracellular traps (NETs), as well as several others all contributing to inhibition of immune defenses (do Vale et al. 2016; Powers and Bubeck Wardenburg 2014; Thammavongsa et al. 2015).

During studies of uropathogenic bacteria in humans, *Escherichia coli*, another major pathogen associated with bovine mastitis, has also been found to modulate neutrophil chemotaxis (Loughman and Hunstad 2011; Tiwari et al. 2013). Loughman and Hunstad (2011) found that uropathogenic strains of *E. coli* suppressed neutrophil activity. These strains were able to evade neutrophil-mediated killing and reduce the antimicrobial response. Analysis of changes to human PMN gene expression in response to uropathogenic *E. coli* revealed a significant downregulation of genes related to the innate immune response (e.g., proinflammatory cytokines, intermediates of cell-signaling pathways) and chemotaxis (e.g., chemokines, cellular adhesion molecules). These effects are suggested to allow the initial establishment of a bacterial infection by delaying the activities of neutrophils prior to the onset of an inflammatory response (Loughman and Hunstad 2011). The release of Kynurenines by uropathogenic *E. coli* has also been found to inhibit neutrophil chemotaxis. Kynurenines are able to bind the aryl hydrocarbon receptor (AHR) which is translocated into the nucleus to exert mainly immunosuppressive changes to gene transcription (Loughman et al. 2016). Uropathogenic *E. coli* has additionally been found to release $\mathrm{YbcL}_{\mathrm{UTI}}$, a potent inhibitor of neutrophil migration, upon lysis in proximity to bladder epithelial cells. This mechanism was described as an altruistic cooperation whereby the lysis of a small number of bacteria can strongly inhibit neutrophil chemotaxis and thus promote bacterial colonization overall (Lau et al. 2012). These mechanisms have yet to be evaluated in cattle. Given that the bacterial species observed to employ immunosuppressive mechanisms in humans are also commonly associated with bovine mastitis, it would be of interest to evaluate whether the strains infecting the mammary can also exert these effects.

3.5 Effect of Immunosuppression on Health and Productivity

Immunosuppression and the high incidence of disease around parturition are of major concern to dairy producers as both health and productivity are negatively affected. For example, disease and inflammation are associated with decreased milk production (Detilleux et al. 1997; Huzzey et al. 2015; Rajala and Grohn 1998) and impaired fertility (Sheldon et al. 2009; Williams et al. 2008a), which can lead to short-term economic loss and increased involuntary culling (Simenew and Wondu 2013). Unfortunately, in some cases the effects of disease can span subsequent lactations as well, such as the irreversible mammary damage from mastitis reducing subsequent milk production (Mehrzad et al. 2005; Zhao and Lacasse 2008), leading to significant long-term losses. Additionally, impaired immunity poses a concern to calf health in production systems where calves receive maternal colostrum and milk (Mallard et al. 1998).

3.5.1 Disease Susceptibility

Neutrophils are a significant part of the innate immune system and play a critical role in the timely resolution of bacterial infections. The mammary and uterus are highly susceptible to bacterial infection during the periparturient period and this is mainly attributed to neutrophil dysfunction (Mallard et al. 1998; Sheldon and Dobson 2004). Typically, neutrophils migrate from the circulation into the site of infection in response to certain mediators, such as chemoattractants, and subsequently phagocytose and kill bacteria through an oxidative burst (Sordillo and Streicher 2002). As previously discussed, many of these key features are impaired around parturition leading to increased incidence of uterine and mammary infections.

Immunosuppression has additionally been shown to play a significant role in the pathogenesis of RP. For the normal expulsion of the placenta there must be degradation of the cotyledon-caruncle attachment which separates the placental membrane from the maternal tissue. This process requires an immune response against the fetal membranes which is repressed throughout gestation in order to maintain the pregnancy. Reduced recruitment of neutrophils as well as reduced neutrophil oxidative burst activity has been observed in cows with RP (Kimura et al. 2002; LeBlanc 2008; Mordak and Stewart 2015).

3.5.2 Productivity

Periparturient diseases have a lasting impact on the productivity of dairy cattle through lower milk production and fertility. Most early studies of immunosuppression seem to be in connection to increased mastitis susceptibility around calving. This is for good reason as mastitis remains among the most common and costly of all dairy diseases (Cha et al. 2011; Ingvartsen and Moyes 2015) despite years of research. Mastitis negatively impacts profitability through decreased milk production, increased milk discard, premature culling, reduced reproductive performance, and increased costs (Hamadani et al. 2013; Ruegg 2012; Schrick et al. 2001). Reduced milk production represents the primary source of economic loss associated with mastitis, accounting for roughly 70%. A large portion of this occurs from irreversible mammary tissue damage as this lowers productivity even after the infection has been resolved (Zhao and Lacasse 2008).

Increased susceptibility to uterine disease is also of particular concern in the dairy industry as uterine infection is the most common cause of infertility. Some of the mechanisms by which uterine infection contributes to infertility include prolonged luteal phase which prevents ovulation, impaired function of the hypothalamus and pituitary, and impaired steroidogenesis (Williams et al. 2008b). Subsequently, affected animals experience significantly reduced reproductive performance in terms of increased days open, delayed first service, lower conception rates, and increased involuntary culling due to reproductive failure (Gilbert et al. 2005).

3.5.3 Calf Health

Maternal immune status during the periparturient period is critical to calf health. Throughout gestation there is a separation of maternal and fetal blood supplies meaning that calves are born with naïve immunity and require colostrum to absorb maternal antibodies (Godden 2008; Mallard et al. 1998). It follows that the concentration of antibodies in the maternal circulation will be lower in animals with suppressed humoral immune responses and thus would not be able to incorporate higher concentrations of antibodies during colostrogenesis. Colostrum management has been described as the single most important factor for determining the health and survival of dairy calves. High quality colostrum is not only associated with improved health and survival of calves but also several long-term benefits such as improved reproductive performance and increased milk production early in life (Godden 2008). These effects, however, only apply when calves are receiving natural colostrum as supposed to colostrum replacer. Additionally, immunosuppression is closely associated with increased susceptibility to mastitis (Mallard et al. 1998) and it has been found that calves fed high SCC colostrum have lower serum IgG concentrations and compromised health status during the first 42 days of age as well as depressed weaning weight gain (Ferdowsi Nia et al. 2010).

3.6 Systems Veterinary Contribution to Immunosuppression

There is immense potential of systems veterinary to expand and deepen our understanding of periparturient dairy cow immunity. These technologies give us the ability to investigate underlying changes occurring not only within whole tissues but also within specific cells during the onslaught of changes at this time, providing additional information into why these changes occur. Expanding from this increased understanding we are then able to implement strategies to combat these issues, namely selective breeding and correlational studies.

3.6.1 Genome-Wide Association Studies

A new area of study with some groundbreaking results includes genome-wide association studies (GWAS) in animals with varying immune responsiveness (Thompson-Crispi et al. 2014). In combining GWAS with a patented system for identifying cows with superior cell-mediated and antibody-mediated immune responses developed at the University of Guelph, it is possible to determine global genetic differences in high responding cows and identify candidate genes for selective breeding. Several thousand genes regulate immunity in mammals and selection for cattle that are more able to mount an immune response may dramatically increase profitability of dairy farms through reduced disease incidence, improved milk quality, improved calf health, and increased longevity (Godden 2008; Mallard et al. 2015; Ruegg 2012). The success of genetic selection depends on heritability of these candidate genes and some immune response traits used to determine estimated breeding value (EBV) for antibody-mediated and cell-mediated immunity have heritability estimates within a similar range as production traits and better than most reproduction traits. In addition to reducing disease incidence around parturition, evidence indicates that selection for increased immune responsiveness can also improve colostrum quality and reproductive traits such as increased pregnancy rate (Mallard et al. 2015).

Thompson-Crispi et al. (2014) recently performed a genome-wide association study comparing high and low responding Canadian Holstein dairy cows for cell-mediated and antibody-mediated responses. Significant differences were found between high and low responding cows for both types of immune response indicating the potential to identify animals genetically superior for disease resistance. Candidate genes included bovine leukocyte antigen (BoLA) which is analogous with the major histocompatibility complex (MHC), interleukin 17 and its receptor IL17RA, TNF, genes associated with the classical complement pathway, as well as several others (Thompson-Crispi et al. 2014). A genome-wide association study was also performed for clinical mastitis traits in three breeds of dairy cattle utilizing high-density SNPs. Several candidate genes were identified although no specific genes or polymorphisms could be identified as causal factors (Sahana et al. 2014).

3.6.2 Functional Genomics and Insight into Mechanisms of Immunosuppression

Bovine functional genomics is a science that is still being developed but provides the opportunity for researchers to connect variation in genetic expression to differing physiological states, such as during the transition to lactation, or caused by a specific treatment (Pareek et al. 2011). Studies have been performed in cattle to identify changes in genetic expression within immune cells during the transition to lactation. Madsen et al. (2002) compared the genetic expression of bovine neutrophils from primiparous Holstein cows in mid-gestation and at several time points from 14 days prepartum to 7 days postpartum. The most significant finding of this study was the decreased expression of the genes for cytochrome *b*, a critical component of cellular respiration, and ribosomal protein S15, required for normal formation of the ribosomal complex for translation of mRNA to form proteins, which were nearly 50% lower postpartum. Interestingly, a correlation was observed between the significant drop in progesterone at and after parturition and the repression of these genes, possibly providing further evidence of the role of reproductive hormones in immunosuppression. Decreased expression of genes for DNA binding proteins and enzymes of the citric acid cycle was also observed (Madsen et al. 2002). A similar study by Burton et al. (2001) analyzed the effects of parturition on gene expression within blood leukocytes utilizing cDNA microarray analysis. Comparison of leukocyte RNA from a high producing Holstein cow at 14 days prepartum and 6 h postpartum revealed that 18 genes were repressed after calving. Repressed genes included one for the β-chain of MHC class II, 11 associated with cell growth, metabolism and responsiveness including 2 involved in gene transcription, and 6 that have yet to be identified.

Genomics can also provide deeper insight into effects of specific treatments that have previously been observed while the underlying mechanisms remain a mystery. As described previously, studies have been able to show that glucocorticoid treatment impairs certain functions of bovine neutrophils but the underlying mechanisms have so far remained speculative. Burton et al. (2005) investigated the impact of glucocorticoids on genetic expression in neutrophils. Results from this study, interestingly, indicated that glucocorticoids may play a critical role in immune defense and tissue repair during parturition. It was observed that during parturition blood neutrophils increased expression of pro-survival mechanisms while downregulating proapoptotic factors, indicating a shift towards cell survival (Burton et al. 2005). A subsequent study analyzing the effect of Dex treatment on the bovine neutrophil transcriptome further substantiated these findings as gene expression profiling and phenotyping results indicated that Dex delayed neutrophil apoptosis, prolonging cell survival (Weber et al. 2006). Correlations were observed between alterations in gene expression and not only the increase in cortisol but the marked drop in progesterone as well around the time of parturition. Blood cortisol levels were additionally correlated with the significant downregulation of genes for chemotaxis, and regulation of ROS production and redox status of cells between the day of calving and the first day postpartum (Burton et al. 2005).

In combination with the findings by Madsen et al. (2002) previously discussed it has been observed that parturition impairs bactericidal function of neutrophils by impairing cellular respiration which generates ROS needed for the oxidative burst. Burton et al. (2005) have speculated that this may redirect overall neutrophil function at parturition towards tissue remodeling at parturition in favor of breaking down the extracellular matrix and their evaluation of changes to tissue remodeling genes in neutrophils were in support of this hypothesis although heightened tissue degradation activity by neutrophils was not confirmed. This would further support the association of immune dysfunction and RP in cattle as the degradation of cotyledon-caruncle attachment is required for proper placental expulsion, while providing new information that the reduced oxidative burst activity may be a natural shift towards remodeling as supposed to an inherent dysfunction (Burton et al. 2005; Kimura et al. 2002). However, there are conflicting results regarding the role of glucocorticoids in extracellular matrix degradation as the expression of matrix metalloproteinase-9 (MMP-9) which was upregulated during parturition according to the findings by Burton et al. (2005) was downregulated during Dex treatment in a subsequent study (Weber et al. 2006) indicating a need to continue investigations in this area.

Functional genomics has also been used to identify the role of metabolic changes in immunosuppression. A study utilizing models for different levels of negative energy balance (NEB; mild or severe) identified alterations in gene expression in the endometrium of early lactation cows to evaluate the effect of NEB on uterine repair and inflammation. It was hypothesized that failure to adapt to the metabolic changes at parturition would delay uterine involution and promote chronic inflammation, thus possibly contributing to infertility. Animals in a state of severe NEB remained in a state of immune activation longer than those with mild NEB as indicated by elevated levels of antimicrobial genes, notably S100A8, S100A9, and S100A12. Evidence of increased oxidative stress was present as well as indicated by altered expression of genes related to Nrf-2-mediated oxidative stress. Severe NEB was also associated with the significant downregulation of genes for cell proliferation and communication, such as NTRK2, CCNB1, MYB, and NOV. Upregulation of IL-1, a proinflammatory cytokine, and IL-8, the previously described chemotactic factor, was observed in severe NEB cows as well as many other genes for immune factors including cytokines, chemokines and their receptors, adhesion molecules, and interferons. Upregulation of tissue repair genes, such as the aforementioned MMP-9, during severe NEB suggests prolonged tissue remodeling in comparison with mild NEB cows. Overall, results of this study suggested that animals with severe NEB had impaired mechanisms of tissue repair leading to delayed uterine involution and were in a state of chronic inflammation suggesting less efficient bacterial clearance (Wathes et al. 2009). Delayed uterine involution and impaired bacterial clearance can promote the development of uterine infection in cattle and subsequently leads to impaired fertility (Sheldon and Dobson 2004).

An additional application of genomic techniques is to identify differences in gene expression of target tissues during a disease state. For example, Walker et al. (2015) looked at genome-wide DNA methylation and gene expression in the endometrium of dairy cows with subclinical endometritis and these results were further correlated to bacteriology within the uterus. Interestingly, no differences in bacteriology were found between healthy cows and those with endometritis while the latter were in a state of immune activation. Specific findings indicated that endometritis may be caused by an inability to resolve inflammation once contaminating bacteria have been cleared. Additionally, the dysfunction of macrophages was suggested as another possible contributing factor as monocyte chemotactic factors were upregulated in cows with endometritis yet no differences in macrophage counts were observed (Walker et al. 2015). In identifying the overall changes occurring during a disease state we can potentially gain critical insight into the mechanisms underlying these diseases.

Lastly, as previously described it is possible that bacterial toxins are an external factor able to modulate the immune response in cattle. Loughman and Hunstad (2011) utilized microarrays to evaluate global changes in genetic expression of human PMN in response to uropathogenic *E. coli*. The results of this study indicated a significant downregulation of genes for the immune response and chemotaxis. These techniques could similarly be applied in dairy cattle, utilizing pathogenic strains which cause mastitis, to determine if the immunosuppressive effects observed in humans during infection of the urinary tract are also present during infection of the bovine mammary. This represents an interesting new approach to future research of immunosuppression utilizing systems biology techniques.

3.6.3 Application of Proteomics

Proteomics also has clinical applications that benefit research into the health and welfare of dairy cattle. New techniques in mass spectrometry allow for a shotgun proteomics approach that can simultaneously detect, identify, and quantify a complex protein mixture free from the bias of looking only at specific protein families (Lippolis et al. 2006; Yates 2004). A proteomic analysis of neutrophils during the periparturient period and those from Dex-treated cows revealed that protein expression significantly changes around parturition and during Dex treatment. However, although a portion of periparturient immunosuppression has been attributed to the effects of glucocorticoids, it was discovered that not all differences in protein expression were similar between peripartum neutrophils and those treated with Dex. For example, myeloperoxidase, an enzyme that contributes to antimicrobial activity, was downregulated in the membrane of both Dex-treated and periparturient neutrophils while others moved in opposing directions or some differed in only one of the two groups (Lippolis et al. 2006).

A subsequent proteomic study evaluated changes to plasma proteins at calving in animals with or without subclinical mastitis after calving. Upregulation of positive APPs, including Hp and SAA, were observed at parturition for both healthy and mastitic animals indicating a state of immune activation. Although the periparturient changes to these serum proteins were found to be similar between healthy and mastitic cows, results found that cows with subclinical mastitis had a prolonged elevation of APPs compared with healthy cows indicating a failure to resolve the inflammation naturally occurring at this time (Yang et al. 2012). These results may serve as a confirmation of the findings by Wathes et al. (2009) discussed previously where animals that fail to appropriately adapt to transition are in a state of chronic inflammation. Additionally, findings by Burton et al. (2005) also described a similar dysfunction whereby natural shifts, such as that of glucocorticoids possibly shifting the function of neutrophils towards tissue repair, can become problematic and promote disease when prolonged.

3.7 Overall Contribution of Omics Sciences

While approaches to understanding periparturient immunosuppression have classically focused on a reductionist view, the outcome has only come to show that this phenomenon is produced by a multifactorial etiology. Initially, infectious and metabolic diseases were thought to be separate while further investigation proved this is not the case. Evidence of the interconnectivity between nutrition, oxidative stress, endocrine changes, and the immune system has shown that the unsuccessful transition to lactation can significantly impair the immune system leading to increased disease susceptibility. While previous studies have been able to identify the ability of multiple factors to impact immunity, the underlying mechanisms have remained largely unknown. In taking a systems biology approach we have the ability to uncover underlying mechanisms of immunosuppression by investigating global changes in gene and protein expression of specific tissues and cells during transition, treatments, and disease states. Recent studies utilizing omics to investigate periparturient immunosuppression have in some cases identified changes that explain observed effects, and in others found novel or conflicting results. The beauty of these studies is that by taking a broad look at what changes are occurring we may be able to develop highly effective techniques to improve disease susceptibility. In addition, genome-wide association studies are being applied to identify genetic markers of superior immune responsiveness in cattle for application into selective breeding programs. This development shows promise in not only reducing disease incidence but also improving milk production, fertility, and calf health as well. Overall, the systems biology approach to transition immunity will benefit our understanding of transition and greatly increase profitability of the dairy industry.

References

Abuelo A, Hernandez J, Benedito JL, Castillo C (2015) The importance of the oxidative status of dairy cattle in the periparturient period: revisiting antioxidant supplementation. J Anim Physiol Anim Nutr 99(6):1003–1016

Aleri JW, Hine BC, Pyman MF, Mansell PD, Wales WJ, Mallard B, Fisher AD (2016) Periparturient immunosuppression and strategies to improve dairy cow health during the periparturient period. Res Vet Sci 108:8–17

Burton JL, Kehrli ME Jr, Kapil S, Horst RL (1995) Regulation of L-selectin and CD18 on bovine neutrophils by glucocorticoids: effects of cortisol and dexamethasone. J Leukoc Biol 57(2):317–325

Burton JL, Madsen SA, Yao J, Sipkovsky SS, Coussens PM (2001) An immunogenomics approach to understanding periparturient immunosuppression and mastitis susceptibility in dairy cows. Acta Vet Scand 42(3):407–424

Burton JL, Madsen SA, Chang LC, Weber PS, Buckham KR, van Dorp R, Hickey MC, Earley B (2005) Gene expression signatures in neutrophils exposed to glucocorticoids: a new paradigm to help explain "neutrophil dysfunction" in parturient dairy cows. Vet Immunol Immunopathol 105(3–4):197–219

Busato A, Faissle D, Kupfer U, Blum JW (2002) Body condition scores in dairy cows: associations with metabolic and endocrine changes in healthy dairy cows. J Vet Med A Physiol Pathol Clin Med 49(9):455–460

Cha E, Bar D, Hertl JA, Tauer LW, Bennett G, Gonzalez RN, Schukken YH, Welcome FL, Grohn YT (2011) The cost and management of different types of clinical mastitis in dairy cows estimated by dynamic programming. J Dairy Sci 94(9):4476–4487

Chacin MFL, Hansen PJ, Drost M (1990) Effects of stage of the estrous cycle and steroid treatment on uterine immunoglobulin content and polymorphonuclear leukocytes in cattle. Theriogenology 34(6):1169–1184

Coico R, Sunshine G (2015) Immunology: a short course, 7th edn. Wiley, Chichester

Dann HM, Litherland NB, Underwood JP, Bionaz M, D'Angelo A, McFadden JW, Drackley JK (2006) Diets during far-off and close-up dry periods affect periparturient metabolism and lactation in multiparous cows. J Dairy Sci 89(9):3563–3577

Dervishi E, Zhang G, Hailemariam D, Dunn SM, Ametaj BN (2015) Innate immunity and carbohydrate metabolism alterations precede occurrence of subclinical mastitis in transition dairy cows. J Anim Sci Technol 57:46

Dervishi E, Zhang G, Hailemariam D, Dunn SM, Ametaj BN (2016a) Occurrence of retained placenta is preceded by an inflammatory state and alterations of energy metabolism in transition dairy cows. J Anim Sci Biotechnol 7:26

Dervishi E, Zhang G, Hailemariam D, Goldansaz SA, Deng Q, Dunn SM, Ametaj BN (2016b) Alterations in innate immunity reactants and carbohydrate and lipid metabolism precede occurrence of metritis in transition dairy cows. Res Vet Sci 104:30–39

Detilleux JC, Grohn YT, Eicker SW, Quaas RL (1997) Effects of left displaced abomasum on test day milk yields of Holstein cows. J Dairy Sci 80(1):121–126

Dosogne H, Burvenich C, Freeman AE, Kehrli ME Jr, Detilleux JC, Sulon J, Beckers JF, Hoeben D (1999) Pregnancy-associated glycoprotein and decreased polymorphonuclear leukocyte function in early post-partum dairy cows. Vet Immunol Immunopathol 67(1):47–54

Drackley JK (1999) ADSA foundation scholar award. Biology of dairy cows during the transition period: the final frontier? J Dairy Sci 82(11):2259–2273

Ferdowsi Nia E, Nikkhah A, Rahmani HR, Alikhani M, Mohammad Alipour M, Ghorbani GR (2010) Increased colostral somatic cell counts reduce pre-weaning calf immunity, health and growth. J Anim Physiol Anim Nutr 94:628–634

Gilbert RO, Shin ST, Guard CL, Erb HN, Frajblat M (2005) Prevalence of endometritis and its effects on reproductive performance of dairy cows. Theriogenology 64(9):1879–1888

Godden SM (2008) Colostrum management for dairy calves. Vet Clin N Am Food Anim Pract 24(1):19–39

Griffin JF (1989) Stress and immunity: a unifying concept. Vet Immunol Immunopathol 20(3):263–312

Grinberg N, Elazar S, Rosenshine I, Shpigel NY (2008) Beta-hydroxybutyrate abrogates formation of bovine neutrophil extracellular traps and bactericidal activity against mammary pathogenic Escherichia coli. Infect Immun 76(6):2802–2807

Hamadani H, Khan AA, Banday MT, Ashraf I, Handoo N, Shah AB, Hamadani A (2013) Bovine mastitis—a disease of serious concern for dairy farmers. Int J Livest Res 3(1):42–55

Hillreiner M, Flinspach C, Pfaffl MW, Kliem H (2016) Effect of the ketone body Beta-Hydroxybutyrate on the innate defense capability of primary bovine mammary epithelial cells. PLoS One 11(6):e0157774

Hoeben D, Burvenich C, Massart-Leen AM, Lenjou M, Nijs G, Van Bockstaele D, Beckers JF (1999) In vitro effect of ketone bodies, glucocorticosteroids and bovine pregnancy-associated glycoprotein on cultures of bone marrow progenitor cells of cows and calves. Vet Immunol Immunopathol 68(2–4):229–240

Huzzey JM, Mann S, Nydam DV, Grant RJ, Overton TR (2015) Associations of peripartum markers of stress and inflammation with milk yield and reproductive performance in Holstein dairy cows. Prev Vet Med 120(3–4):291–297

Ingvartsen KL, Moyes KM (2015) Factors contributing to immunosuppression in the dairy cow during the periparturient period. Jpn J Vet Res 63(1):S15–S24

Kehrli ME Jr, Nonnecke BJ, Roth JA (1989a) Alterations in bovine lymphocyte function during the periparturient period. Am J Vet Res 50(2):215–220

Kehrli ME Jr, Nonnecke BJ, Roth JA (1989b) Alterations in bovine neutrophil function during the periparturient period. Am J Vet Res 50(2):207–214

Kimura K, Goff JP, Kehrli ME Jr, Reinhardt TA (2002) Decreased neutrophil function as a cause of retained placenta in dairy cattle. J Dairy Sci 85(3):544–550

Kindahl H, Kornmatitsuk B, Konigsson K, Gustafsson H (2002) Endocrine changes in late bovine pregnancy with special emphasis on fetal well-being. Domest Anim Endocrinol 23(1–2):321–328

Lamote I, Meyer E, Duchateau L, Burvenich C (2004) Influence of 17β-estradiol, progesterone, and dexamethasone on diapedesis and viability of bovine blood polymorphonuclear leukocytes. J Dairy Sci 87(10):3340–3349

Lamote I, Meyer E, De Ketelaere A, Duchateau L, Burvenich C (2006) Influence of sex steroids on the viability and CD11b, CD18 and CD47 expression of blood neutrophils from dairy cows in the last month of gestation. Vet Res 37(1):61–74

Lau ME, Loughman JA, Hunstad DA (2012) YbcL of uropathogenic Escherichia coli suppresses transepithelial neutrophil migration. Infect Immun 80(12):4123–4132

LeBlanc SJ (2008) Postpartum uterine disease and dairy herd reproductive performance: a review. Vet J 176(1):102–114

LeBlanc S (2010) Monitoring metabolic health of dairy cattle in the transition period. J Reprod Dev 56(35):S29–S35

LeBlanc SJ, Lissemore KD, Kelton DF, Duffield TF, Leslie KE (2006) Major advances in disease prevention in dairy cattle. J Dairy Sci 89(4):1267–1279

Lippolis JD, Peterson-Burch BD, Reinhardt TA (2006) Differential expression analysis of proteins from neutrophils in the periparturient period and neutrophils from dexamethasone-treated dairy cows. Vet Immunol Immunopathol 111(3–4):149–164

Loughman JA, Hunstad DA (2011) Attenuation of human neutrophil migration and function by uropathogenic bacteria. Microbes Infect 13(6):555–565

Loughman JA, Yarbrough ML, Tiemann KM, Hunstad DA (2016) Local generation of Kynurenines mediates inhibition of neutrophil Chemotaxis by Uropathogenic Escherichia coli. Infect Immun 84(4):1176–1183

Madsen SA, Weber PS, Burton JL (2002) Altered expression of cellular genes in neutrophils of periparturient dairy cows. Vet Immunol Immunopathol 86(3–4):159–175

Mallard BA, Dekkers JC, Ireland MJ, Leslie KE, Sharif S, Lacey Vankampen C, Wagter L, Wilkie BN (1998) Alteration in immune responsiveness during the peripartum period and its ramification on dairy cow and calf health. J Dairy Sci 81(2):585–595

Mallard BA, Emam M, Paibomesai M, Thompson-Crispi K, Wagter-Lesperance L (2015) Genetic selection of cattle for improved immunity and health. Jpn J Vet Res 63(1):S37–S44

Maslanka T (2014) Effect of dexamethasone and meloxicam on counts of selected T lymphocyte subpopulations and NK cells in cattle - in vivo investigations. Res Vet Sci 96(2):338–346

Mehrzad J, Desrosiers C, Lauzon K, Robitaille G, Zhao X, Lacasse P (2005) Proteases involved in mammary tissue damage during endotoxin-induced mastitis in dairy cows. J Dairy Sci 88(1):211–222

Mordak R, Stewart PA (2015) Periparturient stress and immune suppression as a potential cause of retained placenta in highly productive dairy cows: examples of prevention. Acta Vet Scand 57(84):015–0175

Nagahata H, Ogawa A, Sanada Y, Noda H, Yamamoto S (1992) Peripartum changes in antibody producing capability of lymphocytes from dairy cows. Vet Q 14(1):39–40

Nonnecke BJ, Burton JL, Kehrli ME Jr (1997) Associations between function and composition of blood mononuclear leukocyte populations from Holstein bulls treated with dexamethasone. J Dairy Sci 80(10):2403–2410

Overton TR, Waldron MR (2004) Nutritional management of transition dairy cows: strategies to optimize metabolic health. J Dairy Sci 87:E105–E119

Pareek CS, Smoczynski R, Pierzchala M, Czarnik U, Tretyn A (2011) From genotype to phenotype in bovine functional genomics. Brief Funct Genomics 10(3):165–171

Powers ME, Bubeck Wardenburg J (2014) Igniting the fire: Staphylococcus aureus virulence factors in the pathogenesis of sepsis. PLoS Pathog 10(2):e1003871

Rajala PJ, Grohn YT (1998) Effects of dystocia, retained placenta, and metritis on milk yield in diary cows. J Dairy Sci 81(12):3172–3181

Ruegg PL (2012) New perspectives in udder health management. Vet Clin North Am Food Anim Pract 28(2):149–163

Sahana G, Guldbrandtsen B, Thomsen B, Holm LE, Panitz F, Brondum RF, Bendixen C, Lund MS (2014) Genome-wide association study using high-density single nucleotide polymorphism arrays and whole-genome sequences for clinical mastitis traits in dairy cattle. J Dairy Sci 97(11):7258–7275

Sato S, Suzuki T, Okada K (1995) Suppression of mitogenic response of bovine peripheral blood lymphocytes by ketone bodies. J Vet Med Sci 57(1):183–185

Scalia D, Lacetera N, Bernabucci U, Demeyere K, Duchateau L, Burvenich C (2006) In vitro effects of nonesterified fatty acids on bovine neutrophils oxidative burst and viability. J Dairy Sci 89(1):147–154

Schrick FN, Hockett ME, Saxton AM, Lewis MJ, Dowlen HH, Oliver SP (2001) Influence of subclinical mastitis during early lactation on reproductive parameters. J Dairy Sci 84(6):1407–1412

Seifi HA, Gorji-Dooz M, Mohri M, Dalir-Naghadeh B, Farzaneh N (2007) Variations of energy-related biochemical metabolites during transition period in dairy cows. Comp Clin Pathol 16(4):253–258

Senger PL (2003) Pathways to pregnancy and parturition, 2nd edn. Current Conceptions, Pullman

Sheldon IM, Dobson H (2004) Postpartum uterine health in cattle. Anim Reprod Sci 83:295–306

Sheldon IM, Price SB, Cronin J, Gilbert RO, Gadsby JE (2009) Mechanisms of infertility associated with clinical and subclinical endometritis in high producing dairy cattle. Reprod Domest Anim 3:1–9

Shi X, Li D, Deng Q, Li Y, Sun G, Yuan X, Song Y, Wang Z, Li X, Liu G (2015) NEFAs activate the oxidative stress-mediated NF-kappaB signaling pathway to induce inflammatory response in calf hepatocytes. J Steroid Biochem Mol Biol 145:103–112

Simenew K, Wondu M (2013) Transition period and immunosuppression: critical period of dairy cattle reproduction. Int J Anim Vet Adv 5(2):44–57

Sordillo LM (2013) Selenium-dependent regulation of oxidative stress and immunity in periparturient dairy cattle. Vet Med Int 154045(10):14

Sordillo LM, Aitken SL (2009) Impact of oxidative stress on the health and immune function of dairy cattle. Vet Immunol Immunopathol 128(1–3):104–109

Sordillo LM, Raphael W (2013) Significance of metabolic stress, lipid mobilization, and inflammation on transition cow disorders. Vet Clin North Am Food Anim Pract 29(2):267–278

Sordillo LM, Streicher KL (2002) Mammary gland immunity and mastitis susceptibility. J Mammary Gland Biol Neoplasia 7(2):135–146

Sordillo LM, Contreras GA, Aitken SL (2009) Metabolic factors affecting the inflammatory response of periparturient dairy cows. Anim Health Res Rev 10(1):53–63

Spears JW, Weiss WP (2008) Role of antioxidants and trace elements in health and immunity of transition dairy cows. Vet J 176(1):70–76

Ster C, Loiselle MC, Lacasse P (2012) Effect of postcalving serum nonesterified fatty acids concentration on the functionality of bovine immune cells. J Dairy Sci 95(2):708–717

Suriyasathaporn W, Heuer C, Noordhuizen-Stassen EN, Schukken YH (2000) Hyperketonemia and the impairment of udder defense: a review. Vet Res 31(4):397–412

Thammavongsa V, Kim HK, Missiakas D, Schneewind O (2015) Staphylococcal manipulation of host immune responses. Nat Rev Microbiol 13(9):529–543

Thompson-Crispi KA, Sargolzaei M, Ventura R, Abo-Ismail M, Miglior F, Schenkel F, Mallard BA (2014) A genome-wide association study of immune response traits in Canadian Holstein cattle. BMC Genomics 15(559):1471–2164

Tiwari JG, Babra C, Tiwari HK, Williams V, Wet SD, Gibson J, Paxman A, Morgan E, Costantino P, Sunagar R, Isloor S, Mukkur T (2013) Trends in therapeutic and prevention strategies for management of bovine mastitis: an overview. J Vaccines Vaccin 4(176):2

do Vale A, Cabanes D, Sousa S (2016) Bacterial toxins as pathogen weapons against phagocytes. Front Microbiol 7:42

Van Dyke TE, Bartholomew E, Genco RJ, Slots J, Levine MJ (1982) Inhibition of neutrophil chemotaxis by soluble bacterial products. J Periodontol 53(8):502–508

Van Merris V, Meyer E, Duchateau L, Burvenich C (2004) Differential effects of steroids and retinoids on bovine myelopoiesis in vitro. J Dairy Sci 87(5):1188–1195

Walker CG, Meier S, Hussein H, McDougall S, Burke CR, Roche JR, Mitchell MD (2015) Modulation of the immune system during postpartum uterine inflammation. Physiol Genomics 47(4):89–101

Wathes DC, Cheng Z, Chowdhury W, Fenwick MA, Fitzpatrick R, Morris DG, Patton J, Murphy JJ (2009) Negative energy balance alters global gene expression and immune responses in the uterus of postpartum dairy cows. Physiol Genomics 39(1):1–13

Weber PS, Madsen-Bouterse SA, Rosa GJ, Sipkovsky S, Ren X, Almeida PE, Kruska R, Halgren RG, Barrick JL, Burton JL (2006) Analysis of the bovine neutrophil transcriptome during glucocorticoid treatment. Physiol Genomics 28(1):97–112

Williams AE (2012) Immunology: mucosal and body surface defences. Wiley, Chichester

Williams EJ, Herath S, England GC, Dobson H, Bryant CE, Sheldon IM (2008a) Effect of *Escherichia coli* infection of the bovine uterus from the whole animal to the cell. Animal 2(8):1153–1157

Williams EJ, Sibley K, Miller AN, Lane EA, Fishwick J, Nash DM, Herath S, England GC, Dobson H, Sheldon IM (2008b) The effect of *Escherichia coli* lipopolysaccharide and tumour necrosis factor alpha on ovarian function. Am J Reprod Immunol 60(5):462–473

Yang YX, Wang JQ, Bu DP, Li SS, Yuan LY, Yang JH, Sun P (2012) Comparative proteomics analysis of plasma proteins during the transition period in dairy cows with or without subclinical mastitis after calving. Czech J Anim Sci 57(10):481–489

Yates JR 3rd (2004) Mass spectral analysis in proteomics. Annu Rev Biophys Biomol Struct 33:297–316

Zarrin M, Wellnitz O, van Dorland HA, Bruckmaier RM (2014) Induced hyperketonemia affects the mammary immune response during lipopolysaccharide challenge in dairy cows. J Dairy Sci 97(1):330–339

Zhang G, Hailemariam D, Dervishi E, Deng Q, Goldansaz SA, Dunn SM, Ametaj BN (2015) Alterations of innate immunity reactants in transition dairy cows before clinical signs of lameness. Animals 5(3):717–747

Zhang G, Hailemariam D, Dervishi E, Goldansaz SA, Deng Q, Dunn SM, Ametaj BN (2016) Dairy cows affected by ketosis show alterations in innate immunity and lipid and carbohydrate metabolism during the dry off period and postpartum. Res Vet Sci 107:246–256

Zhao X, Lacasse P (2008) Mammary tissue damage during bovine mastitis: causes and control. J Anim Sci 86(13 Suppl):57–65

Systems Biology and Ruminal Acidosis

4

Morteza H. Ghaffari, Ehsan Khafipour, and Michael A. Steele

Abstract

Dairy cattle are commonly transitioned to a diet that is rich in rapidly fermentable carbohydrates (high-grain) to increase milk production in early lactation. When the shift in diet is too abrupt or the grain level too high, the rate of ruminal fermentation exceeds the rate of ruminal absorption and buffering, making the cow susceptible to ruminal acidosis. Most investigations of ruminal acidosis have taken a reductionist approach and focused on the impact on the rumen without considering other organs. However, the impact of grain-induced ruminal acidosis involves a whole animal inflammatory response and it is becoming increasingly evident that other gut compartments, namely the lower gut and visceral tissue such as the liver, also play an important role in the etiology of whole animal, multi-organ inflammatory response. Over the past decade, characterizing the gastrointestinal tract and whole animal inflammatory response to grain-induced ruminal acidosis has been initiated using a systems biology approach. To accomplish this, combinations of high-throughput omics-data (i.e., genomics, metabolomics, transcriptomics, and proteomics) have produced unique and meaningful advances in our understanding of the etiology of ruminal acidosis. This chapter will focus on the application of systems biology relating to ruminal acidosis, which provides meaningful insight into the ruminal microbial ecology, metabolome, and host gene expression changes during ruminal acidosis.

M.H. Ghaffari • M.A. Steele, Ph.D. (✉)
Department of Agricultural, Food and Nutritional Science, University of Alberta, Edmonton, AB, Canada
e-mail: morteza1@ualberta.ca; masteele@ualberta.ca

E. Khafipour, Ph.D.
Departments of Animal Science and Medical Microbiology, University of Manitoba, Winnipeg, MB, Canada
e-mail: ehsan.khafipour@umanitoba.ca

B.N. Ametaj (ed.), *Periparturient Diseases of Dairy Cows*,
DOI 10.1007/978-3-319-43033-1_4

4.1 Introduction

Intensive dairy production systems rely on large amounts of grain in diets to increase energy intake to support high levels of milk production. However, feeding excessive amounts of rapidly fermentable dietary carbohydrate to ruminants causes a shift in the rumen microbiota which leads to an accumulation of short-chain fatty acids (SCFA) and a depression of ruminal pH. These changes lead to the digestive disorder termed ruminal acidosis (Khafipour et al. 2009a). When ruminal pH drops to 5.6 or below, the microbial populations in the rumen shift toward the production of lactic acid, which further decreases rumen pH. Lactic acid has a much lower pKa compared to volatile fatty acids (pKa 3.9 vs.4.9) and it is less protonated compared to SCFA at pH 5.0. As a result, lactic acid accumulates in the rumen and leads to a downward spiral in ruminal pH (Nagaraja and Titgemeyer 2007).

The subacute form of ruminal acidosis (SARA) is defined as prolonged periods of ruminal pH depression below 5.6 for at least 3 h per day (Gozho et al. 2005) and is most commonly found in dairy cattle. It has been estimated that 20% of cows in early lactation suffer from SARA (Plaizier et al. 2008). The disorder is associated with a decrease in dry matter intake (Plaizier et al. 2008; Gozho et al. 2005), ruminal fiber digestion (Plaizier et al. 2001), and milk fat percentage (Khafipour et al. 2009b). It has also been connected to changes in richness and diversity of ruminal microbial populations, alterations in ruminal biohydrogenation of unsaturated fatty acids, and milk fatty acid profiles (Kleen et al. 2009; Colman et al. 2013). Moreover, it compromises the health of the animal by causing diarrhea, gastrointestinal damage, liver abscesses, and a whole animal inflammatory response (Plaizier et al. 2008; Emmanuel et al. 2008). Such decreases in production and health make SARA—a topic of critical importance in dairy cow nutrition.

Cows are at greater risk of ruminal acidosis in early lactation because their diet switches from a high-forage ration (dry cow) to high-concentrate ration (fresh cow) over a short period (often overnight). The capacity of absorption of SCFA across the rumen is also limited during this period as ruminal papillae need 2–3 weeks to increase in size to handle the SCFA load from a high-concentrate diet (Dieho et al. 2016). The rapid accumulation of acid alters the activity and abundance of many bacterial species within the rumen, such that cellulolytic bacteria decline and acid tolerant bacteria proliferate (Khafipour et al. 2009c). There are also prominent alterations in morphological and histological characteristics of the rumen epithelium (RE) during ruminal acidosis, which strongly suggest impaired barrier function, and, in turn, compromised cow health (Steele et al. 2011a, b, 2016).

Most studies related to SARA have focused primarily on the rumen, and characterizing the changes in ruminal fermentation and how they related to cow performance and behavior (Plazier et al. 2008). However, the disorder of SARA involves several biological systems that are dynamic. By taking a reductionist approach and focusing on a single organ such as the rumen, we may be ignoring larger impairments occurring in other organs that may equally contribute to the state of depressed health. A novel approach for characterizing these changes is Systems Biology—an emerging interdisciplinary research field combining biology, bioinformatics, statistics, mathematics, and computational science to obtain, integrate, and analyze complex data sets from multiple experimental sources in order to arrive at a better understanding of the underlying biology (Woelders et al. 2011; Loor et al. 2013). Over the past decade, new techniques related to systems biology involving whole genome, transcriptome, proteome, and metabolome analysis have been applied in ruminal acidosis experiments and produced meaningful advances which have helped researchers gain greater insight into the mechanisms between different systems in the dairy cow during SARA. The goal of this chapter is to summarize the new developments in ruminal microbial ecology, metabolome, and host gene expression changes during SARA arising from the use of high-throughput technologies.

4.2 Ruminal Acidosis and the Rumen Microbiota

The earliest work describing the microbial etiology of ruminal acidosis characterized the ruminal changes associated with dietary shift from a high-forage to a high-grain diet through sets of culture-based experiments (Hungate et al. 1952; Hungate 1966). Some significant findings were derived from the culture-based experiments such as how imbalance between lactate producing (*S. bovis*) and lactate utilizing bacteria (*Magasphaera elsdenii*) in the rumen might play a role in the onset of acidosis (Russell et al. 1979, 1981). Although culture-based techniques such as roll tubes or continuous culture systems were useful for studying specific microbes, they were underestimating the diversity of the microbial community and the interactions between microorganisms in their natural environment (McSweeney et al. 2007). This limitation cannot be fully addressed using quantitative PCR techniques as they were targeting specific microorganisms within the community and not the community as a whole (Tajima et al. 2000; Khafipour et al. 2016). In the early 2000s, a range of DNA fragmentation techniques, such as Denaturing Gradient Gel Electrophoresis (DGGE), Temperature Gradient Gel Electrophoresis (TGGE), Ribosomal Intergenic Spacer Analysis (RISA), Automated Ribosomal Intergenic Spacer Analysis (ARISA), Restriction Fragment Length Polymorphism (RFLP), and Terminal Restriction Fragment Length Polymorphism (TRFLP), became available for studying microbial diversity.

One of the first studies investigating the bacterial community structure during SARA using TRFLP was described by Khafipour et al. (2009c), during which authors demonstrated that grain- and alfalfa pellet-induced SARA challenges differentially impact rumen microbial community and animal health regardless of similarity in rumen pH, SCFA and free lipopolysaccharide (LPS; endotoxin) concentrations. Authors reported that both dietary challenge models reduced richness, evenness, and diversity of rumen microbial community, hence transforming these microbiotas into a less functional state. In particular, mild and severe grain-induced SARA challenges resulted in a significant decline in members of gram-negative *Bacteroidetes* phylum (i.e., *Prevotella albensis, Prevotella brevis*, and *Prevotella ruminicola*); however, the magnitude of depression was smaller in mild compared to severe grain-induced SARA groups. In contrast, SARA induced by feeding alfalfa pellets maintained a greater proportion of *Bacteroidetes*, in particular, *P. albensis* and *P. ruminicola,* compared with mild or severe grain-induced SARA groups. Such differential effects on the composition and functionality of the rumen microbiome may provide the opportunity for pathogenic and opportunistic members of the community to proliferate in the rumen under one challenge condition but not the other. For example, high-grain feeding that induced severe acidosis was associated with increases in populations of pathogenic *E. coli* and *Clostridium perfringens* in the rumen (Khafipour et al. 2009c; Plaizier et al. 2016). This was an important discovery, for although the outcome of SARA in terms of rumen fermentation conditions and free rumen LPS content were the same, the rumen bacterial communities were different and only grain-induced SARA provoked an inflammatory response.

Since the mid-2000s, the rapid advances in sequencing technologies have offered low-cost molecular-based techniques for investigating microbial communities as a whole. These new techniques are either based on high-throughput sequencing of hypervariable regions of highly conserved and universal 16S rRNA genes for bacterial and archaeal communities (Woese and Fox 1977; Pace et al. 1985) and 18S or internal transcribed spacer (ITS) region of rRNA gene for fungal and protozoal communities (Firkins and Yu 2015). Broad sequencing of total DNA (metagenomics) or RNA (metatranscriptomics) has also allowed the functional capacity of the microbiome to be more comprehensively studied (Desai et al. 2012). These methodologies have become commonly employed in recent studies to characterize microbial communities in the dairy cow's gastrointestinal tract to determine changes in the microbial community composition and function under high-grain feeding.

When microbial communities are described using omics methodologies, where microorganisms are not directly evaluated, the term "operational taxonomic units" (OTUs) must be used instead of "species" (Khafipour et al. 2016). An OTU is a cluster of sequence reads with a given similarity that is expected to be assigned to a taxonomical level (e.g., sequences with a 97% similarity approximately correspond to species) (Khafipour et al. 2016). In a recent comprehensive study of the rumen microbiome, 742 samples of foregut (reticulum, rumen, and omasum) microbial communities were collected from 32 species across 35 countries from different geographical regions (Henderson et al. 2015). The results from that study showed that the genera *Prevotella* (phylum *Bateroidetes*), followed by *Butyrivibrio* and *Ruminococcus* (both from phylum *Firmicutes*), are the most relatively abundant taxa and are most likely to make up the core microbiome of all ruminants.

Several recent studies using high-throughput sequencing have provided more insight into the changes in microbial community structure during acidosis (summarized in Table 4.1). In general, the results consistently show that excessive grain feeding decreases microbiota richness and diversity in the rumen (Khafipour et al. 2011; Petri et al. 2012; Mao et al. 2013). While the majority of this research has been conducted on rumen fluid samples to assess the rumen microbiota, several recent studies have characterized the microbial composition of rumen solids and those attached to the RE (rumen epimural) (Petri et al. 2013a; McCann et al. 2016).

Table 4.1 Summary of bovine rumen microbiome after animals were exposed to an acidotic challenge

Title	Sequencing platform	Major finding	Reference
"Rumen microbial population dynamics during adaptation to a high-grain diet"	ABI 3700	– Feeding high-grain diets increased *M. elsdenii, S. bovis, S. ruminantium,* and *P. bryantii* while decreased *fibrolytic* bacteria	Fernando et al. (2010)
"Microbial ecology of the rumen evaluated by 454 GS FLX pyrosequencing is affected by starch and oil supplementation of diets"	454 GS-FLX	– High starch treatment significantly increased the relative abundance of *Olsenella, Barnesiella,* and *Oribacterium* genera, but decreased the relative abundances of *Butyrivibrio-Pseudobutyrivibrio* and *Rikenellaceae_RC9*	Zened et al. (2013)
"Impact of subacute ruminal acidosis (SARA) adaptation of rumen microbiota in dairy cattle using pyrosequencing"	454 GS-FLX	– Feeding grain-induced SARA increased the abundances of *Bifidobacterium* and unclassified *Clostridiales,* and decreased the abundances of *Prevotella, Acinetobacter, Treponema, Anaeroplasma, Papillibacter,* and unclassified *Lentisphaerae* compared with the control group	Mao et al. (2013)
"Characterization of the core rumen microbiome in cattle during transition from forage to concentrate as well as during and after an acidotic challenge"	454 GS-FLX	– The relative abundances of *Prevotella, Streptococcus, Lactobacillus,* and *Acetitomaculum* increased in rumen of heifers suffering from clinical vs. subclinical acidosis	Petri et al. (2013b)
"Changes in the rumen epimural bacterial diversity of beef cattle as affected by diet and induced ruminal acidosis"	454 GS-FLX	– Relative abundances of *Atopobium, Desulfocurvus, Fervidicola, Lactobacillus,* and *Olsenella* increased in the high-grain diet compared with the forage diet during acidosis. In addition, *Solobacterium, Atopobium, cc142, RC39, Succiniclasticum, Lactobacillus, Sharpea, Olsenella, and Syntrophococcus* were particularly prevalent during acidosis	Petri et al. (2013a)
"Impact of subacute ruminal acidosis on the diversity of liquid and solid-associated bacteria in the rumen of goats"	454 GS-FLX	– Pyrosequencing analysis showed that SARA indication increased percentage of *Firmicutes* and decreased the proportion of phylum *Bacteroidetes* in the liquid and solid fractions	Huo et al. (2014)

(continued)

Table 4.1 (continued)

Title	Sequencing platform	Major finding	Reference
"High-grain feeding causes strong shifts in ruminal epithelial bacterial community and expression of Toll-like receptor genes in goats"	454 GS-FLX	– At the genus level, feeding high-grain diets increased the relative abundance of the relative abundance of taxa *Butyrivibrio*, unclassified *Clostridiales*, *Mogibacterium*, unclassified *Anaerolineaceae*, *Succiniclasticum*, and decreased the proportion of unclassified *Ruminococcaceae*, unclassified *Rikenellaceae*, *Howardella*, unclassified *Erysipelotrichaceae*, and unclassified *Neisseriaceae* in goats	Liu et al. (2015)
"Microbiome–metabolome analysis reveals unhealthy alterations in the composition and metabolism of ruminal microbiota with increasing dietary grain in a goat model"	454 GS-FLX	– As the proportion of grain increased, the percentage of *Succiniclasticum, Lysinibacillus, Prevotella, Thalassospira,* and *Papillibacter,* as well as some unclassified bacteria, including unclassified *Ruminococcaceae* unclassified *Bacteroidales,* and unclassified *Prevotellaceae,* linearly decreased. Conversely, the proportions of *Acetitomaculum, Butyrivibrio, Mogibacterium,* and unclassified *Anaerolineaceae* linearly increased with the increase in grain proportion	Mao et al. (2016)
"Induction of subacute ruminal acidosis affects the ruminal microbiome and epithelium"	MiSeq Illumina	– SARA induction on relative abundances of *Firmicutes* decreased and *Bacteroidetes* increased in the solid fraction; however, the relative abundance of *Bacteroidetes*, representing more than 60% of the sequences, tended to increase within the liquid fraction	McCann et al. (2016)

Table 4.1 (continued)

Title	Sequencing platform	Major finding	Reference
"Metagenomic analysis of rumen microbial population in dairy heifers fed a high-grain diet supplemented with dicarboxylic acids or polyphenols"	Illumina sequencing	"Polyphenols and organic acid significantly increased the family *Christenenellaceae* and decreased *Prevotella brevis*compared to control in dairy heifers fed high-grain diets. Polyphenol supplementation increased the abundance of many taxa belonging to *Tenericutes*, *Firmicutes*, and *Bacteroidetes* phyla due to a potential antimicrobial activity of flavonoids"	De Nardi et al. (2016)
"Grain-rich diets altered the colonic fermentation and mucosa associated bacterial communities and induced mucosal injuries in rumen bacterial communities can be acclimated faster to high-concentrate diets than currently implemented feedlot programs goats"	Illumina MiSeq sequencing	"Feeding high-grain diets to goat increased the abundance of genus *Blautia* and decreased in the abundance of genera *Bacillus*, *Enterococcus*, and *Lactococcus* in the colonic mucosal bacterial communities"	Ye et al. (2016)

The rumen epimural microbiome is different from rumen fluid and the particle-associated microbiome (Li et al. 2012). Moreover, they have specific functions, such as recycling epithelial tissue, oxygen scavenging, and hydrolysis of urea (Petri et al. 2013a). Petri et al. (2013b) suggest that the core epimural rumen microbiome in cattle is probably stable during acidotic challenges and that alterations in rumen microbiota community structure appear to quickly recover during the recovery period.

In dairy cattle, the main source of free LPS is the gastrointestinal (GI) tract, which contains a high proportion of gram-negative bacteria (Eckel and Ametaj 2016). Members of phylum *Proteobacteria* together with the phylum *Bacteroidetes* and *Fibrobacteres* are major contributors to the free LPS pool in the rumen digesta. These endotoxins are part of the outer membrane of the gram-negative bacterial cell wall, which in their free form act as immunogenic compounds (Hurley 1995). They are extensively shed during the logarithmic and stationary phases of bacterial growth and also released following cell disintegration and lysis (Nagaraja et al. 1978a, b; Hurley 1995; Plaizier et al. 2012). Previous reports indicated that the LPS content of the rumen digesta increase during high-grain feeding and is partly due to a release of LPS during the logarithmic growth phase, which results in a higher absolute number of gram-negative bacteria in the rumen (Plaizier et al. 2008, 2012). This increase, however, might not necessarily translate into a higher proportion of these bacteria in the community. For example, Li et al. (2012) reported that large increases in grain feeding increased the abundance of *Proteobacteria*, whereas Khafipour et al. (2009b) and Plaizier et al. (2016) did not observe any such effect. Regardless, it is important to keep in mind that LPS potency varies among gram-negative species, and therefore subsequent proinflammatory effects of LPS may differ. For instance, in the studies of Khafipour et al. (2009a, b, c), rumen LPS content during a grain-based SARA challenge was close to that of alfalfa pellet-induced SARA, but rumen *E. coli* numbers were significantly higher during the grain-based SARA (Khafipour et al. 2011). Since *E. coli* LPS is substantially more potent than major gram-negative bacteria in the rumen, that might have contributed to the differences in the inflammatory response observed between the two challenge models.

High-throughput approaches to studying the microbiome have undoubtedly improved our understanding of how the rumen community structures shift during acute acidosis and SARA. Still, our understanding of the functionality of the rumen microbiome as a whole requires further research. Application of systems biology tools, such as metagenomics, metatranscriptomics, and metabolomics, will provide more insights into the functional capacity of the microbiome and its interactions with the host.

4.3 Metabolomics as a Novel Approach to Understanding Ruminal Acidosis

The key changes in ruminal metabolites during a grain-induced SARA are well documented and consist of an increase in total ruminal SCFA, namely propionate and butyrate (Lettat et al. 2010). For the past two decades, researchers' attention has been focused on the changes in SCFA since they represent the metabolites at the largest concentrations in the rumen. With this approach, we neglect the hundreds of other metabolites that exist in the rumen at lower concentrations that may have bioactive properties of relevance to SARA. With the advancement of new techniques in the field of metabolomics, it is now possible to characterize hundreds of metabolites in a sample simultaneously, rather than just SCFA. A more detailed characterization of the ruminal metabolome using metabolomics could improve our understanding of microbiota–metabolome–host interactions during SARA.

The methodologies to study the metabolome are predominately based on two platforms: nuclear magnetic resonance (NMR) spectroscopy and gas chromatography-mass spectroscopy (GC-MS), allowing for the detection of hundreds of metabolites in parallel. The first ruminal acidosis metabolomics study employed NMR and GC-MS metabolomics analysis (Ametaj et al. 2010). In this study, it was found that increasing the grain in the total diet by 30 and 45% increased the concentration of several rumen metabolites including methylamines, N-nitroso-dimethylamine, dimethylamine, glucose, alanine, maltose, uracil, propionate, fumarate, butyrate, valerate, xanthine, ethanol, and phenylacetate, valine, leucine, lysine, nicotinate, glycerol, as well as phenyl acetyl glycine, but decreased the level of phosphatidylcholines and 3-phenylpropionate in the rumen of cows. In another study by Saleem et al. (2012), rumen fluid was profiled based on a combination of NMR and GC-MS spectroscopy, which identified and quantified a total of 93 metabolites in the rumen of dairy cows fed high-grain diets. Data from that study showed that feeding high-grain diets (>30% replacement of standard lactating cow diet) to dairy cows resulted in increased ruminal fluid biogenic amines (i.e., histamine, tyramine and tryptamine, cadaverine, and putrescine) in ruminal fluid derived from the decarboxylation of amino acids (arginine, lysine, and arginine/ornithine) by certain types of rumen bacteria (e.g., carbohydrate fermenting bacteria such as *Lactobacillus* spp.). At low concentrations, biogenic amines are required for the normal pattern of cell growth and differentiation (Bardócz et al. 1995), but they may cause a toxic effect for livestock in large quantities (i.e., 1.4 g per day) resulting in cell death (Fusi et al. 2004). Although there are limited studies in metabolomics of rumen fluid during acidosis, it has developed more hypotheses and some of these metabolites, albeit in lower concentrations, may change during SARA and may signal to the dairy cow to change gastrointestinal function.

4.4 Transcriptomic and Proteomic Analysis of Ruminal Tissue in Animals Fed High-Grain Diets

Recent advances in transcriptomics and proteomics techniques have provided researchers the ability to investigate a vast number of RNA transcripts and protein abundances simultaneously in a biological sample. In review, transcriptomics is the measurement and study of RNA transcripts (i.e., mRNAs, miRNA, noncoding RNAs, and small nuclear RNAs) and provides the signature of what genes are responsive to grain-induced SARA. Quantitative real-time PCR (qRT-PCR), microarrays, and RNA-sequencing are three commonly used methods for characterizing the transcriptome. On the other hand, proteomics provides information about protein abundance and protein posttranslational modifications which is of importance to their biological activity. The laboratory techniques of proteomics include various gel-based and mass-spec-based techniques that allow high-throughput analyses for proteome profiling. Changes in gene expression patterns do not necessarily reflect by similar changes in corresponding protein expression because of posttranslational modifications and variation in gene regulation of translation (Shebl et al. 2010).

To date, functional genomic characterization of cows with SARA has focused on RNA expression of rumen tissue with very few research groups investigating protein abundances with high-throughput technologies. Nevertheless, our understanding of how metabolism, growth, and barrier function of rumen tissue during a grain challenge has improved using these functional genomic approaches. In this section, the main developments in our understanding of the metabolism, growth, and barrier function of the RE during an acidosis are summarized. The focus will be on findings over the past 5 years from high-throughput approaches to investigating the RE transcriptome and proteome.

4.4.1 Metabolism

Although once considered only as a protective barrier, it is now well known that the RE is metabolically active and is the greatest consumer of energy of the total viscera (Huntington 1990). The RE plays an important role in whole-animal energy energetics as it is responsible for ruminal SCFA transport and metabolism that ultimately make up the majority of metabolizable energy available for the dairy cow (Baldwin 1998). The RE is a stratified squamous epithelium consisting of four distinct strata with multiple functions (stratum basale, stratum spinosum, stratum granulosum, and stratum corneum). The stratum basale and spinosum is adjacent to the basal lamina and the stratum granulosum, and has significant numbers of fully functional mitochondria and other organelles that contribute most to the metabolic properties of the RE, i.e., ketogenesis (Steele et al. 2011a). Consequently, stratum basale and spinosum are important cells of the rumen with regard to energy metabolism (Baldwin et al. 2004).

The transport of SCFA through the RE is regulated by several isoforms of sodium/hydrogen exchangers, monocarboxylate transporters and anion exchanger transporters (Gaebel and Martens 1988). Although some targeted studies have shown that sodium/hydrogen exchangers increase during a grain challenge in goats (Yang et al. 2009), there are very little changes in these transporters as detected by high-throughput transcriptomic studies in dairy cattle. There has been only one proteomic study in sheep fed high-grain diets which detected a reduction in carbonic anhydrase isoform 1 protein abundance, which is involved in RE pH regulation and ion transport (Bondzio et al. 2011). Although limited differential expression has been found using high-throughput techniques, it has been very effective for understanding what specific isoforms of each exchanger and transporter to scope future more targeted studies.

Once transported into the cell, SCFA can make its way to circulation or be metabolized. Of the SCFA, butyrate is the preferred energy source for rumen epithelial tissue and most of the ruminal absorbed butyrate is metabolized to beta-hydroxy butyrate (BHBA) in the mitochondria (Bergman 1990). The enzymes Acetyl-CoA acetyl transferase 1 (*ACAT1*) and 3-hydroxy, 3-methylglutaryl CoA synthase2 (*HMGCS2*) are considered to be rate-limiting enzymes in the pathway of ketogenesis in RE (Lane et al. 2002). The expression of *ACAT1*, *HMGCS2,* and other ketogenic genes is controlled by a transcriptional factor termed the peroxisome proliferator activated receptor alpha *(PPARα)* transcriptional factor (Kinoshita et al. 2002). Although it is common to have elevated levels of BHBA in blood during a grain challenge, the RE transcriptome and proteome have not detected differences in ketogenic enzyme expression (Penner et al. 2009, 2011; Steele et al. 2011a). Although no changes in ketogenic genes were detected in the RE of cows with SARA, large changes in the pathway of cholesterol biosynthesis which is controlled by the *PPARα* and sterol regulatory element binding proteins have been identified (*SREBP*). The downregulation of genes involved in cholesterol biosynthesis during a grain challenge that induced SARA was not expected. However, when one considers that the amount of available SCFA to be metabolized is higher during a grain challenge, it can be expected that the synthesis of cholesterol—which is downstream of some of the most major SCFA metabolic pathways in the RE—is tightly controlled. Although the biological significance of this finding is still in question, the regulation of cholesterol may be essential to reduce the potential negative impact that arise from elevated levels of cholesterol being produced with RE cells, which may be involved in maintaining cell membrane function or controlling inflammation (Steele et al. 2011a, b, 2012). The systems biology approach was very useful for studying the RE adaptation during SARA as it led to the discovery of new pathways that may be of biological relevance that was not being considered in the past.

4.4.2 Growth

It is well known that when switching a diet to a more rapidly fermentable source of carbohydrate the RE proliferates to increase the surface area for SCFA absorption, yet molecular mechanisms that control growth and differentiation of the RE are not clearly defined. As stated previously, feeding of readily fermentable carbohydrates increases the proportions of propionate and butyrate in the rumen, which are thought to stimulate the growth of the rumen papillae directly and indirectly to increase the overall surface area for absorption. Since the SCFA do not cause a response in RE proliferation in-vitro, it is likely that growth is controlled indirectly via hormones (Baldwin 1999). Among the hormones promoting RE growth, increasing attention has been directed toward the role of insulin-like growth factor-1 (*IGF-1*) and IGF binding proteins (Baldwin 1999; Penner et al. 2011).

In one of the first studies evaluating RE gene expression in goats fed high-energy diets, it was determined that RE proliferation was associated with increased IGF-1 receptors in RE cells and IGF-1 concentrations in the plasma (Shen et al. 2004). The involvement of IGF-1 axis in RE adaptation under conditions of SARA in dairy cows was validated in recent gene transcriptomic-based studies. In a study by Steele et al. (2011b), the short-term and prolonged effects of SARA on the mRNA expression of candidate genes involved in the RE proliferation in lactating dairy cattle were evaluated using microarray technology. Results of that study showed the relative mRNA expression of insulin-like growth factor-binding proteins 5 (*IGFBP5*) was upregulated, whereas the expression of *IGFBP3* and *IGFBP6* were downregulated in the RE when cattle were transitioned to a high-concentrate. This result is interesting because it is well known that *IGFBP3* opposes and *IGFBP5* potentiates *IGF-1* mediated cellular events; therefore, the downregulation of *IGFBP3* and the upregulation of *IGFBP5* in cattle fed high-grain diet may stimulate cellular growth in the RE in addition to circulating IGF-1 concentrations (Steele et al. 2011b). A similar expression profile of IGFBP was also detected in a study evaluating transition cow RE gene expression from the dry period (3 weeks prior to calving) to early lactation (weeks 1 and 6 of lactation) (Steele et al. 2015). However, the results contrast a recent study by McCann et al. (2016) where a short-term SARA induction did not affect the relative mRNA expression of *IGFBP3* and *IGFBP5* in lactating Holstein cows (McCann et al. 2016). To date, the expression of genes related to the

IGF-Axis has not been validated at the protein level and represents the next logical step to validate their potential role in RE proliferation.

In a recent study in dairy cattle that biopsied rumen tissue during the transition to lactation, additional growth factors that may be involved in proliferation and differentiation of the RE during the shift to a more rapidly fermentable diet were uncovered. For example, transforming growth factor-beta (*TGFB*) is a multifunctional peptide that regulates growth and differentiation in the RE and Steele et al. (2015) reported that the gene expression of *TGFB1* and *TGFB2* and their subfamily of bone morphogenetic proteins were upregulated in early lactation. In the same study, it was also noted that epidermal growth factor receptor (*EGFR*) and growth hormone receptor (*GHR*) were downregulated in early lactation when cows were fed a high-energy diet. Although these findings are preliminary and need to be validated at the protein level, the collection of high-throughput research of the RE during acidosis has pinpointed some key genes that require further investigation.

4.4.3 Barrier Function

Not only does the reticulo-rumen play an important role in metabolism, it also acts as a barrier to protect the host from the harsh ruminal environment. Preventing the translocation of bacteria and toxins into the portal circulation and also maintaining the concentration gradients required for ion absorption are the responsibility of RE. It is known that ruminal acidosis can impair RE barrier function mainly associated with a decrease in ruminal pH (Aschenbach et al. 2011) and an increase in ruminal osmolality (Penner et al. 2010). With regard to the RE, the stratum granulosum and corneum play an important role in barrier function. Endotoxins such as LPS and lipoteichoic acid can compromise the epithelial barrier of gastrointestinal tract (Korhonen et al. 2002; Singh et al. 2007). Metzler-Zebeli et al. (2013) noted that the multilayered stratified squamous RE may be less permeable to allow the transmigration of LPS compared with the single layer of the columnar epithelium in the lower gut.

Recent research using high-throughput gene expression analyses have been used to uncover genes related to cell adhesion and epithelial structure that may be responsive to elevated levels of grain in the diet. In the first microarray study conducted in SARA cows, it was revealed that the barrier function in the rumen was compromised due to ruminal acidosis and was associated with downregulated desmoglein 1 expression in rumen tissue (Steele et al. 2011a)—a key component of desmosomes which are evident in the granulosum layer of the epithelium. This data has been recently confirmed by McCann et al. (2016) further suggesting its role in RE function during a grain challenge. In the same study, it was shown that the expression of genes related to barrier function (claudin1 and claudin4) in ruminal epithelium was upregulated for SARA cows. In addition, transcriptome analysis by Mackey (2013), using RNA-sequencing technologies, revealed that genes involved in the molecular pathway of homophilic cell adhesion (protocadherin betas4 (PCDHB4), PCDHB15, PCDHB14, and PCDHB7) were upregulated to oppose the formation of a weakened permeability barrier in RE of dairy cows in response to SARA. Although these findings have been important from a discovery point of view, further studies are needed to elucidate the extent of their control on RE barrier function in addition to their responsiveness at the RNA level.

4.5 Other Organs: A Systems Biology Approach in Understanding Ruminal Acidosis

Although most research findings about ruminal acidosis have focused on the rumen, there is growing evidence that other organs such as the lower gut and liver play a key role in the etiology of SARA (Fig. 4.1). Feeding high-grain diets increases the starch bypass to the small intestine (Li et al. 2012) and excessive flow of readily fermentable carbohydrates to the small intestine promotes fermentative acidosis with an accumulation of SCFA, including lactic acid (Gressley et al. 2011). Previous research has shown that the richness, evenness, and diversity of the microbiota of the hindgut (cecum and colon) and feces are decreased as a result of feeding high-grain diets to cattle (Mao et al. 2013; Li et al. 2012). Feeding high-grain diets has also been reported to increase populations of pathogenic *Clostridium perfringens* and *Escherichia coli* (LPS-producing bacteria) in the hindgut (Plaizier et al. 2016). However, there is limited research—especially research using high-throughput sequencing—examining the lower gut microbiome of ruminants fed high-grain diets.

Accumulation of SCFA and increasing osmotic pressure of digesta in the lower gut may lead to damage of the gut barrier and allow bacteria, endotoxins, or amines into the systemic circulation (Plaizier et al. 2008). In contrast to the multilayered stratified squamous epithelium in the rumen, the lower gut has a simple epithelium where the relative proportion of specialized cell type change dramatically (Lavker and Motoltsy 1970). Feeding lactating cows diets containing elevated proportions of concentrate results in an increased release of bacterial endotoxin (i.e., LPS) in the rumen or hind gut (Gozho et al. 2006; Plaizier et al. 2016). The toxic effects exerted by endotoxins on the gastrointestinal tract epithelial cells might be responsible for the local inflammatory reaction in the early phase (Thibault et al. 2010), as well as the majority of the proinflammatory cytokine production and acute phase proteins (APP) by translocation of endotoxins across the gastrointestinal tract into circulating blood (Gozho et al. 2006; Khafipour et al. 2009b; Emmanuel et al. 2007).

Acute phase proteins including haptoglobin (Hp), serum Amyloid-A (SAA), and LPS-binding protein (LBP) synthesis from the hepatocytes are biomarkers for determining inflammation (Emmanuel et al. 2008; Eckersall and Bell 2010). Gozho et al. (2007) showed that an increase of Hp and SAA levels activated a systemic inflammatory response in the blood of cows fed a high-grain diet. The blood concentration of LBP is also often used as an indicator for LPS-induced systemic inflammation. QRT-PCR analysis demonstrated that the mRNA expression of 13 different genes involved in immune response, as well as TLR4 protein in the liver, was upregulated in goats fed high-grain diets compared to those fed low-grain diets,

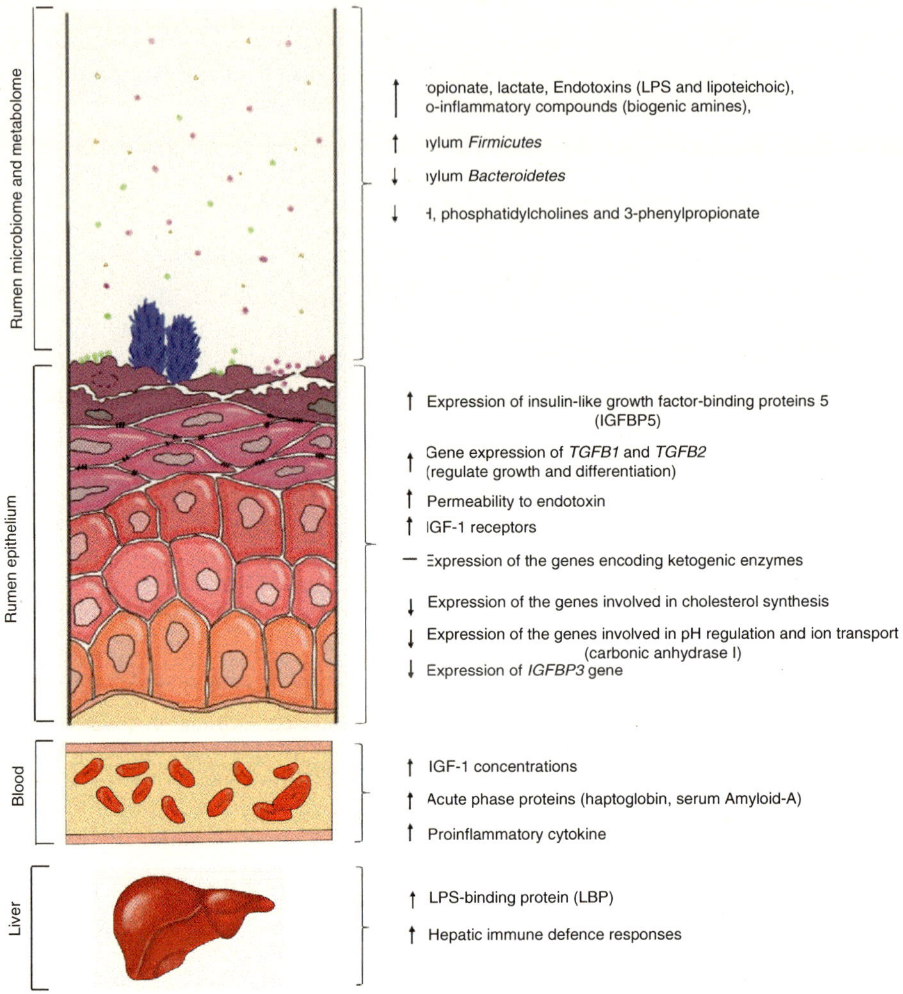

Fig. 4.1 Illustration of diet–microbiota–host interactions in cattle fed high-grain diets

suggesting that increased entry of endotoxin into the liver exacerbated hepatic immune defense responses (Chang et al. 2015). These results showed that SARA had an important impact on immune defense responses in liver.

In the context of systems biology, more research is needed to better understand the link between microorganisms in the gastrointestinal tract and the systemic response of animals fed high-grain diets. Although many studies have been published over the past 5 years using "omics" and bioinformatics tools in livestock nutrition, there has been limited integration of the different omics platforms and little attention devoted to understanding how feeding high-grain diets can affect the function of the animal.

Conclusions

Investigation of dairy cow health challenges such as ruminal acidosis has traditionally taken a reductionist approach, focusing on the rumen to diagnose and provide treatments and preventions. The term "ruminal acidosis" however is misleading, as the lower gut may play a significant role in the etiology of whole animal inflammation during SARA. Now that we know that SARA is a whole-animal disorder impacting multiple organs, and that many of these organs and biofluids have not yet been characterized during SARA, it would be advantageous to utilize a high-throughput, systems biology approach to expand our knowledge base. The recent advances in systems biology, including genomics, transcriptomics, metabolomics, and proteomics, have provided novel insights into the biology of SARA as well as a new approach to study multiple biological targets in a high-throughput manner. Recent progress using high-throughput sequencing methodologies showed that the richness, diversity, and functionality of the microbiome in the rumen and lower gut can decrease with excessive grain feeding and acidotic conditions, thereby affecting animal production and health. These analyses also revealed that the expression levels of multiple genes associated with rumen epithelium proliferation and metabolism, gastrointestinal tract barrier function, and hepatic enzyme activity were altered by feeding high-grain diets to cattle. An integrated systems biology approach is required to fully uncover the causes and mechanisms driving the interaction among the gastrointestinal microbes and host during this SARA challenge. As such, future systems biology studies in ruminal acidosis should focus on—identifying reliable biomarkers for molecular diagnosis and early detection of ruminal acidosis which may improve transition cow health.

References

Ametaj BN, Zebeli Q, Saleem F et al (2010) Metabolomics reveals unhealthy alterations in rumen metabolism with increased proportion of cereal grain in the diet of dairy cows. Metabolomics 6:583–594

Aschenbach JR, Penner G, Stumpff BF et al (2011) Ruminant nutrition symposium: role of fermentation acid absorption in the regulation of ruminal pH. J Anim Sci 89:1092–1107

Baldwin RL (1998) Use of isolated ruminal epithelial cells in the study of rumen metabolism. J Nutr 128:293S–296S

Baldwin RL (1999) The proliferative actions of insulin, insulin like growth factor-I, epidermal growth factor, butyrate and propionate on ruminal epithelial cells in vitro. Small Rumin Res 32:261–268

Baldwin RL, McLeod IKR, Klotz JL et al (2004) Rumen development, intestinal growth and hepatic metabolism in the pre- and postweaning ruminant. J Dairy Sci 87(Suppl E):E55–E65

Bardócz S, Duguid TJ, Brown DS et al (1995) The importance of dietary polyamines in cell regeneration and growth. Br J Nutr 73:819–828

Bergman EN (1990) Energy contributions of volatile fatty acids from the gastrointestinal tract in various species. Physiol Rev 70:567–590

Bondzio A, Gabler C, Badewien-Rentzsch B et al (2011) Identification of differentially expressed proteins in ruminal epithelium in response to a concentrate-supplemented diet. Am J Physiol Gastrointest Liver Physiol 301:G260–G268

Chang G, Zhang K, Xu T et al (2015) Feeding a high-grain diet reduces the percentage of LPS clearance and enhances immune gene expression in goat liver. BMC Vet Res 11:67

Colman E, Khafipour E, Vlaeminck B et al (2013) Grain-based versus alfalfa-based subacute ruminal acidosis induction experiments: similarities and differences between changes in milk fatty acids. J Dairy Sci 96:4100–4111

De Nardi R, Marchesini G, Li S et al (2016) Metagenomic analysis of rumen microbial population in dairy heifers fed a high grain diet supplemented with dicarboxylic acids or polyphenols. BMC Vet Res 12:29

Desai N, Antonopoulos D, Gilbert JA et al (2012) From genomics to metagenomics. Curr Opin Biotechnol 23:72–76

Dieho K, Dijkstra J, Schonewille JT et al (2016) Changes in ruminal volatile fatty acid production and absorption rate during the dry period and early lactation as affected by rate of increase of concentrate allowance. J Dairy Sci 99:5370–5384

Eckel EF, Ametaj BN (2016) Invited review: role of bacterial endotoxins in the etiopathogenesis of periparturient diseases of transition dairy cows. J Dairy Sci 99(8):5967–5990

Eckersall PD, Bell R (2010) Acute phase proteins: biomarkers of infection and inflammation in veterinary medicine. Vet J 185:23–27

Emmanuel DGV, Madsen KL, Churchill TA et al (2007) Acidosis and lipopolysaccharide from *Escherichia coli* B:055 cause hyperpermeability of rumen and colon tissues. J Dairy Sci 90:5552–5557

Emmanuel DGV, Dunn SM, Ametaj BN (2008) Cover image article feeding high proportions of barley grain stimulates an inflammatory response in dairy cows. J Dairy Sci 91:606–614

Fernando SC, Purvis HT, Najar FZ et al (2010) Rumen microbial population dynamics during adaptation to a high-grain diet. Appl Environ Microbiol 76:7482–7490

Firkins JL, Yu Z (2015) Ruminant nutrition symposium: how to use data on the rumen microbiome to improve our understanding of ruminant nutrition. J Anim Sci 93:1450–1470

Fusi E, Rossi L, Rebucci R et al (2004) Administration of biogenic amines to Saanen kids: effects on growth performance, meat quality and gut histology. Small Rumin Res 53:1–7

Gaebel G, Martens H (1988) Reversibility of acid induced changes in absorptive function of sheep rumen. Zentralbl Vet A 35:157–160

Gaebel G, Martens H, Suendermann M (1987) The effect of diet, intraruminal pH and osmolarity on sodium, chloride and magnesium absorption from the temporarily isolated and washed reticulo-rumen of sheep. Q J Exp Physiol 72:501–511

Gozho GN, Plaizier JC, Krause DO et al (2005) Subacute ruminal acidosis induces ruminal lipopolysaccharide endotoxin release and triggers an inflammatory response. J Dairy Sci 88:1399–1403

Gozho GN, Krause DO, Plaizier JC (2006) Rumen lipopolysaccharide and inflammation during grain adaptation and subacute ruminal acidosis in steers. J Dairy Sci 89:4404–4413

Gozho GN, Krause DO, Plaizier JC (2007) Ruminal lipopolysaccharide concentration and inflammatory response during grain-induced subacute ruminal acidosis in dairy cows. J Dairy Sci 90:856–866

Gressley TF, Hall MB, Armentano LE (2011) Ruminant nutrition symposium: productivity, digestion, and health responses to hindgut acidosis in ruminants. J Anim Sci 89:1120–1130

Henderson G, Cox F, Ganesh S et al (2015) Rumen microbial community composition varies with diet and host, but a core microbiome is found across a wide geographical range. Sci Rep 5:14567

Hungate RE (1966) The rumen and its microbes. Academic, New York

Hungate RE, Dougherty RW, Bryant MP, Cello RM (1952) Microbiological and physiological changes associated with acute indigestion in sheep. Cornell Vet 42:423–449

Huntington GB (1990) Energy metabolism in the digestive tract and liver of cattle: influence of physiological state and nutrition. Reprod Nutr Dev 30:35–47

Huo W, Zhu W, Mao S (2014) Impact of subacute ruminal acidosis on the diversity of liquid and solid-associated bacteria in the rumen of goats. World J Microbiol Biotechnol 30:669–680

Hurley JC (1995) Endotoxemia: methods of detection and clinical correlates. Clin Microbiol Rev 8:268–292

Khafipour E, LiS THM (2016) Effects of grain feeding on microbiota in the digestive tract of cattle. Anim Front 6:13–19

Khafipour E, Krause DO, Plaizier JC (2009a) Alfalfa pellet induced subacute ruminal acidosis in dairy cows increases bacterial endotoxin in the rumen without causing inflammation. J Dairy Sci 92:1712–1724

Khafipour E, Krause DO, Plaizier JC (2009b) A grain-based subacute ruminal acidosis challenge causes translocation of lipopolysaccharide and triggers inflammation. J Dairy Sci 92:1060–1070

Khafipour E, Li S, Plaizier JC et al (2009c) Rumen microbiome composition determined using two nutritional models of subacute ruminal acidosis. Appl Environ Microbiol 75:7115–7124

Khafipour E, Plaizier JC, Aikman PC et al (2011) Population structure of rumen *Escherichia coli* associated with subacute ruminal acidosis (SARA) in dairy cattle. J Dairy Sci 94:351–360

Kinoshita M, Suzuki Y, Saito Y (2002) Butyrate reduces colonic paracellular permeability by enhancing PPAR activation. Biochem Biophys Res Commun 293:827–831

Kleen JL, Hooijer GA, Rehage J et al (2009) Subacute ruminal acidosis in Dutch dairy herds. Vet Rec 164:681–684

Korhonen R, Korpela R, Moilanen E (2002) Signalling mechanisms involved in the induction of inducible nitric oxide synthase by *Lactobacillus rhamnosus* GG, endotoxin, and lipoteichoic acid. Inflammation 26:207–214

Lane MA, Baldwin VIRL, Jesse BW (2002) Developmental changes in ketogenic enzyme gene expression during sheep rumen development. J Anim Sci 80:1538–1544

Lavker RM, Matoltsy AG (1970) The fate of cell organelles and differentiation products in ruminal epithelium. J Cell Biol 44:501–512

Lettat A, Nozière P, Silberberg M et al (2010) Experimental feed induction of ruminal lactic, propionic, or butyric acidosis in sheep. J Anim Sci 88:3041–3046

Li S, Khafipour E, Krause DO et al (2012) Effects of subacute ruminal acidosis challenges on fermentation and endotoxins in the rumen and hindgut of dairy cows. J Dairy Sci 95:294–303

Liu JH, Bian G, Zhu W et al (2015) High-grain feeding causes strong shifts in ruminal epithelial bacterial community and expression of toll-like receptor genes in goats. Front Microbiol 6:167

Loor JJ, Bionaz M, Drackley JK (2013) Systems physiology in dairy cattle: nutritional genomics and beyond. Annu Rev Anim Biosci 1:365–392

Mackey, E (2013) Effects of ruminal acidosis on rumen papillae transcriptome. Thesis, The University of Delaware

Mao SY, Zhang RY, Wang DS (2013) Impact of subacute ruminal acidosis (SARA) adaptation on rumen microbiota in dairy cattle using pyrosequencing. Anaerobe 24:12–19

Mao S, Huo W, Zhu WY (2016) Microbiome–metabolome analysis reveals unhealthy alterations in the composition and metabolism of ruminal microbiota with increasing dietary grain in a goat model. Environ Microbiol 18:525–541

McCann JC, Luan S, Cardoso FC et al (2016) Induction of subacute ruminal acidosis affects the ruminal microbiome and epithelium. Front Microbiol 7:701

McSweeney CS, Denman SE, Wright A-DG et al (2007) Application of recent DNA/RNA-based techniques in rumen ecology. Asian Aust J Anim 20:283–294

Metzler-Zebeli BU, Schmitz-Esser S, Klevenhusen F et al (2013) Grain-rich diets differently alter ruminal and colonic abundance of microbial populations and lipopolysaccharide in goats. Anaerobe 20:65–73

Nagaraja TG, Titgemeyer EC (2007) Ruminal acidosis in beef cattle: the current microbiological and nutritional outlook. J Dairy Sci 90(Suppl 1):E17–E38

Nagaraja TG, Bartley EE, Fina LR et al (1978a) Quantitation of endotoxin in cell-free rumen fluid of cattle. J Anim Sci 46:759–1767

Nagaraja TG, Bartley EE, Fina LR et al (1978b) Evidence of endotoxins in the rumen bacteria of cattle fed hay or grain. J Anim Sci 47:226–234

Pace NR, Stahl DA, Lane DJ et al (1985) Analyzing natural microbial populations by rRNA sequences. ASM News 51:4–12

Penner GB, Taniguchi BM, Guan LL et al (2009) Effect of dietary forage to concentrate ratio on volatile fatty acid absorption and the expression of genes related to volatile fatty acid absorption and metabolism in ruminal tissue. J Dairy Sci 92:2767–2781

Penner GB, Oba M, Gäbel G et al (2010) A single mild episode of subacute ruminal acidosis does not affect ruminal barrier function in the short term. J Dairy Sci 93:4838–4845

Penner GB, Steele MA, Aschenbach JR et al (2011) Molecular adaptation of ruminal epithelia to highly fermentable diets. J Anim Sci 89:1108–1119

Petri R, Forster R, Yang W et al (2012) Characterization of rumen bacterial diversity and fermentation parameters in concentrate fed cattle with and without forage. J Appl Microbiol 112:1152–1162

Petri RM, Schwaiger T, Penner GB et al (2013a) Changes in the rumen epimural bacterial diversity of beef cattle as affected by diet and induced ruminal acidosis. Appl Environ Microbiol 79:3744–3755

Petri RM, Schwaiger T, Penner GB et al (2013b) Characterization of the core rumen microbiome in cattle during transition from forage to concentrate as well as during and after an acidotic challenge. PLoS One 8:e83424

Plaizier J, Keunen J, Walton J et al (2001) Effect of subacute ruminal acidosis on in situ digestion of mixed hay in lactating dairy cows. Can J Anim Sci 81:421–423

Plaizier JC, Krause DO, Gozho GN et al (2008) Subacute ruminal acidosis in dairy cows: the physiological causes, incidence and consequences. Vet J 176:21–31

Plaizier JC, Khafipour E, Li S et al (2012) Subacute ruminal acidosis (SARA), endotoxins and health consequences. Anim Feed Sci Technol 172:9–21

Plaizier JC, Li S, Tun HM et al (2016) Effects of experimentally induced subacute ruminal acidosis (SARA) on the rumen and hindgut microbiome in dairy cows. Front Microbiol In press

Russell JB, Sharp WM, Baldwin RL (1979) The effect of pH on maximum bacterial growth rate and its possible role as a determinant of bacterial competition in the rumen. J Anim Sci 48(2):251–255

Russell JB, Cotta MA, Dombrowski DB (1981) Rumen bacterial competition in continuous culture: *Streptococcus bovis* versus *Megasphaera elsdenii*. Appl Environ Microbiol 41(6):1394–1399

Saleem F, Ametaj BN, Bouatra S et al (2012) A metabolomics approach to uncover the effects of grain diets on rumen health in dairy cows. J Dairy Sci 95:6606–6623

Shebl FM, Pinto LA, García-Piñeres A et al (2010) Comparison of mRNA and protein measures of cytokines following vaccination with human papillomavirus-16 L1 virus-like particles. Cancer Epidemiol Biomark Prev 19:978–981

Shen Z, Seyfert H, Lo B et al (2004) An energy-rich diet causes rumen papillae proliferation associated with more IGF type 1 receptors and increased plasma IGF-1 concentrations. J Anim Sci 90:307–317

Singh AK, Jiang Y, Gupta S (2007) Effects of bacterial toxins on endothelial tight junction in vitro: a mechanism-based investigation. Toxicol Mech Methods 17:331–347

Steele MA, Croom J, Kahler M et al (2011a) Bovine rumen epithelium undergoes rapid structural adaptations during grain-induced subacute ruminal acidosis. Am J Physiol Regul Integr Comp Physiol 300:R1515–R1523

Steele MA, Vandervoort G, AlZahal O et al (2011b) Rumen epithelial adaptation to high-grain diets involves the coordinated regulation of genes involved in cholesterol homeostasis. Physiol Genomics 43:308–316

Steele MA, Greenwood SL, Croom J et al (2012) An increase in dietary non-structural carbohydrates alters the structure and metabolism of the rumen epithelium in lambs. Can J Anim Sci 92:123–130

Steele MA, Schiestel C, AlZahal O et al (2015) The periparturient period is associated with structural and transcriptomic adaptations of rumen papillae in dairy cattle. J Dairy Sci 98:2583–2595

Steele MA, Penner GB, Chaucheyras-Durand F et al (2016) Development and physiology of the rumen and the lower gut: targets for improving gut health. J Dairy Sci 99:4955–4966

Tajima K, Arai S, Ogata K et al (2000) Rumen bacterial community transition during adaptation to high-grain diet. Anaerobe 6(5):273–284

Thibault R, Blachier F, Darcy-Vrillon B et al (2010) Butyrate utilization by the colonic mucosa in inflammatory bowel diseases: a transport deficiency. Inflamm Bowel Dis 16:684–695

Woelders H, Te Pas MF, Bannink A et al (2011) Systems biology in animal sciences. Animal 5:1036–1047

Woese CR, Fox GE (1977) Phylogenetic structure of the prokaryotic domain: the primary kingdoms. Proc Natl Acad Sci 74(11):5088–5090

Yang W, Martens H, Shen Z (2009) Effects of energy intake and ruminal SCFA on mRNA expression of Na+/H+ exchangers in rumen epithelium of goats. In: Proceedings of 11th international symposium of ruminant physiology, Clermont-Ferrand, France, pp 412–413

Ye H, Liu J, Feng P et al (2016) Grain-rich diets altered the colonic fermentation and mucosa-associated bacterial communities and induced mucosal injuries in goats. Sci Rep 6:20329

Zened A, Combes S, Cauquil L et al (2013) Microbial ecology of the rumen evaluated by 454 GS FLX pyrosequencing is affected by starch and oil supplementation of diets. FEMS Microbiol Ecol 83:504–514

Cattle Gastrointestinal Tract Microbiota in Health and Disease

5

André Luiz Garcia Dias and Burim N. Ametaj

Abstract

Rumen is the most significant organ of the ruminant animal containing trillions of bacteria known also as microbiota. The ruminant host and the microbiota live in a symbiotic relationship where bacteria are provided shelter and nutrients and the host benefits from many essential nutrients released by bacterial fermentation activities, otherwise not available to the ruminant animal because of the lack of specific enzymes to do so. In nature the ruminants usually graze and have almost no grain as part of their daily diet. However, under intensive husbandry dairy cows producing large amounts of milk need very large amounts of energy to support the high demand for milk production. The usual form of energy fed to ruminants in intensive dairy farms is grain whether in the form of corn (maize), barley, wheat, or any other species of cereals. Feeding large amounts of grain in the ration, especially immediately after parturition, has been reported to influence majorly composition of gastrointestinal (GI) tract microbiota and their metabolite composition. Besides releasing large amounts of beneficial and essential nutrients bacterial activity has been reported to be associated with the release of multiple harmful compounds including bacterial endotoxins. Translocation of those compounds into important organs like mammary or uterus or into the systemic circulation is associated with multiple deleterious effects on health status of the postpartum cow. Gastrointestinal microbiota, factors that influence its composition, and harmful effects on the host will be discussed in detail in this chapter.

A.L.G. Dias, Ph.D. (✉) • B.N. Ametaj, D.V.M., Ph.D., Ph.D.
Department of Agriculture, Food and Nutritional Science, University of Alberta, Edmonton, AB, T6R 2J6, Canada
e-mail: garciadi@ualberta.ca; burim.ametaj@ualberta.ca

B.N. Ametaj (ed.), *Periparturient Diseases of Dairy Cows*,
DOI 10.1007/978-3-319-43033-1_5

5.1 Introduction

Animal microbiologists and ruminant health scientists have focused on three main organs in regard to studying composition of microbiota in dairy cows including rumen, udder, and uterus. These three organs are very important from the physiological point of view but also from the health prospective. In this chapter we will focus on the microbiota of the gastrointestinal tract (GIT) as this is considered currently as one of the most important "organs" of the ruminant animal. The total number of bacteria in the rumen of an adult cow might be more than 10^{12}. The stomach of the cow is divided into four sections which include rumen, reticulum, and omasum, known also as forestomachs, and the real stomach or abomasum. The rumen is a distinct organ that is specific to ruminants, and its primary function is fermentation of feedstuff materials by microbiota and absorption of large amounts of nutrients otherwise not available to the host. Plant fibrous materials are digested by the microbiota, and further converted into essential nutrients for maintenance, milk production, growth of the fetus, or meat production. Fermentation of the feedstuff fibrous materials from the microbiota provides essential nutrients to the host in the form of fatty acids, lipids, carbohydrates, and vitamins.

An important lesson learned from the recent intensive research on rumen composition and bacterial activities is that rumen microbiota, besides being a "friend" to the ruminant host, due to their contribution to digestion of the feedstuff, can sometimes become a "foe" as they are a source of various pathogenic compounds, for example, endotoxins, which have been reported to trigger various pathologies in dairy cows.

In this chapter we will focus on the new knowledge generated by system's biology approach in regard to the microbiota of the ruminant GI tract specific to dairy cows and its relation to health-disease states. First we will discuss rumen and intestinal core microbiota as related to factors that affect their bacterial composition. Then, we will focus on the potential role of the microbiome in the pathobiology of diseases in dairy cows.

5.2 Rumen Microbiota

5.2.1 Rumen Bacterial Core

Rumen is an anaerobic organ that houses trillions of microorganisms. Bacteria live, digest, replenish, interact, compete or fight with each other, and also assist or harm the host depending on the conditions established within the surrounding environment. Ruminal microbiota are complex, and characterized by high density, diversity, and complexity of interactions. Moreover, rumen content is a distinct ecosystem that includes bacteria, protozoa, fungi, and archaea. It should be noted that bacteria are the most abundant and diverse taxa representing approximately 95% of the total microbiota community in the rumen, and we will focus our discussion in this chapter only on bacteria (Brulc et al. 2009).

Ruminal bacteria have been classified into three main groups including those that are free in the rumen liquid, those that are attached to feed particles, and those adherent to the rumen epithelium or epimural bacteria. Interestingly, the function of the first group of bacteria is to degrade and use feed materials that are soluble or insoluble in the rumen fluid. These types of bacteria are very mobile because they move along with digesta into the other sections of the GI tract. The second group of bacteria attached to the rumen solid particles function to degrade and use feedstuff materials spread in the rumen in the form of particles. The third group of bacteria (i.e., epimural ones) accounts for 1–2% of all rumen bacteria adherent to the rumen epithelium. They play significant roles in oxygen scavenging (i.e., 25–50% of epimural bacteria are facultative anaerobes), hydrolysis of urea, and tissue recycling (Cheng et al. 1979; Wallace et al. 1979). It has been reported that the composition of this group of bacteria is different from those that are free in the rumen liquid or attached to solid particles. Given the fact that the epithelial layer of cells is shed every 1–2 days, these bacteria also fall in the rumen digesta, and move along to other sections of the GI tract. Most bacteria that move to the abomasum are digested or destroyed by the acidic conditions and are used as a source of nutrients from the host.

Rumen microbiota live in a milieu that is continuously changing in relation to various environmental factors; therefore, characterization of these bacteria has been difficult. There are several factors that have been shown to influence the composition of rumen microbiota including diet, age, breed, stage of lactation, supplementation of antibiotic compounds like monensin, health of the host, and finally geographical location and season (Stewart et al. 1997; Kocherginskaya et al. 2001). Moreover, changes in the microbial community are associated with alterations in both the structure of the microbes and population levels.

In regard to free ruminal fluid bacteria Brulc et al. (2009) analyzed ruminal fluid samples with different metagenomic libraries, and reported three dominant phylotypes present including Bacteroidetes, Firmicutes, and Proteobacteria which represent between 91.3 and 95.2% of all phyla present. General DNA sequencing of rumen samples was also conducted by Jing et al. (2014), who reported that Firmicutes and Bacteroidetes represented 59.3 and 34.3% of all phyla, or in total around 93.6% of all phyla analyzed, confirming the results of Brulc et al.'s (2009) study.

A study by Petri et al. (2013) investigated rumen epimural core microbiome in beef cattle and found that phyla Firmicutes (73%), Proteobacteria (11%), Bacteroidetes (10%), and Actinobacteria (3%) were the most abundant ones. Moreover, Jami and Mizrahi (2012) reported that 98% of all rumen microbiota, in the Holsteins fed a diet composed of 30% roughage and 70% concentrate, were comprised of phyla Bacteroidetes (51%), Firmicutes (42%), and Proteobacteria (5.21%). Actinobacteria and Tenericutes also were present in lesser abundances at 0.9 and 0.7%, respectively. The microbiome was also analyzed at the genus level, and they demonstrated a total of 32 genera, with genus *Prevotella* accounting for almost 52% of all rumen bacterial genera. *Prevotella* are Gram-negative bacteria that thrive in anaerobic conditions. Besides their roles in degradation of feed particles, they are also associated with multiple diseases in humans including sinusitis, otitis, dental infection, peritonsillar abscess, aspiration pneumonia, lung abscesses, peritonitis, pelvic inflammatory disease, vulvovaginal and perianal infections, endocarditis, meningitis, arthritis, osteomyelitis, and many others (Stubbs et al. 1996; Made et al. 1998; Brook and Frazier 2003; Hayashi et al. 2007; Ulrich et al. 2010; Field et al. 2010; Nadkarni et al. 2012; Ruan et al. 2015). Moreover, *Prevotella* have been shown to be involved in uterine disease (Sheldon et al. 2008), periodontitis (Borsanelli et al. 2015), and foot rot in cattle (Kennan et al. 2011; Kaler et al. 2012).

A recent study published by Henderson et al. (2015) evaluated 742 samples from 32 different ruminant animal species in 35 countries in seven global regions. The authors reported that 30 of the most abundant rumen bacterial families were found in over 90% of samples, comprising 89.4% of all sequenced data. The seven most abundant bacterial groups comprised 67.1% of all bacterial sequence data including *Prevotella*, *Butyrivibrio*, and *Ruminococcus* as well as unclassified *Lachnospiraceae*, *Ruminococcaceae*, *Bacteroidetes*, and *Clostridiales*. The authors indicated that these families were the most dominant rumen bacteria and can be considered as a "core bacterial microbiome" at the genus level or higher because they are present in a large selection of ruminants. Henderson et al. (2015) also observed that there was an effect of host and diet on rumen microbiota. They concluded that bacterial communities from ruminants fed forage were similar to each other, those from ruminants fed concentrate also were similar to each other, but distinct from those in the animals fed forage, and intermediate from ruminants fed mixed diets. Unclassified *Bacteroidales* and *Ruminococcaceae* were more abundant in all animals fed forages. In contrast, members of *Prevotella* and unclassified *Succinivibrionaceae* were more abundant in animals fed diets containing concentrate.

5.2.2 Factors Modifying Rumen Bacterial Core

It is well known that various factors including host (genotype), environment (ration and medication), and bacteria (proportion of different species, interactions among them, their adhesion capacity as well as enzymatic and metabolic activities) play an important role in determining the GI microbiota in mammalian species. However, the most common factor that influences composition of ruminal microbiota in dairy cows is the quantity of grain or concentration of the diet, especially at the beginning of lactation. Transitioning from the dry off period to early lactation is associated with major changes in the diet composition. During the dry off period cows are commonly fed diets containing low energy rations (mainly hay and forage) for maintenance and pregnancy needs. However, immediately after calving dairy cows are fed high amounts of energy in the form of grains or concentrates to cope with the high demands for milk production. The amount of grain or concentrate in the ration can reach up to 45–60% of dry matter (DM) in the form of grain and 40–55% as forage.

5.2.2.1 Effect of Grain Feeding on Rumen Microbiota

High proportions of grain in the diet has been shown to be associated with significant changes in the diversity and structure of rumen bacterial populations compared with diets containing high amounts of forage. These changes occur due to the increase in fermentable substrates present in high concentrate diets, which favor growth of amylolytic and other starch-digesting bacterial species.

Fernando et al. (2010) compared beef steers receiving prairie hay or diets with increasing proportions of concentrate, and reported significant changes in the rumen microbiota composition. The authors observed that the most dominant phylotypes present in the rumen of steers fed prairie hay were Firmicutes (35.4%), Bacteroidetes (23.8%), Fibrobacteres (3.2%), Proteobacteria (2.5%), and Spirochaetes (2.5%). There were also around 32.6% of unclassified operational taxonomic units (OTUs) in the samples analyzed. Moreover, bacterial phyla in the rumen fluid of steers fed high concentrate diets were Bacteroidetes (44.9%), Firmicutes (40.1%), Spirochaetes (3.6%), Proteobacteria (1.0%), Fibrobacteres (0.4%), and 10% unclassified OTUs (Fernando et al. 2010).

Additionally, at the genera level Fernando et al. (2010) reported that steers fed a prairie hay ration had increased bacterial genera of *Xylella* sp., *Xanthomonas* sp., *Mycoplasma* sp., *Thauera* sp., and *Bacteroides* sp. in the rumen. Moreover, animals on 20:80 hay:concentrate ration had increased OTUs of *Prevotella* sp., *Actinobacillus* sp., *Alcanivorax* sp., *Arthrobacter* sp., *Bacteroides* sp., *Methylobacillus* sp., *Megasphaera* sp., and *Taylorella* sp. All bacteria in the phylum Fibrobacteres belonged to the genus *Fibrobacter*, whereas most species in the phylum of Bacteroidetes belonged to genus *Prevotella*. There were no differences in the Firmicutes; however, analyses of families within the phylum indicated a greater number of bacteria from families Clostridiaceae and Acidaminococcaceae. The same authors also observed an increase in *Megasphaera elsdenii*, *Streptococcus bovis*, *Selenomonas ruminantium*, and *Prevotella bryantii* and a decrease in the *Butyrivibrio fibrisolvens* and *Fibrobacter succinogenes* species in animals receiving the high concentrate diet.

5.2.2.2 Role of Genotype on Rumen Microbiota

There is a discussion on which factor determines the most the composition of rumen microbiota, the host genotype or the dietary composition. Most of the studies published to date suggest that the most dominant factor in shaping the GI tract microbiota of cows is the ration (Khafipour et al. 2009; Ramirez et al. 2012). However, it should be pointed out that in some individual animals ruminal microbiota is more resistant to change (Weimer et al. 2010a; Mohammed et al. 2012).

For instance, in a study involving three steers, ruminal microbiota analysis showed that there was individuality with respect to rumen microbiota although they were fed the same ration of medium-quality grass-legume hay. The phylogenetic distribution of the microbiota in this case was strikingly different for the fiber-adherent microbiota of one of the steers, where two-thirds (i.e., 63%) of genes sequenced fell into the Gammaproteobacteria class, whereas in the other two steers the class of Gammaproteobacteria represented only 3.5–8% of sequences (Brulc et al. 2009).

In another study conducted by Li et al. (2009) it was reported lower similarities in microbial diversity among three Holstein cows fed a similar ration, ranging from 74.4 to 89.9%. These observations suggest that the host genotype might have a significant role in the taxonomy of ruminal microbial structure. However, the links between the genotype of the host, composition of microbiota, and metabolic activities of the rumen bacteria are not firmly established, and warrants further research.

Mohammed et al. (2012) analyzed alterations in the composition of rumen bacterial community during severe acidosis during the transition period, and observed that the composition of bacteria was not related to dietary treatment. The shift in bacterial populations was greater in individual cows, which was not related to the severity of acidosis. This suggests that each individual animal has a specific rumen bacterial community that is usually persistent during different lactation cycles.

It has also been observed that alterations in the microbial populations change in response to dietary modifications, however, those changes return to the original level once the dietary factor is removed. For instance, Weimer et al. (2010b) triggered a large change in the ruminal microbiota composition after the exchange of all rumen content between different cows; however, the original microbiota community re-established within 9 weeks of the intervention. Moreover, Li et al. (2012) altered rumen bacterial composition by infusion of butyrate, and also observed that the microbiota composition returned to the original status 168 h post-treatment. Additionally, Blanch et al. (2009) reported that the microbial profile of beef heifers returned to the initial composition 4 days after induced ruminal acidosis. These results also suggest that the animal genotype may in fact play a significant role in the microbiota composition of the rumen.

5.2.2.3 Feeding Time and Rumen Site

Feeding time is known to be a farm management practice that might promote alteration of rumen bacterial community. Ruminal pH varies considerably during the course of the day, mostly due to the amount of fermentable carbohydrates in each meal. Shifts from 0.5 to 1.0 pH unit within a day are common; however, these small shifts can represent a five to tenfold change in the hydrogen ion concentration of the rumen (Nocek et al. 2002). Acidification of the rumen environment is one of the most important factors altering the rumen microbial population.

Mullins et al. (2013) evaluated ruminal content during a total rumen evacuation, 2 h before and 4 h after feeding. They found that the total rumen bacterial pool was greater in samples taken after feeding compared with those before feeding. These results have been observed in multiple studies, and are related to the availability of nutrients to the microbial community. The authors also reported that some bacterial populations responded differently in relation to time after feeding. Mullins et al. (2013) showed that *Fibrobacter succinogenes* was more abundant after feeding than before feeding, whereas the population of *Ruminococcus albus* was lower after feeding.

Interestingly, the time relative to feeding was not reported as a factor that triggers alteration of rumen microbiota in a study performed by Li et al. (2009). Similarly, no differences of bacterial composition were detected in samples collected from different locations in the rumen. This suggests that a larger cohort of animals as well as a larger number of studies need to be conducted in order to learn more about the effect of these factors on rumen microbiome.

5.2.2.4 Rumen Solid and Liquid Fractions

Various factors might influence the correlation between the amount of bacteria found in the solid and liquid phases of the rumen. This is important because the methodologies used to determine microbial populations have been different depending on the type of samples collected. It is known that the microorganisms attached to undigested feed particles comprise a major proportion of total ruminal microorganisms (Craig et al. 1987). The microbial population present in the liquid portion has less contact with the substrate and normally moves along with the digesta into the reticulum, omasum, and abomasum.

In a study by Mullins et al. (2013) they found differences in the relative abundance of several species of bacteria between the solid and liquid fractions of rumen content. The authors reported that an average of 92% of total bacterial DNA was found in the solid fraction. Differences between the bacterial populations in the liquid and solid phases were also observed by Welkie et al. (2010), with approximately 14% of the detected reads observed in the rumen liquid, and around 2% in the solid particles. All other bacterial reads in the rumen samples were found in both liquid and solid phases. Interestingly, alterations in the bacterial composition over the course of the feeding cycle were smaller for the solid-adherent fraction of bacteria versus those present in the liquid phase.

Pitta et al. (2014) also studied bacterial phyla present in the solid and liquid fractions of the rumen content, and reported that the liquid fraction contained 55 lineages with 20% belonging to Bacteroidetes and 40% to Firmicutes. In the solid fraction, results were similar with 22% of bacterial population corresponding to Bacteroidetes, and 49% to Firmicutes.

5.3 Intestinal Microbiota

Studies on the composition of intestinal microbiota in ruminants are scarce and warrant further research. In this section we will discuss all the reported data on the composition of intestinal microbiota of dairy cows.

Bacterial population that makes up the intestines of ruminant animals has been reported to be diverse and numerous. It is commonly accepted that animals and humans alike have a core microbiome of the intestines. This community of bacteria is influenced by multiple factors among which nutrition has been highlighted as the most important factor. In ruminant animals, most of the fermentation and digestion of the feed materials occurs in the rumen and partially in other forestomachs. Feed materials that are able to bypass forestomachs are further digested by the intestines. It should be noted that the number of bacteria in the upper portion of the intestinal tract is lower compared to the rumen and large intestines. This is understandable given that the function of the small intestines is to digest and absorb nutrients.

Moreover, presence of bacteria in the small intestines would beat the purpose because they would lower the quantity of absorbable nutrients as they would compete for nutrients with the host.

There are various mechanisms that restrict bacterial growth in the proximal GI tract, including chemical inhibition (e.g., bile), highly competitive rate of nutrient absorption (large absorptive surface and active transport), high passage rate of digesta (washout of free bacteria), continuous sloughing of the epithelial cells and mucus (washout of adhered bacteria), and the immunological defense mechanisms (i.e., production of secretory immunoglobulin A) (Peterson et al. 1998; Thomas and Versalovic 2010; Frank et al. 2011; Pflughoeft and Versalovic 2012).

Similar with the rumen bacterial core, microbiota present in the intestines shows three predominant phyla including Firmicutes, Bacteroidetes, and Proteobacteria (Shanks et al. 2011; Dowd et al. 2008; Mao et al. 2015). For example, Mao et al. (2015) showed that Bacteroidetes phylum was significantly less abundant in the digesta of intestines compared with the rumen samples. The authors reported that the most abundant phyla in the intestinal digesta of dairy cows were Firmicutes (~70% of all reads), Proteobacteria (~15%), and Bacteroidetes (~3%).

Moreover, Shanks et al. (2011) reported that fecal samples of dairy cows were compromised mainly of Firmicutes (73.3%), Bacteroidetes (13.3%), Cyanobacteria (3.33%), Tenericutes (3.33%), and Actinobacteria (0.67%). The most abundant taxa included *Ruminococcaceae* (6.9%), *Prevotella* spp. (3.9%), *Lachnospiraceae* (3.2%), *Peptostreptococcaceae* (1.3%), and *Turicibacter* spp. (1.1%). The authors also analyzed samples individually and found that 10 phyla in the fecal samples belonged to Firmicutes (55.3%), Bacteroidetes (25.4%), Tenericutes (2.9%), and Proteobacteria (2.5%), representing more than 86% of the total bacterial phyla in the fecal samples.

In another study, Dowd et al. (2008) analyzed fecal samples from 20 Holsteins and found that all samples were highly prevalent of *Clostridium* (19.0% of the total population across all cows), *Porphyromonas* (7.34%), *Bacteroides* (9.26%), *Ruminococcus* (3.57%), *Alistipes* (6.61%), *Lachnospira* (3.73%), and *Prevotella* (5.47%).

5.3.1 Factors That Influence Intestinal Bacterial Core

There are several factors that have been reported to modify the intestinal microbiota of dairy cows including diet, particularly the shift from high forage to high concentrate ration, use of antibiotics, stressful periods for the host, and immune status or occurrence of diseases. An additional factor in ruminants is that some metabolites released during ruminal fermentation can trigger alterations in the composition of gut microbiota. Some of the nutrients, like starch, are able to bypass the forestomachs without being degraded, inducing fluctuations of the bacterial community in the gut. More details about factors that influence intestinal bacterial microbiota are discussed below.

5.3.1.1 Amount of Grain in the Ration

Beef cattle fed grain or primarily forage had a 2.4-fold increase of Bacteroidetes in the fecal samples in steers that received grain compared with those that were fed mainly forage (Shanks et al. 2011). Additionally, a closer examination of the Bacteroidetes at a family level revealed a relative abundance of the family of Prevotellaceae. For example, steers that were fed grain contained tenfold greater Prevotellaceae than the steers fed forage. Moreover, the bacterial communities from steers fed grain exhibited a severe reduction in community richness, which the authors attributed largely to the emergence of the Prevotellaceae family.

Shanks et al. (2011) also reported an increase in the relative abundance of Bacteroidetes and a decrease of Firmicutes as the amount of starch increased in the fecal matter. This is not surprising considering that members of Bacteroidetes phylum are known to be involved in digestion of complex carbohydrates.

5.3.1.2 Intestinal Section

Each segment of the intestinal tract has different physiologic functions and consequently the bacterial communities differ between the small and large intestines. Analyzing 10 different sites in the GI tract of dairy cows, Mao et al. (2015) found 21 bacterial phyla in the digesta samples belonging to Firmicutes (64.81%), Bacteroidetes (15.06%), and Proteobacteria (13.29%). The rumen and abomasum sheltered most of the phyla (19 phyla), whereas a lowest number of phyla were observed in the cecum (12 phyla). When the authors compared the bacterial composition regionally, Firmicutes dominated all communities along the GI tract, with the exception of the duodenum, where Proteobacteria (45.6%) were the most dominant phylum. The second most prevalent phylum in the forestomachs was Bacteroidetes, whereas Proteobacteria ranked the second most prevalent phylum in the jejunum, ileum, cecum, and colon. Analysis at the genus level showed that the unclassified family of Enterobacteriaceae was the predominant one in the small intestine, cecum, and colon, whereas a large proportion of *Acetitomaculum*, *Ruminococcus*, and unclassified Lachnospiraceae were dominant in the jejunum. Data also showed that genus *Butyrivibrio* was greater in the duodenum and jejunum than the other regions of the intestines. Unclassified *Peptostreptococcaceae* and *Turicibacter* were increased in the ileum and large intestines, whereas *Clostridium* was more present in the large intestine than in the small intestine (Mao et al. 2015).

5.3.1.3 Digesta and Mucosal Bacteria

There is a significant difference between the diversity and composition of bacterial communities in the mucosa of the intestines and digesta. For example, the proportion of Firmicutes in the digesta of the jejunum and large intestines was greater than their corresponding mucosal tissues (Mao et al. 2015). Interestingly, Bacteroidetes were found to be lower in the digesta of the duodenum, jejunum, and large intestines than their corresponding mucosal tissues. Mucosal tissues of the rectum had greater proportion of Proteobacteria than the digesta samples, whereas digesta of duodenum showed a comparatively higher proportion of Proteobacteria.

5.4 Gastrointestinal Microbiota and Disease

Clostridium spp. is a broad genus ubiquitous in the GI tract. Clostridia can both positively and negatively influence ruminants related to the individual *Clostridium* species involved. For example, species of *C. perfringens*, *C. tetani*, *C. botulinum*, and *C. difficile* can cause a significant impact on milk yield. Conversely, some *Clostridium* spp. like *Clostridium kluyveri* may be beneficial and improve digestion of complex organic matter such as cellulose and even act as beneficial probiotics (Widyastuti et al. 1992).

Bacteroidetes are known intestinal bacteria that can be both beneficial and harmful, and also participate in natural genetic transfer of antimicrobial resistance genes (Shoemaker et al. 1991).

It is interesting to note that the activity of rumen bacteria is associated with the release and production of various compounds. Some of those compounds might be involved in the pathobiology of periparturient diseases. Indeed, several recent metabolomics studies have identified various metabolites originating from digestion of feed materials or bacterial activity that might potentially be involved in the pathobiology of different diseases of transition dairy cows (Ametaj et al. 2010a, b; Saleem et al. 2012).

5.4.1 Grain Feeding, Microbiota Alterations, and Disease

Rumen houses a variety of anaerobic microorganisms, the majority of which are involved in the breakdown and digestion of dietary compounds consumed. Breeds of dairy cows have been selected over several decades for their milk production, however, this was not accompanied by similar selection for increased feed intake. The selection for high milk production has altered how dairy cows were fed over the years, and an increase in the proportion of feed with high level of energy became necessary. High concentrate diets are commonly associated with subclinical or clinical ruminal acidosis (i.e., SARA or ARA), which is characterized by pH below 5.8 in the rumen fluid (Cooper and Klopfenstein 1996).

The consumption of fermentable carbohydrates is associated with the release of VFAs including acetate, propionate, and butyrate, and other acidic products like lactate that subsequently lower the pH of the rumen fluid (Steele et al. 2012). The decrease in rumen pH influences composition of microbiota with lactic acid producers, such as *Streptococcus bovis*, outnumbering lactate utilizers such as *Megasphaera elsdenii* and *Selenomonas ruminantium*, leading to accumulation of lactate in the rumen fluid. Moreover, if rumen pH continues to decrease, *lactobacillus* spp. start surpassing *S. bovis* triggering an excessive accumulation of lactate, leading to ruminal acidosis (Russell and Hino 1985). Due to the low pK_a (i.e., 3.7) of lactate compared to pK_a of the major VFA (i.e., 4.8–4.9), ruminal pH is lowered to acidic values (Nocek 1997). Other changes in the microbiota related to high grain feeding include fiber-degrading bacteria. The majority of the bacterial species that degrade fiber such as *Fibrobacter succinogenes*, *Ruminococcus albus*, and *Ruminococcus flavefaciens* are sensitive to the rumen acidic pH and start decreasing in numbers (Russell and Wilson 1996).

Why is the low rumen pH harmful to the cow? Ruminal acidosis has been clearly associated with major changes in the population of rumen microbiota, ruminal short chain fatty acids fermentation patterns, altered gastrointestinal function, feed intake, milk production and composition, liver abscesses, and consequently might be an etiological factor for a number of other diseases (Nocek 1997; Plaizier et al. 2012; Enemark 2008; Steele et al. 2011).

Ruminal acidosis is also associated with swelling of epithelial layers because rumen motility, during acidosis, decreases and there is stimulation of mucopolysaccharide production by *S. bovis*, an acid-resistant bacterium, that increases the viscosity of rumen content (Khafipour et al. 2009).

Moreover, the acidic pH of the rumen digesta has a negative impact on the integrity of the rumen wall. Repeated assaults promoted by fermentation acids might cause atrophy of the rumen papillae, acute or chronic injury of the diffusion areas, scars resulting from severe local ruminitis, drilling and mucomycoses associated with pain, discomfort and dysfunction in feed intake and alteration of rumen functions (Enemark 2008).

Large numbers of Gram-negative bacteria inhabit the rumen and intestines of dairy cows, playing important roles in the digestive process. The rumen epithelium works as a barrier preventing translocation of bacteria and their products such as lipopolysaccharide (LPS) or other harmful compounds into the bloodstream or lymphatic circulation. Failure of the barrier functions has been implicated in translocation of LPS into the systemic circulation. Once in the blood circulation, LPS interacts with various immune and non-immune cells including macrophages, polymorphonuclear granulocytes, thrombocytes as well as endothelial and smooth muscle cells, stimulating production of pro-inflammatory mediators such as cytokines, which trigger activation of the acute phase response (reviewed in Eckel and Ametaj 2016). Depending on the severity of the barrier failure, the LPS structure, and the magnitude of LPS translocation, the immune response may remain local or expand throughout the body and initiate health issues like hypotension, shock, critical organ failure, respiratory arrest, and death (Jacobsen et al. 2004).

Presence of LPS in the rumen, at acidic pH values, increases more than sixfold the permeability of the rumen tissue and more than fivefold the permeability of the colon tissue to nondigestible compounds (Emmanuel et al. 2007), suggesting that translocation of harmful compounds from rumen or intestines might be easily increased in the presence of LPS.

The mechanism by which endotoxins damage barrier functions of the rumen is related to induction of cell apoptosis, disruption of tight junction protein zonula occludens-1, and enhancement of epithelial permeability in a dose and time dependent manner (Chin et al. 2006). Alterations in the expression and/or location of the structure and function of tight junction proteins might occur through increased expression of inducible nitric oxide synthase, which in turn increases concentrations of nitric oxide (Singh et al. 2007). Nitric oxide lowers the activity of Na^+ pumps and results in the swelling and dysregulation of tight junctional protein expression, and in barrier failure (Han et al. 2004). In addition to disruption of tight junction proteins, nitric oxide has been shown to induce enterocyte apoptosis, potentially by inhibition of cellular respiration and activation of apoptotic cascades (Potoka et al. 2002), which increases tissue damage and its permeability. Excessive amount of nitric oxide combined with superoxide anions, form toxic peroxynitrites, which trigger epithelial cell death or apoptosis (Bossy-Wetzel and Lipton 2003).

Rumen epithelium has a stratified squamous structure and a keratinous layer, known as stratum corneum, which acts as a protective barrier. High grain diets have been reported to cause a dramatic increase in the proliferation of the epithelium resulting in premature transition of cells into the keratinous layer, known as parakeratosis (Steele et al. 2011). Interestingly, an earlier report has shown that LPS alone can cause abnormal keratin production, leading to parakeratosis (Singh et al. 2007), which might explain the frequent incidence of parakeratosis and hyperkeratosis during SARA.

Rumen epithelium changes to adapt to high grain diets by increasing the size of the papillae, through proliferation of stratified squamous rumen epithelium, maximizing the surface area for VFA absorption (Odongo et al. 2006). Steele et al. (2011) demonstrated that feeding dairy cows high grain diets after one week resulted in dramatic changes in regard to rumen epithelium associated with deterioration of the cellular junctions, sloughing, and increased space between the cells, creating conditions for microbial translocation into the blood circulation.

The increase in the concentration of LPS in the rumen and its translocation into the systemic circulation is harmful not only to the GI tract epithelium but it can also modify the structure and composition of ruminal bacterial communities. In a recent study, intravenous infusion of LPS in Holstein lactating cows was associated with alteration of composition of rumen microbiota (Jing et al. 2014). Administration of LPS triggered increased abundance of Firmicutes and linearly lowered the percentages of Bacteroidetes, Tenericutes, Spirochaetes, Chlorobi, and Lentisphaerae. At the genus level one of the major changes observed in this study were the decrease of *Prevotella* reads and unclassified Bacteroidales and an increase of unclassified *Christensenallaceae* in cows treated with LPS compared with controls.

The mechanism of how intravenous administration of LPS changes the composition of rumen bacterial community is still unknown. Jing et al. (2014) suggested that infusion of LPS might indirectly change the physiology of the animals. They suggested that the effects of LPS infusion on ruminal bacterial communities were related to the decrease of rumen pH due to enhanced accumulation of VFA, a decrease in salivation rate, and increased rectal temperature in the cows infused. Another reason might be that the inflammation triggered by intravenous administration of LPS might affect GI tract barrier functions and host–microbiome interactions and a more aggressive response of the host towards bacterial community.

Indeed, another study demonstrated intravenous infusion of LPS lowered rumination activity and salivation in dairy calves (Borderas et al. 2008). It is possible that LPS lowers cow's capacity to buffer rumen acids, and as a consequence there is accumulation of VFA and other acids in the rumen. These findings suggest that translocation of LPS during ruminal acidosis may further aggravate the disease, influence microbiota composition, and exacerbate cow's health.

The damage to the rumen epithelium is triggered by ruminal acidosis or rumenitis, when cows are fed high grain diets, resulting in the epithelium becoming susceptible to invasion and colonization by ruminal pathogenic bacteria (Nagaraja and Lechtenberg 2007; Reinhardt and Hubbert 2015; Amachawadi and Nagaraja 2016). Once colonization of ruminal tissues has occurred, the invading bacteria penetrate rumen wall and subsequently enter into portal circulation. Bacteria from the portal circulation are entrapped in the portal capillary system of the liver, leading to infection and establishment of abscesses. Liver abscesses have shown to be a multimicrobial infection; however, the most predominant bacteria have been shown to be *Fusobacterium necrophorum* and *Actinomiyces pyogenes* (Reinhardt and Hubbert 2015; Amachawadi and Nagaraja 2016).

Although liver abscesses are considered to be a more frequent disease in beef cattle due to the excessive amounts of grain in finishing rations, there are also reports of Holstein feedlot steers being affected by liver abscesses. Moreover, it is interesting to note that the proportion of liver abscesses and liver abscesses with adhesion to the diaphragm and visceral organs is greater in Holstein feedlot steers than in other beef breeds (Vogel and Parrott 1994; Duff and McMurphy 2007). Doré et al. (2007) demonstrated that liver abscesses in dairy cows were associated with other diseases such as peritonitis, vagal indigestion, traumatic reticuloperinonitis, abomasal displacement, pneumonia, and enteritis. On the other hand, Rezac et al. (2014) examined cows at slaughterhouses and observed that approximately 32% of all Holstein cows had liver abscesses.

Ruminal acidosis has also been associated with laminitis, with its severity depending on the gravity and duration of ruminal acidosis (Nocek 1997). During SARA, endotoxins like lipopolysaccharides (LPS) are released due to lysis of Gram-negative bacterial cells and enter into the bloodstream to initiate an inflammatory response in the hoof (Emmanuel et al. 2007, 2008). Moreover, there has been reports that acid-resistant bacterium *Allisonella histaminiformans,* a *Gram-negative* bacterium, isolated from dairy cows fed only grain but not hay is able to produce histamine as an exclusive metabolite of histidine decarboxylation (Garner et al. 2004). Histamine has been implicated in the etiopathology of laminitis in dairy cows (Mgassa et al. 1984; Nocek 1997).

Disturbances of rumen microbiota might also indirectly contribute to the pathobiology of mastitis. It has been suggested that endotoxins translocated into systemic circulation may play a role in increasing the incidence of mastitis by delaying the migration of neutrophils into the mammary gland during the initial infection (Eckel and Ametaj 2016). Endotoxins are associated with increased migration of neutrophils from bone marrow into circulation (i.e., neutrophilia), but there is evidence that migration of neutrophils out of the circulation is delayed, potentially, by nitric oxide induced by endotoxins (Wagner and Roth 1999).

Fatty liver is another disease that is potentially associated with increased concentration of endotoxin in the rumen. A possible mechanism for development of fatty liver in association with endotoxins is through the lipoprotein pathway.

Triglyceride-rich lipoproteins (Lp) bind and neutralize LPS, and liver hepatocytes remove Lp-LPS complexes quickly from circulation through the process of endocytosis. The preferential shunting of Lp-associated LPS into hepatocytes in large quantities may cause large accumulation of triglycerides (Ametaj et al. 2010b; Eckel and Ametaj 2016). Among others, translocated LPS stimulates production of TNF, and injection of TNF alone in late-lactation triggered storage of triglycerides in the liver of dairy cows (Bradford et al. 2009).

5.4.2 Potential Pathologies by Other Bacterial Compounds Produced in the Rumen

In a study conducted by us (Ametaj et al. 2010a) we reported that feeding dairy cows increasing amounts of barley grain is associated with major changes in the metabolite signatures of the rumen fluid. In a later publication, a more in-depth targeted metabolomics analysis of the same rumen fluid samples revealed major metabolite alterations between cows fed lower amounts of barley grain (0 or 15% dry matter basis) versus those fed greater amounts (30 and 45% dry matter basis) (Saleem et al. 2012). Some metabolites of interest increased during feeding of high grain diets included ethanol, ethanolamine, 3-hydroxybutirate, methylamine, dimethylamine, N-nitrosodimethylamine, alanine, uracil, xanthine, phenylacetylglycine, phenylacetate, putrescine, urea, 4-aminobutyrate, methionine, phenylalanine, and threonine. Additionally, the high grain diets decreased concentrations of 1,3-dihydroxyacetone, 3-phenylpropionate, hydrocinnamic acid, and phosphatidylcholines. The potential pathogenic effects of some of the aforementioned rumen fluid metabolites will be discussed in more detail below.

Several of the metabolites reported to be increased in the rumen such as methylamine, and other methylated amines such as putrescine, cadaverine, and dimethylamine derive from decarboxylation of amino acids by the activity of certain types of rumen bacteria (Rice and Koehler 1976). Saleem et al. (2012) suggested that there might be health related issues to dairy cows if methylated amines are absorbed into the blood circulation and catabolized further by semicarbazide-sensitive amine oxidases. Some of the intermediary metabolites including hydrogen peroxide or formaldehyde are toxic and have been associated with diseases like diabetes, vascular disorders, heart failure, and Alzheimer's disease in humans (Yu et al. 2006). There are no reports regarding potential health effects of methylated amines in dairy cows. This would be a very interesting area of research by dairy cow health scientists in the future.

Phenylacetate, enhanced during high grain diets, has been identified as an antifungal substance produced by some *Streptomyces* spp. strains (Hwang et al. 2001) and might have similar effects in the rumen environment. The authors reported that phenylacetate inhibits the growth of some important fungal and yeast species in the rumen such as *Phytophthora capsici*, *Rhizoctonia solani*, *Pseudomonas syringae*, and *Saccharomyces cerevisiae*. Since *S. cerevisiae* has been used as a direct-fed microbial in dairy cows, due to its positive effects on improving ruminal fermentation and dairy cow performance, enhanced phenylacetate during high grain diets might lower its beneficial effects.

One of the intermediates of dimethylamine, N-nitrosodimethylamine, is known as a potent carcinogen (Mitch and Sedlak 2002). Increases in N-nitrosodimethylamine have been reported to be associated with kidney, liver, and lung tumors in humans (Magee and Barnes 1962). Additionally, it was reported that N-nitrosodimethylamine induces tumors in a variety of animal species such as rats, mice, rabbit, fish, and birds and in different organs such as bladder, kidney, liver, esophagus, and stomach (Peto et al. 1984). Human exposure to this metabolite has increased the risk of gastric, bladder, and colon cancer (Knekt et al. 1999). N-nitrosodimethylamine is produced endogenously, in the gastrointestinal tract, or exogenously, with food consumption (Krul et al. 2004). Although there are no studies reported with regard to negative effects of N-nitrosodimethylamine in dairy cows the fact that this compound was increased during feeding of high grain diets (Saleem et al. 2012) and it triggers multiple cancers in humans warrants further investigation.

Putrescine metabolism in animals results in aldehyde and hydrogen peroxide production (Yamashita et al. 1993), which are extremely toxic compounds for various eukaryotic cells resulting in oxidative stress. This may also contribute to periparturient disorders as well as other metabolic diseases (Ronchi et al. 2000) and might be very toxic to ruminal protozoa and rumen epithelial cells (Willard and Kodras 1967). Therefore, elevated rumen putrescine when feeding diets high in grain should be a metabolite of focus in future studies in relation with health of transition dairy cows.

Ethanolamine is a nutrient derived from phospholipids in the membrane of shed enterocytes, and its release in the ruminal fluid might happen due to a change in the turnover of epithelial cells and cell lysis of ruminal microbiota (Nagaraja et al. 1978). Pathogenic Gram-negative bacteria such as enterohemorrhagic *Escherichia coli* and *Salmonella enterica* can use ethanolamine as a nitrogen source, conferring a growth advantage over other commensal microbiota (Bertin et al. 2011; Thiennimitr et al. 2011). These studies suggest that the release of such a metabolite in the rumen during feeding of high grain diets can be critical for multiplication of certain pathogenic bacteria with relevance to both animal health and food safety. Similarly, maltose also can provide a competitive advantage to pathogenic *E. coli* in the intestine (Jones et al. 2008) due to a specific maltose-binding protein located in the cell wall of *E. coli*.

It was reported that concentration of urea in the rumen fluid of cows fed high grain diets was increased (Saleem et al. 2012). Presence of urea can be quite problematic due to its rapid hydrolysis to NH_3 by microbial enzymes in the rumen (Highstreet et al. 2010). Absorption of NH_3 into the blood circulation is toxic. High levels of NH_3 in the blood have been shown to cause dyspnea, excessive salivation, frothing, ataxia, weakness, abdominal pain, violent struggling, and bellowing in ruminants (Blood and Henderson 1963). The aforementioned symptoms are common in cows suffering from acute rumen acidosis.

A summary of all bacteria phyla in various section of the GI tract of cattle are given in Fig. 5.1.

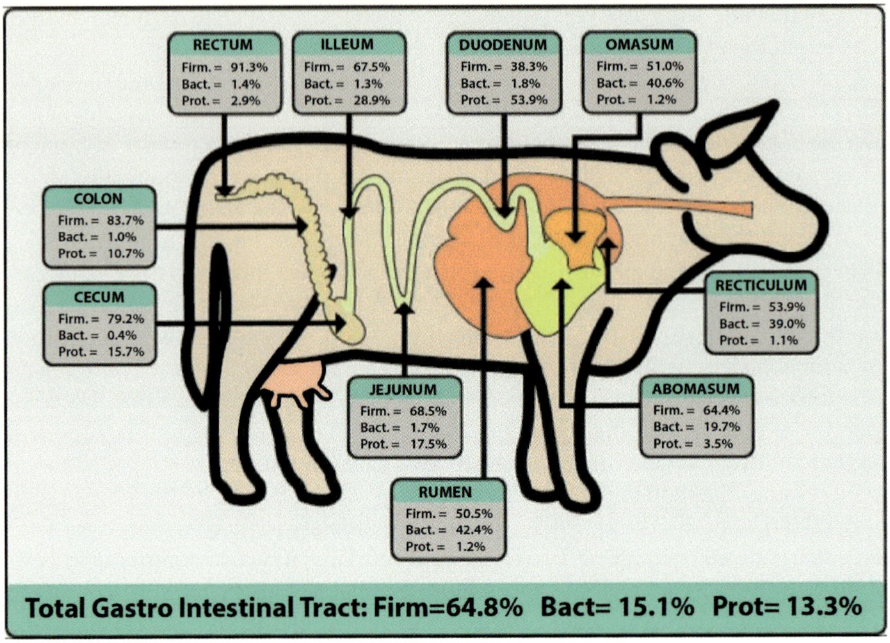

Fig. 5.1 Bacteria phyla in the GI tract of cattle [*Firm.* Firmicutes, *Prot.* Proteobacteria, *Bact.* Bacteroidetes]

Conclusions

Ruminal and intestinal microbiota are extremely important for dairy cows due to their functions in fermenting and utilizing plant components ingested by the host, and producing VFA and other compounds that are utilized as nutrients. Major changes in the GI tract microbiota are observed in dairy cows when fed high proportions of grain, especially after calving, to support milk production. In addition, major changes in rumen fluid metabolites and other compounds have been reported during feeding of high grain diets. Feeding diets with high starch content might support certain groups of bacteria against others; however, such alterations might contribute to the health status of the cow. Although not very much is known about the specific metabolites released during fermentation processes in the GI tract of ruminants by specific bacteria, recent studies are reporting a few compounds with potential health consequences. Translocation of these metabolites into the bloodstream might be associated with the pathobiology of multiple diseases including rumen acidosis and parakeratosis, laminitis, fatty liver, and liver abscesses. Future studies in dairy cows are warranted to better understand host–microbiome interrelationships in health and disease.

References

Amachawadi RG, Nagaraja TG (2016) Liver abscesses in cattle: a review of incidence in Holsteins and of bacteriology and vaccine approaches to control in feedlot cattle. J Anim Sci 94:1620–1632

Ametaj BN, Zebeli Q, Iqbal S (2010a) Nutrition, microbiota, and endotoxin-related diseases in dairy cows. Rev Bras Zootec 39:433–444

Ametaj BN, Zebeli Q, Saleem F et al (2010b) Metabolomics reveals unhealthy alterations in rumen metabolism with increased proportion of cereal grain in the diet of dairy cows. Metabolomics 6:583–594

Bertin Y, Girardeau JP, Chaucheyras-Durand F et al (2011) Enterohaemorrhagic *Escherichia coli* gains a competitive advantage by using ethanolamine as a nitrogen source in the bovine intestinal content. Environ Microbiol 13:365–377

Blanch M, Calsamiglia S, DiLorenzo N et al (2009) Physiological changes in rumen fermentation during acidosis induction and its control using a multivalent polyclonal antibody preparation in heifers. J Anim Sci 87:1722–1730

Blood DC, Henderson JA (1963) Veterinary medicine, 2nd edn. Baillière, Tyndall & Cassell, London, p 107

Borderas TF, Passillé AM, Rushen J (2008) Behavior of dairy calves after a low dose of bacterial endotoxin. J Anim Sci 86:2920–2927

Borsanelli AC, Gaetti-Jardim Júnior E, Schweitzer CM et al (2015) Presence of *Porphyromonas* and *Prevotella* species in the oral microflora of cattle with periodontitis. Pesqui Vet Bras 35:829–834

Bossy-Wetzel E, Lipton SA (2003) Nitric oxide signaling regulates mitochondrial number and function. Cell Death Differ 10:757–760

Bradford BJ, Mamedova LK, Minton JE et al (2009) Daily injection of tumor necrosis factor-α increases hepatic triglycerides and alters transcript abundance of metabolic genes in lactating dairy cattle. J Nutr 139:1451–1456

Brook I, Frazier EH (2003) Immune response to *Fusobacterium nucleatum* and *Prevotella intermedia* in the sputum of patients with acute exacerbation of chronic bronchitis. Chest J 124:832–843

Brulc JM, Antonopoulos DA, Miller MEB et al (2009) Gene-centric metagenomics of the fiber-adherent bovine rumen microbiome reveals forage specific glycoside hydrolases. PNAS 106:1948–1953

Cheng KJ, Bailey CBM, Hironaka R et al (1979) A technique for depletion of bacteria adherent to the epithelium of the bovine rumen. Can J Anim Sci 59:207–209

Chin AC, Flynn AN, Fedwick JP et al (2006) The role of caspase-3 in lipopolysaccharide-mediated disruption of intestinal epithelial tight junctions. Can J Physiol Pharmacol 84:1043–1050

Cooper R, Klopfenstein T (1996) Effect of Rumensin and feed intake variation on ruminal pH. In: Scientific Update on Rumensin/Tylan/Mycotil for the Professional Feedlot Consultant. Elanco Animal Health, Indianapolis, pp A1–A14

Craig WM, Broderick GA, Ricker DB (1987) Quantitation of microorganisms associated with the particulate phase of ruminal ingesta. J Nutr 117:56–62

Doré E, Fecteau G, Hélie P, Francoz D (2007) Liver abscesses in Holstein dairy cattle: 18 cases (1992–2003). J Vet Intern Med 21:853–856

Dowd SE, Callaway TR, Wolcott RD et al (2008) Evaluation of the bacterial diversity in the feces of cattle using 16S rDNA bacterial tag-encoded FLX amplicon pyrosequencing (bTEFAP). BMC Microbiol 8:125

Duff GC, McMurphy CP (2007) Feeding Holstein steers from start to finish. Vet Clin North Am Food Anim Pract 23:281–297

Eckel EF, Ametaj BN (2016) Invited review: Role of bacterial endotoxins in the etiopathogenesis of periparturient diseases of transition dairy cows. J Dairy Sci 99:1–24

Emmanuel DGV, Madsen KL, Churchill TA et al (2007) Acidosis and lipopolysaccharide from *Escherichia coli* B:055 cause hyperpermeability of rumen and colon tissues. J Dairy Sci 90:5552–5557

Emmanuel DGV, Dunn SM, Ametaj BN (2008) Feeding high proportions of barley grain stimulates an inflammatory response in dairy cows. J Dairy Sci 91:606–614

Enemark JMD (2008) The monitoring, prevention and treatment of sub-acute ruminal acidosis (SARA): a review. Vet J 176:32–43

Fernando SC, Purvis HT, Najar FZ et al (2010) Rumen microbial population dynamics during adaptation to a high grain diet. Appl Environ Microbiol 76:7482–7490

Field TR, Sibley CD, Parkins MD et al (2010) The genus *Prevotella* in cystic fibrosis airways. Anaerobe 16:337–344

Frank D, Zhu W, Sartor R, Li E (2011) Investigating the biological and clinical significance of human dysbioses. Trends Microbiol 19:427–434

Garner MR, Gronquist MR, Russell JB (2004) Nutritional requirements of *Allisonella histaminiformans*, a ruminal bacterium that decarboxylates histidine and produces histamine. Curr Microbiol 49:295–299

Han X, Fink MP, Yang R, Delude RL (2004) Increased iNOS activity is essential for intestinal epithelial tight junction dysfunction in endotoxemic mice. Shock 21:261–270

Hayashi H, Shibata K, Sakamoto M et al (2007) *Prevotella copri* sp. *nov*. and *Prevotella stercorea* sp. *nov*., isolated from human faeces. Int J Syst Evol Microbiol 57:941–946

Henderson G, Cox F, Ganesh S et al (2015) Rumen microbial community composition varies with diet and host, but a core microbiome is found across a wide geographical range. Sci Rep 5:145–167

Highstreet A, Robinson PH, Robison J, Garrett JG (2010) Response of Holstein cows to replacing urea with a slowly rumen released urea in a diet high in soluble crude protein. Livest Sci 129:179–185

Hwang BK, Lim SW, Kim BS et al (2001) Isolation and in vivo and in vitro antifungal activity of phenylacetic acid and sodium phenylacetate from *Streptomyces humidus*. Appl Environ Microbiol 67:3739–3745

Jacobsen S, Andersen PH, Toelboell T, Heegaard PMG (2004) Dose dependency and individual variability of the lipopolysaccharide-induced bovine acute phase protein response. J Dairy Sci 87:3330–3339

Jami E, Mizrahi I (2012) Composition and similarity of bovine rumen microbiota across individual animals. PLoS One 7:e33306

Jing L, Zhang R, Liu Y et al (2014) Intravenous lipopolysaccharide challenge alters ruminal bacterial microbiota and disrupts ruminal metabolism in dairy cattle. Br J Nutr 112:170–182

Jones SA, Jorgensen M, Chowdhury FZ et al (2008) Glycogen and maltose utilization by *Escherichia coli* O157:H7 in the mouse intestine. Infect Immun 76:2531–2540

Kaler J, Wani SA, Hussain I et al (2012) A clinical trial comparing parental oxytetracycline and enrofloxacin on time to recovery in sheep lame with acute or chronic footrot in Kashmir, India. BMC Vet Res 8:1–12

Kennan RM, Han X, Porter CJ, Rood JI (2011) Pathogenesis of ovine footrot. Vet Microbiol 153:59–66

Khafipour E, Li S, Plaizier JC, Krause DO (2009) Rumen microbiome composition determined using two nutritional models of subacute ruminal acidosis. Appl Environ Microbiol 75:7115–7124

Knekt P, Jarvinen R, Dich J, Hakulinen T (1999) Risk of colo-rectal and other gastrointestinal cancers after exposure to nitrate, nitrite and N-nitroso compounds: a follow-up study. Int J Cancer 80:852–856

Kocherginskaya SA, Aminov RI, White BA (2001) Analysis of the rumen bacterial diversity under two different diet conditions using denaturing gradient gel electrophoresis, random sequencing, and statistical ecology approaches. Anaerobe 7:119–134

Krul CAM, Zeilmaker MJ, Schothorst RC, Havenaar R (2004) Intragastric formation and modulation of N-nitrosodimethylamine in a dynamic in vitro gastrointestinal model under human physiological conditions. Food Chem Toxicol 42:51–63

Li M, Penner GB, Hernandez-Sanabria E et al (2009) Effects of sampling location and time, and host animal on assessment of bacterial diversity and fermentation parameters in the bovine rumen. J Appl Microbiol 107:1924–1934

Li RW, Wu S, Baldwin RLVI et al (2012) Perturbation dynamics of the rumen microbiota in response to exogenous butyrate. PLoS One 7:e29392

Made N, Okamoto M, Kondo K et al (1998) Incidence of *Prevotella intermedia* and *Prevotella nigrescens* in periodontal health and disease. Microbiol Immunol 42:583–589

Magee PN, Barnes JM (1962) Induction of kidney tumors in the rat with dimethylnitrosamine (N-nitrosodimethylamine). J Pathol Bacteriol 84:19–31

Mao S, Zhang M, Liu J, Zhu W (2015) Characterising the bacterial microbiota across the gastrointestinal tracts of dairy cattle: membership and potential function. Sci Rep 5:16116

Mgassa MN, Amaya-Posada G, Hesselholt M (1984) Pododermitis aseptica diffusa (laminitis) in free-range beef cattle in tropical Africa. Vet Rec 115:413

Mitch WA, Sedlak DL (2002) Formation of N-nitrosodimethylamine (NDMA) from dimethylamine during chlorination. Environ Sci Technol 36:588–595

Mohammed R, Stevenson DM, Weimer PJ et al (2012) Individual animal variability in ruminal bacterial communities and ruminal acidosis in primiparous Holstein cows during the periparturient period. J Dairy Sci 95:6716–6730

Mullins CR, Mamedova LK, Carpenter AJ et al (2013) Analysis of rumen microbial populations in lactating dairy cattle fed diets varying in carbohydrate profiles and *Saccharomyces cerevisiae* fermentation product. J Dairy Sci 96:5872–5881

Nadkarni MA, Browne GV, Chhour K et al (2012) Pattern of distribution of *Prevotella* species/phylotypes associated with healthy gingiva and periodontal disease. Eur J Clin Microbiol Infect Dis 31:2989–2999

Nagaraja TG, Lechtenberg KF (2007) Liver abscesses in feedlot cattle. Vet Clin North Am Food Anim Pract 23:351–369

Nagaraja TG, Bartley EE, Fina LR, Anthony HD (1978) Relationship of rumen gram-negative bacteria and free endotoxin to lactic acidosis in cattle. J Anim Sci 47:1329–1337

Nocek JE (1997) Bovine acidosis: implications on laminitis. J Dairy Sci 80:1005–1028

Nocek JE, Allman JG, Kautz WP (2002) Evaluation of an indwelling ruminal probe methodology and effect of grain level on diurnal pH variation in dairy cattle. J Dairy Sci 85:422–428

Odongo NE, AlZahal O, Lindinger MI et al (2006) Effects of mild heat stress and grain challenge on acid-base balance and rumen tissue histology in lambs. J Anim Sci 84:447–455

Peterson G, Allen C, Holling C (1998) Ecological resilience, biodiversity, and scale. Ecosystems 1:6–18

Peto R, Gray R, Brantom P, Grasso P (1984) Nitrosamine carcinogenesis in 5210 rodents: chronic administration of sixteen different concentrations of NDEA, NDMA, NPYR and NPIP in the water of 4440 inbred rats, with parallel studies on NDEA alone of the effect of age of starting (3,6 or 20 weeks) and on species (rats, mice or hamsters). In: O'Neill IK, Von Borstel RC, Miller CT, Long J, Bartsch H (eds) N-nitroso compounds: occurrence, biological effects and relevance to human cancer. IARC Sci. Publ. No. 84, Lyon, pp 627–665

Petri RM, Schwaiger T, Penner GB et al (2013) Changes in the rumen epimural bacterial diversity of beef cattle as affected by diet and induced ruminal acidosis. Appl Environ Microbiol 79:3744–3755

Pflughoeft K, Versalovic J (2012) Human microbiome in health and disease. Annu Rev. Pathol 7:99–122

Pitta DW, Parmar N, Patel AK et al (2014) Bacterial diversity dynamics associated with different diets and different primer pairs in the rumen of Kankrej cattle. PLoS One 9:e111710

Plaizier JC, Khafipour E, Li S et al (2012) Subacute ruminal acidosis (SARA), endotoxins and health consequences. Anim Feed Sci Technol 172:9–21

Potoka DA, Nadler EP, Upperman JS, Ford HR (2002) Role of nitric oxide and peroxynitrite in gut barrier failure. World J Surg 26:806–811

Ramirez HAR, Nestor K, Tedeschi LO et al (2012) The effect of brown midrib corn silage and dried distillers' grains with soluble on milk production, nitrogen utilization and microbial community structure in dairy cows. Can J Anim Sci 92:365–380

Reinhardt CD, Hubbert ME (2015) Control of liver abscesses in feedlot cattle: A review. Prof Anim Sci 31:101–108

Rezac DJ, Thomson DU, Siemens MG et al (2014) A survey of gross pathologic conditions in cull cows at slaughter in the Great Lakes region of the United States. J Dairy Sci 97:4227–4235

Rice SL, Koehler PE (1976) Tyrosine and histidine decarboxylase activities of *Pediococcus cerevisiae* and *Lactobacillus* specie and the production of tyramine in fermented sausages. J Milk Food Technol 39:166–169

Ronchi B, Bernabucci U, Lacetera N, Nardone A (2000) Oxidative and metabolic status of high yielding dairy cows in different nutritional conditions during the transition period. In: Proceedings of 51st Annual Meeting EAAP, Vienna, Austria. Wagningen Pers., Wageningen, The Netherlands, p 125

Ruan Y, Shen L, Zou Y et al (2015) Comparative genome analysis of *Prevotella intermedia* strain isolated from infected root canal reveals features related to pathogenicity and adaptation. BMC Genomics 16:1–22

Russell JB, Hino T (1985) Regulation of lactate production in *Streptococcus bovis*: a spiraling effect that contributes to rumen acidosis. J Dairy Sci 68:1712–1721

Russell JB, Wilson DB (1996) Why are ruminal cellulolytic bacteria unable to digest cellulose at low pH? J Dairy Sci 79:1503–1509

Saleem F, Ametaj BN, Bouatra S et al (2012) A metabolomics approach to uncover the effects of grain diets on rumen health in dairy cows. J Dairy Sci 95:6606–6623

Shanks OC, Kelty CA, Archibeque S et al (2011) Community structures of fecal bacteria in cattle from different animal feeding operations. Appl Environ Microbiol 77:2992–3001

Sheldon M, Williams EJ, Miller ANA et al (2008) Uterine diseases in cattle after parturition. Vet J 176:115–121

Shoemaker NB, Anderson KL, Smithson SL et al (1991) Conjugal transfer of a shuttle vector from the human colonic anaerobe *Bacteroides uniformis* to the ruminal anaerobe *Prevotella* (Bacteroides) *ruminicola* B(1)4. Appl Environ Microbiol 57:2114–2120

Singh AK, Jiang Y, Gupta S (2007) Effects of bacterial toxins on endothelial tight junction in vitro: A mechanism-based investigation. Toxicol Mech Methods 17:331–347

Steele MA, Croom J, Kahler M et al (2011) Bovine rumen epithelium undergoes rapid structural adaptations during grain-induced subacute ruminal acidosis. Am J Physiol Regul Integr Comp Physiol 300:1515–1523

Steele MA, AlZahal O, Walpole ME, McBride BW (2012) Short communication: Grain-induced subacute ruminal acidosis is associated with the differential expression of insulin-like growth factor-binding proteins in rumen papillae of lactating dairy cattle. J Dairy Sci 95:6072–6076

Stewart CS, Flint HJ, Bryant MP (1997) The rumen bacteria. In: Hobson PN, Stewart CS (eds) The rumen microbial ecosystem, 2nd edn. Springer, New York, pp 10–72

Stubbs S, Lewis MAO, Waddington RJ, Embery G (1996) Hydrolytic and depolymerising enzyme activity of *Prevotella intermedia* and *Prevotella nigrescens*. Oral Dis 2:272–278

Thiennimitr P, Winter SE, Winter MG et al (2011) Intestinal inflammation allows *Salmonella* to use ethanolamine to compete with the microbiota. Proc Natl Acad Sci U S A 108:17480–17485

Thomas C, Versalovic J (2010) Probiotics-host communication: modulation of signaling pathways in the intestine. Gut Microbes 1:148–163

Ulrich M, Beer I, Braitmaier P et al (2010) Relative contribution of *Prevotella intermedia* and *Pseudomonas aeruginosa* to lung pathology in airways of patients with cystic fibrosis. Thorax 65:978–984

Vogel GJ, Parrott JC (1994) Mortality survey in feed yards: the incidence of death from digestive, respiratory and other causes in feed yards on the Great Plains. Compend Contin Educ Pract Vet 16:227–234

Wagner JG, Roth RA (1999) Neutrophil migration during endotoxemia. J Leukoc Biol 66:10–24

Wallace RJ, Cheng KJ, Dinsdale D, Orskov ER (1979) An independent microbial flora of the epithelium and its role in the ecomicrobiology of the rumen. Nature 279:424–426

Weimer PJ, Stevenson DM, Mantovani HC, Man SLC (2010a) Host specificity of the ruminal bacterial community in the dairy cow following near-total exchange of ruminal contents. J Dairy Sci 93:5902–5912

Weimer PJ, Stevenson DM, Mertens DR (2010b) Shifts in bacterial community composition in the rumen of lactating dairy cows under milk fat-depressing conditions. J Dairy Sci 93:265–278

Welkie DG, Stevenson DM, Weimer PJ (2010) ARISA analysis of ruminal bacterial community dynamics in lactating dairy cows during the feeding cycle. Anaerobe 16:94–100

Widyastuti Y, Lee SK, Suzuki K, Mitsuoka T (1992) Isolation and characterization of rice-straw degrading *Clostridia* from cattle rumen. J Vet Med Sci 54:185–188

Willard FL, Kodras R (1967) Survey of chemical compounds tested in vitro against rumen protozoa for possible control of bloat. Appl Microbiol 15:1014–1019

Yamashita Y, Bowen WH, Burne RA, Kuramitsu HK (1993) Role of the *Streptococcus mutans gtf* genes in caries induction in the specific-pathogen-free rat model. Infect Immun 61:3811–3817

Yu L, Wan F, Dutta S et al (2006) Autophagic programmed cell death by selective catalase degradation. Proc Natl Acad Sci U S A 103:4952–4957

A Systems Biology Approach to Dairy Cattle Subfertility and Infertility

6

Fabrizio Ceciliani, Domenico Vecchio, Esterina De Carlo, Alessandra Martucciello, and Cristina Lecchi

Abstract

Omics techniques have been widely applied to veterinary science, including reproductive disease and infertility, although their application is still far behind as compared to human medicine. This chapter presents the most recent achievement in the application of postgenomic techniques, such as transcriptomics, proteomics, and metabolomics, and how their application advanced the knowledge on genes and proteins, as well as the pathways they are involved in, related to the field of dairy cow subfertility. Focus will be on both female and male apparatus. System biology approaches to the study of gametes, including spermatozoa and seminal fluid, and ovaries, and follicular fluid, as well as the specific tissues where the gametes have origin from, namely epididymis and ovaries, will be presented. OMICS technologies were also applied to the characterization of maternal environment, such as the oviduct, uterus, and placenta, and the results were also described hereby. Omics applications to biomarker discovery and their potential for subfertility diagnostics in veterinary medicine are also highlighted.

F. Ceciliani, D.V.M., Ph.D. (✉) • C. Lecchi, D.V.M., Ph.D.
Department of Veterinary Medicine, Università degli Studi di Milano, Milano, Italy
e-mail: fabrizio.ceciliani@unimi.it; cristina.lecchi@unimi.it

D. Vecchio, Ph.D. • E. De Carlo, Ph.D. • A. Martucciello, Ph.D.
Istituto Zooprofilattico Sperimentale del Mezzogiorno—Centro di Referenza Nazionale sull'Igiene e le Tecnologie dell'Allevamento e delle Produzioni Bufaline—Sezione di Salerno, Salerno, Italy
e-mail: domenico.vecchio@izsmportici.it; esterina.decarlo@cert.izsmportici.it; alessandra.martucciello@cert.izsmportici.it

B.N. Ametaj (ed.), *Periparturient Diseases of Dairy Cows*,
DOI 10.1007/978-3-319-43033-1_6

Abbreviations

2D/MALDI-TOF	2D gel electrophoresis/matrix-assisted laser desorption ionization-time of flight
2D/MS	2D gel electrophoresis/mass spectrometry
2-DGE	2 Dimensional gel electrophoresis
2-DIGE	2-D fluorescence difference gel electrophoresis
2-DIGE/MS	2-D fluorescence difference gel electrophoresis/mass spectrometry
LC-MS/MS	Liquid chromatography-mass spectrometry
MALDI-TOF	Matrix-assisted laser desorption ionization-time of flight
NGS	Next generation sequencing

6.1 Introduction

Since early 1980s, fertility in dairy cows has steadily declined, resulting in the increase of culling and decrease of the longevity of the animals. In high-yielding cows, the reproductive disorder named as "subfertility" defines a delayed and irregular fertility or conception including conditions such as anovulatory estrus, anestrus, cystic ovarian disease, repeat breeder syndrome, endometritis, and delayed ovulation. Economic losses of stillbirth to the dairy industry in the USA caused by losses of replacement heifers due to subfertility only were estimated to be $125 million per year (Galligan 1999; Royal et al. 2000; Lucy 2001).

Current available assessment of fertility relies on morphological features of oocytes, sperm, and embryos and includes the tissue environments where they are produced and developed, namely epididymis, ovaries, oviduct, uterus, and placenta. This approach is too narrowly focused, and somehow subjective. New diagnostic and therapeutic tools are therefore required. Omics technologies such as transcriptomics, proteomics, and metabolomics are increasing the knowledge in the field of reproduction and fertility.

Next generation sequencing (NGS) and gel-free high throughput proteomics stand at the cutting edge of the techniques currently used for data acquisition in molecular pathogenesis studies and biomarker search and are probably poised to replace microarrays (Fig. 6.1) and 2-dimensional gel electrophoresis (2-DGE), although so far, microarrays and 2-DGE are still the workhorse among omics applied to bovine reproductive pathophysiology. Figure 6.1 and 6.2 graphically present the current transcriptomics and proteomics techniques applied to investigation of reproduction diseases at omics level.

This chapter is aimed to provide the reader with the most recent achievement of omics technologies, namely proteomics, transcriptomics, and metabolomics applied to the study of subfertility and infertility in dairy cattle, and how their application advanced the knowledge on genes and proteins, and related pathways and functions.

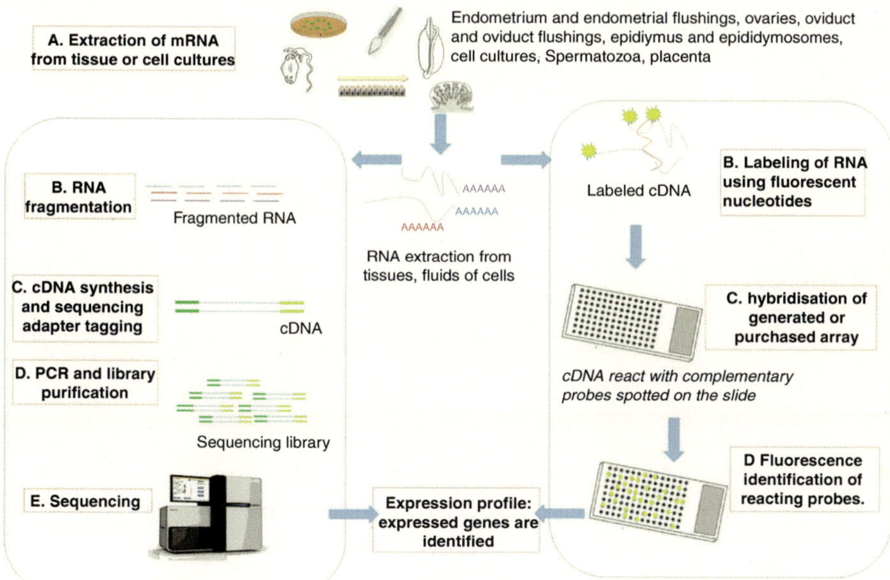

Fig. 6.1 *Transcriptomic workflow*. RNA-seq workflow: The workflow begins with poly-A-mRNA purification using poly-T beads. The RNA is cleaved into fragments of 100–200 bp by enzymatic reaction or by chemical hydrolysis. The fragmented RNA is converted into a double-stranded cDNA library. RNA fragments are hybridized and ligated to an adapter mixture using the RNA ligase. The adaptors' linked RNA is converted to single strand cDNA using reverse transcriptase and purified. The cDNA library is finally enriched with PCR and then purified. During the PCR step it is also possible to introduce specific short DNA sequences acting as barcodes to identify different samples. The final product consists of dsDNA molecules of 200–300 bp containing the copies of the RNAs present in the original sample surrounded by adapters and creates the final cDNA library. Microarray workflow: The workflow starts with RNA purification and retrotranscription to double-stranded cDNA. After purification, cDNA are fluorescently labeled with distinct fluorescent dyes—such as, Cy-3 (*green*) and Cy-5 (*red*)—and detected by hybridization onto immobilized DNA probes on the microarray. Each sample sequence (target) hybridizes to the complementary strand on the array (probe) allowing confirmation of the presence of a target gene. Multiple DNA probes are spotted on a thin support—such as, silicium, glass, or polymers—with each one being specific for a DNA or RNA target sequence

6.2 Systems Biology Techniques Applied to Gamete Quality and Their Environment

Poor semen quality has long been associated to infertility. Mammalian spermatozoa are produced in the testis and have to transit along the epididymis to acquire their fertilizing capability and their motility. The maturation process is related to the production of the proteins progressively secreted by epididymal sectors and seminal glands. In this chapter we will provide insights into the actual knowledge of omics techniques applied to bovine spermatozoa and their environment, including the epididymal and sexual gland secretions.

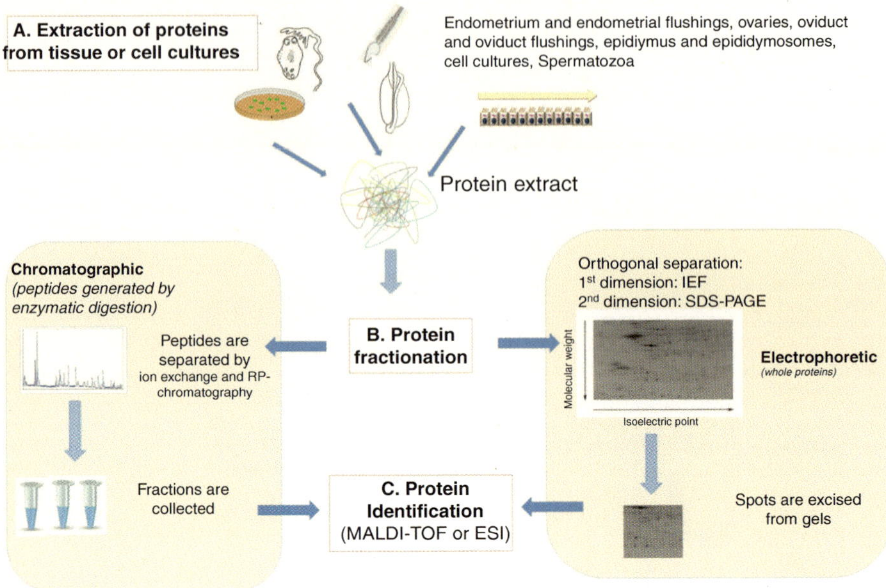

Fig. 6.2 Proteomic workflow. The proteomic workflow starts with the extraction of proteins from cell or tissues, or from part of them after subcellular fractionations, such as those carried out in spermatozoa or cell cultures. The protein fractionation may be either electrophoretic (*upper panel*), or chromatography (*lower panel*). The electrophoretic fractionation system former is applied to intact proteins. Conventional 2DE involves separation of protein by means of isoelectric focusing in the first dimension, which is thereafter followed by sodium dodecyl sulfate electrophoresis in the second dimension by means of a polyacrylamide gel matrix, allowing a migration as spots according to their isoelectric point and molecular weight. The resulting spots can be excised directly from the gel for identification by mass spectrometry (MS). The chromatographic separation involves a tryptic digestion of the whole protein extract generating peptides that can be further fractionated by high performance liquid chromatography (HPLC), by means of ionic exchange or /and reverse phase chromatography. The chromatography eluate flows into an ESI-MS (LC-MS). The MS records the mass of analytes and also isolates and fragments peptide ions (MS/MS, or tandem MS) to generate information about structure

6.2.1 Spermatozoa

Spermatozoa are formed in the seminiferous tubules during their transition through the epididymis. They, then, undergo further maturation, namely a change in the composition of membrane lipids and proteins and a cytoskeleton rearrangement. Maturation of spermatozoa is fundamental to acquire motility, oocyte binding, and fertilizing capacity. After maturation is achieved, during epididymal transit, spermatozoa remain in a dormant state in the cauda region until ejaculation (Gatti et al. 2004). A proteomic analysis of high and LF bull spermatozoa identified 125 biomarkers associated with fertility (Peddinti et al. 2008). Proteomic analysis was carried out after trypsin digestion, followed by reverse phase liquid chromatography separation of peptides, which were then identified by electrospray ionization mass spectrometry (ESI) ion trap mass spectrometry. Spermatozoa from HF bulls had an overexpression of proteins involved in spermatogenesis and cell motility, as well as energy metabolism and cell communication. An investigation aiming to identify fertility biomarkers was carried out after separating the spermatozoa protein collected from normal ERCR+ (estimated relative conception rate) and ERCR− bulls, by means of 2-DIGE and mass spectrometry (Soggiu et al. 2013). Alpha enolase was found to be downregulated in the ERCR− group, while two other proteins, isocitrate dehydrogenase and triosephosphate isomerase, were upregulated in ERCR− as compared to comparison to ERCR+. ESI-MS analysis identified other proteins including calmodulin, ATP synthase mitochondrial subunits alpha and delta, malate dehydrogenase, and sperm equatorial segment protein 1 as downregulated in the ERCR− group and 1 protein upregulated. Enolase was also found on a parallel study that compared protein expression profiles of spermatozoa from high and LF bulls by 2-DGE (Park et al. 2012). Beside enolase 1 (ENO1), the study also identified ATP synthase H^+ transporting mitochondrial F1 complex beta subunit, apoptosis-stimulating of p53 protein 2, alpha-2-HS-glycoprotein, and phospholipid hydroperoxide glutathione peroxide as overexpressed in HF bulls, whereas voltage dependent anion channel 2, ropporin-1, and ubiquinol-cytochrome-c reductase complex core protein 2 (UQCRC2), as overexpressed in LF bulls. Three proteins, namely ENO1, VDAC2, and UQCRC2, were found to be significantly correlated with individual fertility.

Subcellular fractionation allows to identify proteins which have eluded studies using whole cell approaches, pinpointing the focus on selected districts and cellular fractions of the spermatozoa that are believed to be of strategic importance, such as acrosome and plasma membrane. An investigation on bull sperm membranes was carried out with the aim to investigate protein markers associated with bull fertility. Samples were collected from Nelore bulls with great, low, and normal fertility. The membrane proteins were separated by 2-DGE, and proteins were identified by MALDI-TOF mass spectrometry. The results highlighted several differences in protein mobility between the three groups, in particular, aSFP amount was 8.5 times greater in the sperm membrane protein profile of the higher fertility groups, whereas BSP-A3 was 2.5 times greater in the lower fertility group (Roncoletta et al. 2006).

Sperm surface proteome has been characterized recently. Subcellular fractionation carried out by nitrogen cavitation provided mature bull sperm surface proteome, allowing to identify 419 proteins after ESI and MALDI mass spectrometry analyses (Byrne et al. 2012). A number of 118 proteins were predicted to be located along the membrane and include proteins involved in cell adhesion, acrosomal exocytosis, vesicle transport, and more in general, fertilization events, but also immunity, such as several complement regulatory proteins. It should be pointed out that none of these results were validated with other techniques, such as Western Blotting, for example.

A study aimed to investigate how spermatozoa motility is related to protein expression pattern was carried out on semen from Brahman bull (Thepparat et al. 2012) after separating the protein extracts from spermatozoa with 2-DGE and identifying the differentially expressed proteins with LC-MS/MS. The results allowed to sort the samples in five groups following the expression pattern of Tektin-4, a protein associated to axonemal microtubules of flagella. Tektin-4 was found to migrate on 2D gel following different electrophoresis pattern, named from A to E. The bull whose tektin-4 migrates following pattern A showed the highest spermatozoa motility. These results suggested the possible use of tektin-4 as a biomarker to identify spermatozoa motility.

A very recent study improved cell fractionation technique, applying a cell surface biotin-labeling paired with differential centrifugation to enrich sperm surface proteins. A total number of 338 proteins was identified by means of nano-LC MS/MS. Subcellular enrichment allowed to identify cell surface proteins that were not yet described on bovine sperm such as, for example, plasma glutamate carboxypeptidase, CPQ, and carboxypeptidase, vitellogenic-like, CPVL, confidently identified in the PM-enriched proteome (Kasvandik et al. 2015).

Although the main function of spermatozoa is to deliver the paternal genome into the oocyte, spermatozoa cargo includes also RNA (Ostermeier et al. 2004) and microRNA (Govindaraju et al. 2012). The role of transcripts during embryonic and fetal development remains still poorly understood. An analysis conducted by microarray hybridization technique investigated the difference between gene expression profiling of spermatozoa collected from high- and low-fertility (LF) bulls, with the aim to identify a "fingerprinting" between the two groups (Feugang et al. 2010). The study revealed that 211 gene transcripts were at least twofold higher in the HF bull spermatozoa, whereas 204 transcripts were detected as at least twofold higher in the LF bull spermatozoa. The validation of the individual transcript was hampered by the absence of proper housekeeping genes (GAPDH and β-actin remained undetected in spermatozoa). Only one transcript, the CD36, an integral membrane protein and a member of the class B scavenger receptors, was found to be decreased in LF bulls.

The population of mRNA was investigated in both spermatids and spermatozoa by using DNA microarrays (Gilbert et al. 2007). The authors showed for the first time that the spermatic RNA population is composed mainly of naturally truncated mRNA, and that most of the transcripts of spermatids are also present in spermatozoa. A comparative transcriptome analysis between a low-fertile and a high-fertile group of fertile bulls with nonreturn rates was carried out (Lalancette et al. 2008). By using a specific technique called suppression-subtractive hybridization, the authors demonstrated that mitochondrial-related category genes were more represented in HF bull spermatozoa, suggesting mitochondria expressed genes as potential markers to identify bulls with desired fertility potential.

Given the background of their impact as translation regulators, and how they become popular as biomarkers for diseases or altered functionality, microRNA also provide a logic source of biomarkers of male infertility.

MicroRNA have been shown to be differentially expressed in spermatozoa collected from HF and LF bulls (Govindaraju et al. 2012). Although several microRNA were found to be differentially expressed between the two groups, only five were validated by quantitative PCR, namely hsa-aga-8197, -6727, -11,796, -14,189, -6125, all of them being more abundant in the LF group as compared with the HF group. Besides providing biomarkers to discriminate LF from HF semen, this study also demonstrated that microRNA, being expressed at higher level in LF spermatozoa, may play an important role in modulating the expression of genes involved in fertilization.

The hypothesis that the expression of microRNA in spermatozoa can be related to fertility rates has been recently confirmed (Fagerlind et al. 2015). The study was not carried out by using omics techniques, but by conventional quantitative PCR. Nonetheless, the number of targets was very wide (178 miRNA), and the authors found that miR-502-5p, miR-1249, miR-320a, miR-34c-3p, miR-19b-3p, miR-27a-5p, and miR-148b-3p were downregulated in bulls with high nonreturn rate as compared with those with moderate return rate.

Application of NGS technique on bovine spermatozoa identified 959 microRNA, of which 8 were novel (Du et al. 2014), and miR10 and let-7 family member being the most abundant. A parallel investigation on sperm microRNAome identified 1582 unique small RNA present in the sperm (Stowe et al. 2014). Both these studies provide a solid starting point to investigate future applications of sperm miRNA profiles as indicators of male infertility. None of the results of these two studies were validated with Real-Time PCR.

In another study bull infertility was investigated by applying 2-DIGE to detergent-extracted sperm proteins collected from bulls with different fertility indexes (D'Amours et al. 2010). Eight proteins were identified by LC-MS. In particular, T-complex protein 1 subunits 3 and q (CCT5 and CCT8), two isoforms of epididymal sperm-binding protein E12 (ELSPBP1), proteasome subunit a type-6, and binder of sperm 1 (BSP1) were more expressed in the LF group than in the high-fertility (HF) group, whereas adenylate kinase isoenzyme 1 (AK1) and phosphatidylethanolamine-binding protein 1 (PEBP1) were more expressed in the HF group than in the LF group. The functions of each of these proteins partially explain the decreased fertilization capabilities of the spermatozoa. Comparative proteomics between Holstein Friesian (a Taurine breed), Tharpaktar (an Indicine breed), and Karan Fries cross-bred (a Holstein Friesian × Tharpaktarand cross breed) was carried out to investigate the origin of the high male infertility in crossbreds (Muhammad Aslam et al. 2015). A comparison by means of 2-DIGE was carried out, and proteins were identified by means of MALDI-TOF analysis. By comparing the Holstein with Crossbred, 17 differentially expressed proteins were detected, among which, 9 were overexpressed in Holstein and 8 were under-expressed. Comparison between Tharpaktar and crossbred demonstrated that 4 proteins were overexpressed and 4 were under-expressed, confirming that proteomics may provide important insight into the understanding the causes of hypofertility. Figure 6.3 summarizes a list of proteins and mRNA transcripts (including also microRNA) upregulated and downregulated in spermatozoa with different fertility.

Abnormal sperm morphology is regarded as one of the main causes of male infertility. Pyriform shape of the spermatozoa is a common sperm abnormality in both human and animal species, which is the major cause of male hypofertility (Rousso et al. 2002). The differential expression of proteins in pyriform bovine sperm was compared with that of morphologically normal sperm by paring 2-DGE

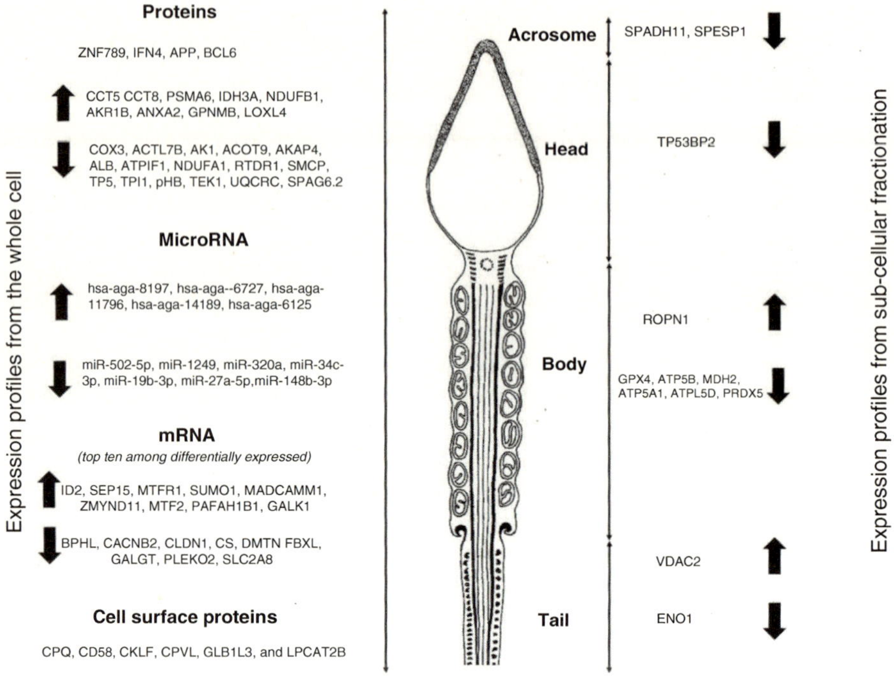

Fig. 6.3 The list of proteins, microRNA, and transcripts (mRNA) differentially regulated in low-fertility spermatozoa. The list of proteins, microRNA, and transcripts (mRNA) upregulated (↑) and downregulated (↓) in low-fertility spermatozoa. Results are summarized from: (Roncoletta et al. 2006; Peddinti et al. 2008; D'Amours et al. 2010; Park et al. 2012; Thepparat et al. 2012; Soggiu et al. 2013; Kasvandik et al. 2015; Muhammad Aslam et al. 2015), transcripts (Feugang et al. 2010), and microRNA (from Govindaraju et al. 2012; Fagerlind et al. 2015). Most of the proteins were found to be expressed in Holstein Friesian, with the exception of Brahman (Thepparat et al. 2012) and Taurine × Indicine crossbred (Muhammad Aslam et al. 2015) for this reference, only the proteins differentially expressed in Holstein breed are listed in the figure. The figure has been drawn after modification from Valour et al. (2015)

with LC-MS/MS (Shojaei Saadi et al. 2013). Results confirmed that the molecular pathogenesis of pyriform sperm is related to oxidative stress, and suggests that proteins which regulate antioxidant activity, such as CLU, GPX5, and PRDX5, were increased in pyriform spermatozoa. Consistently, reactive oxygen species and ubiquitinated proteins were also higher in pyriform sperm. On the contrary, proteins involved in sperm capacitation, sperm–egg interaction, and sperm cytoskeletal structure were decreased in pyriform sperm.

Gender selection is important in cow farming. For example, dairy farmers desire more females for milk production, since meat obtained from dairy breeds usually provides minor economical outcome. The X and Y spermatozoa were differentiated by means of 2-DGE coupled with MALDI-TOF/TOF and LC-MS/MS analysis (Chen et al. 2012). A number of 14 proteins were identified as being differentially expressed between X and Y spermatozoa. These proteins were found to be involved in energy metabolism, stress resistance, cytoskeleton, binding and fusion of sperm/oocytes, and development of the zygotic embryo.

6.2.2 Seminal Plasma

After being produced in the testis, spermatozoa have to transit along the epididymis to acquire their fertilization capability and their motility. Therefore, an exchange of message between the cells lining the epididymal duct and the maturating spermatozoa occurs. A preliminary study identifying 24 proteins on a 2-DGE map was carried out in 1998 (Mortarino et al. 1998). The major proteins already described in bull seminal plasma, like PDC-109 and aSFP, were located and, most importantly, a reference map was established.

The mechanism of transfer of the extracellular proteins is complex, and involves direct secretion of proteins from the epididymal epithelium undergoing apocrine secretion at its apical pole (Hermo and Jacks 2002). These apical blebs disintegrate have a diameter of 50–500 nm, and are named epididymosomes. Proteomic analysis of epididymosomes and their interaction with spermatozoa demonstrated that it was the epididymal origin of epididymosomes that influences which proteins are transferred to sperm (Frenette et al. 2006). These data revealed that the endoplasmin precursor (HSP90B1), reticulocalbin 1, nucleobindin 2 precursor, tumor-associated calcium signal transducer 1, and testis-expressed sequence 101 were present in epididymosomes from the caput epididymis, whereas regucalcin was found in epididymosomes from cauda epididymis. Clusterin, HSPA5, chaperonin-containing TCP-1 subunit 2, beta actin, and aldose reductase were found in epididymosomes from both cauda and caput epididymis. Beside proteins related to spermatozoa fertility, epididymosomes' cargo also include microRNA. MicroRNA (miRNA) belong to a recently discovered class of small, noncoding RNA regulating posttranscriptionally protein expression, including also interfering RNAs (siRNAs) and piwi-interacting RNAs (piRNAs) (Sayed and Abdellatif 2011). Small RNA have recently emerged as regulators of gene expression at the posttranscriptional or translation level in several biological processes, including host–pathogen interaction and spermatogenesis (He et al. 2009). Epididymosomes from two different regions of epididymis, the caput epididymis and the cauda epididymis, were collected and their miRNA content was compared. The miRNA content of epididymal epithelial cells was investigated as well (Belleannée et al. 2013). MiRNA let-7a and miR-200 families as well as miR-26a, miR-103, and miR-191 were those mostly represented in epididymosomes, collected from both regions. A number of 118 miRNA were differentially detected between caput and cauda epididymosomes. Among the others, miR-654, miR-1224, and miR-395 were overexpressed in the epididymosomes from the cauda region, whereas miR-145, miR-143, miR-214, and miR-199 were more abundant in epididymosomes from caput epididymis. Moreover, the microRNA contained in epididymosomes were different as compared with those from the epididymis epithelium. Figure 6.4 presents the list of proteins and microRNA expressed in various sections of epididymis.

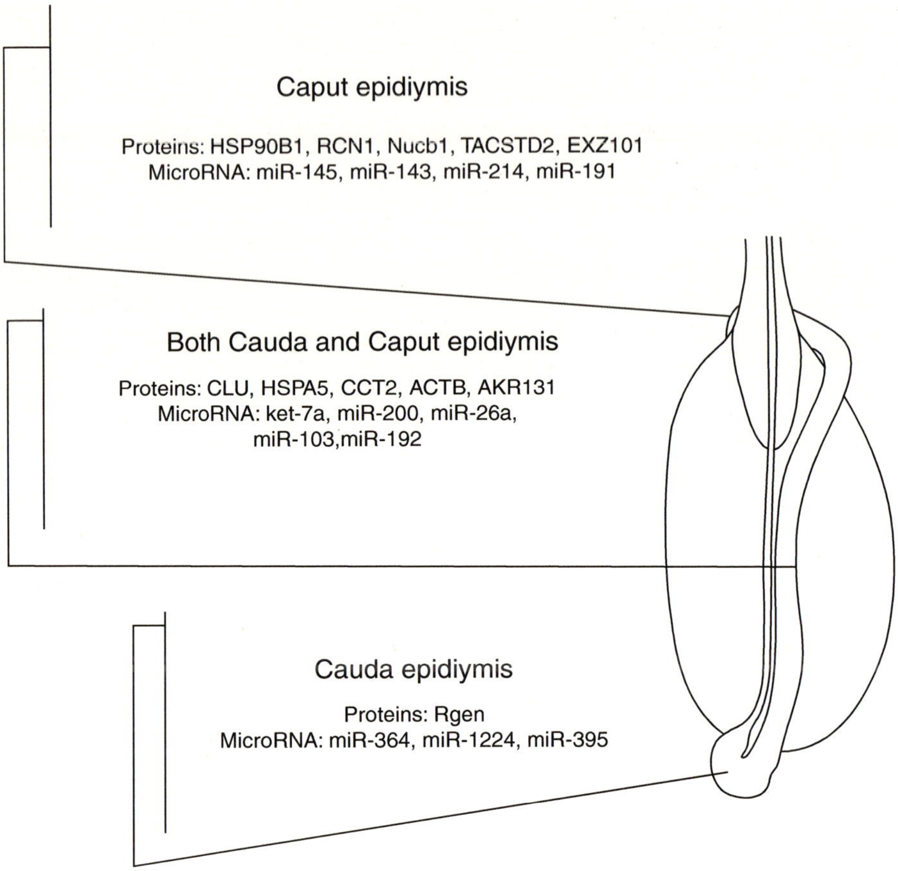

Fig. 6.4 The list of proteins and microRNA expressed in epididymis regions. The list of proteins, microRNA, and mRNA expressed in epididymis regions, as previously reported after collection of samples from different areas of epididymis (Frenette et al. 2006; Belleannée et al. 2013)

A comparative proteomics of bull and other ruminant seminal plasma was recently carried out. Proteins were separated by means of SDS-PAGE and identified using LC-MS/MS. The results identified several major proteins, which were different between species, possibly explaining variation in reproductive capacity (Druart et al. 2013). A proteomic analysis of the accessory sex gland fluid (Moura et al. 2007) identified 13 different proteins involved in various pathways of sperm protection and capacitation as well as antimicrobial peptides. Cauda epididymal fluid, from vas deferens of Holstein bulls, was analyzed by means of 2-DGE. The fluid was collected from vas deferens, and was subjected to 2-D SDS-PAGE. Several spots were identified by LC-MS/MS and MALDI-ToF/ToF (Moura et al. 2010). This study highlighted presence of two new proteins in the cauda epididymal, acidic seminal fluid protein (aSFP) and nucleobindin. ASFP was previously reported as a constituent of the sex gland secretion (Moura et al. 2007). Its role in cauda epididymal was hypothesized to be of protection of sperm from oxidative stress as well as inhibition of motility during storage. The role of nucelobindin, which is a calcium-binding protein, playing a role in activation of intracellular event, is presently unknown.

Luminal and secreted proteins from nine regions of bull epididymis were collected and identified after 2-DGE and LC-MS/MS (Belleannée et al. 2011). The luminome included 172 different proteins, most of which were enzymes, proteases, and protease inhibitors. Other proteins were involved in carbohydrate and glycoprotein metabolism, binding proteins, and proteins involved in response to stress. Differences between the various tracts of epididymis were also identified, since the distribution of luminal proteins was epididymal tract-dependent. Epididymal secretome was also determined, identifying several proteins. Also in this case, secreted proteins were found to be epididymal tract-dependent.

Finally, proteomics analysis as applied to the study of protein patterns, related to different freezability of semen (Magalhães et al. 2016). The authors found that expression of Binder of SPerm 1 (BSP1) protein, which is involved in the fertilization and semen cryopreservation processes, is increased in bulls with low semen freezability and its absence in bulls presenting high semen freezability. 2-DGE also showed that different BSP1 isoforms are present, and that they are possibly related in fertilization and cryopreservation process.

6.2.3 Oocytes and Ovaries

Several studies have focused their attention on bovine oocyte transcriptomics and proteomics in non-pathological status. Table 6.1 provides a list of papers that present the current status of the research on oocyte transcriptomics and proteomics. A review on animal oocyte proteomics studies has been recently published (Virant-Klun and Krijgsveld 2014).

A 2-DGE map of proteins contained in fluid from bovine ovarian follicles at different stages of development and from bovine ovarian cyst was reported (Mortarino et al. 1999). Alpha-1-antitrypsin, albumin, serotransferrin and apolipoprotein A-I and A-IV were located on the map and comparison between protein patterns revealed the differential expression of some spots among follicles of smaller diameter, follicles of larger diameter, and cysts.

A study of bovine ovarian follicular fluid was carried out using 2-DGE followed by identification of proteins with MALDI-TOF analysis (Maniwa et al. 2005). The study identified eight proteins, which were differentially expressed between normal and cystic follicular fluids, namely mitochondrial f1-atpase (BMFA), erythroid associated factor (EAF), methionine synthase (MeS), VEGF-receptor, glyceraldehydes 3-phosphate dehydrogenase (GAPDH), heat shock protein 70 (HSP70), β-lactoglobulin (BLG), and succinate dehydrogenase Ip subunit (SD). It must be said that these results were by no means validated.

A thorough proteomic analysis of preovulatory follicular fluid defined both the bovine preovulatory fluid proteome and how it changes in less fertile cows, as compared to controls, was very recently published (Zachut et al. 2016). Less fertile cows were subjected to follicular aspiration. Protein profile was determined after trypsin digestion and separation of peptides with liquid chromatography. Identification of proteins was done by means of ESI mass spectrometry. The study identified 219 proteins in the follicular fluid of preovulatory follicles, containing serum-derived proteins and other produced by granulosa and thecal cells. The study identified three proteins which were increased in follicular fluid of LF cows, that is metalloproteinase inhibitor 2, inter-alpha-trypsin inhibitor heavy chain H1, and complement component C8 alpha chain. Other seven proteins were found to be decreased in LF follicular fluid, including alpha-1-antiproteinase, basement membrane-specific heparan sulfate proteoglycan core protein, collagen alpha-2(I) chain, prothrombin, alpha-S1-casein, alpha-S1-casein, and one uncharacterized protein. Where antibodies were available, results were validated by means of Western blotting.

Table 6.1 Proteomic and transcriptomics studies on non-pathological cells, fluid and tissues

Tissue/cell/metabolic process	Technique	References
Protein and phosphoprotein profile in oocytes	2-DGE/MALDI-TOF	Bhojwani et al. (2006)
Trophectoderm cell lines	2-DGE/MALDI-TOF/ LC-MS/MS	Talbot et al. (2010)
Early embryo	LC-MS/MS	Deutsch et al. (2014)
Blastocoel fluid and blastocyst cells	LC-MS/MS	Jensen et al. (2014a)
Yolk sac fluid and cells	LC-MS/MS	Jensen et al. (2014b)
Histotroph during pre-implantation	LC-MS/MS	Mullen et al. (2012)
Embryonic development and uterus	2-DGE/MALDI-TOF	Ledgaard et al. (2012)
Embryo-maternal interactions	2-DIGE and LC-ESI-MS/MS	Munoz et al. (2011)
Germinal vesicle oocyte and cumulus	2-DGE/LC-MS/MS	Memili et al. (2007)
Uterine fluid during pre-implantation	LC-MS/MS	Forde et al. (2014)
Blastocoel fluid and blastocyst cells	LC-MS/MS	Jensen et al. (2013)
Morulae and blastocyst	LC-MS/MS	Demant et al. (2015)
Oocyte maturation and early embryos	2-DGE	Massicotte et al. (2006)
Transcriptomic studies on non-pathological cells, fluid and tissues		
Oviductal gene transciptome	Next generation sequencing	Gonella-Diaza et al. (2015)
microRNA in ovary	Sequencing after plasmid DNA preparation	Munir Hossain et al. (2009)
Germinal vesicles, oocytes, embyos and blastocysts	Next generation sequencing	Graf et al. (2014)
MiroRNA in embryo culture media	Next generation sequencing	Kropp and Khatib (2015)
Conceptus cells at the onset of elongation	Microarray	Ribeiro et al. (2016b)
Conceptus development	Microarray	Riberiro et al. (2016a)
microRNA signatures in early pregnancy	Next generation sequencing	Ioannidis and Donade (2016)

6.3 The Role of Maternal Environment: Oviduct, Uterus, and Placenta

Interaction with the maternal environment is of paramount importance for a successful fertilization and embryo development. In this section, the contribution of omics techniques to the knowledge of the support of the maternal environment, including oviduct, uterus, and placenta, is presented.

6.3.1 Oviduct

The role of oviduct during early pregnancy is to support fertilization and the development of the embryo (Besenfelder et al. 2012), and proteomics and transcriptomics profiles of the oviduct in different domestic animals have been previously reviewed (Mondéjar et al. 2012).

Each of the various parts of the oviduct can produce an individual transcriptomic profile (Maillo et al. 2016). In the isthmus, for example, the main upregulated pathways include synthesis of nitrogen, lipids, nucleotides, steroids and cholesterol as well as vesicle-mediated transport, cell cycle, apoptosis, endocytosis, and exocytosis. In ampulla, on the contrary, the main upregulated pathways include cell motion, motility and migration, DNA repair, calcium ion homeostasis, carbohydrate biosynthesis, regulation of cilium movement, and beat frequency. Figure 6.5 presents the pathways activated in isthmus and ampulla of oviduct.

A gene expression study of oviduct epithelial cells comparing estrus with diestrus cycle was carried out by means of cDNA hybridization array (Bauersachs et al. 2004). A total of 37 different genes were highly expressed at estrus, and 40 genes were upregulated during diestrus. Results were partially validated by means of

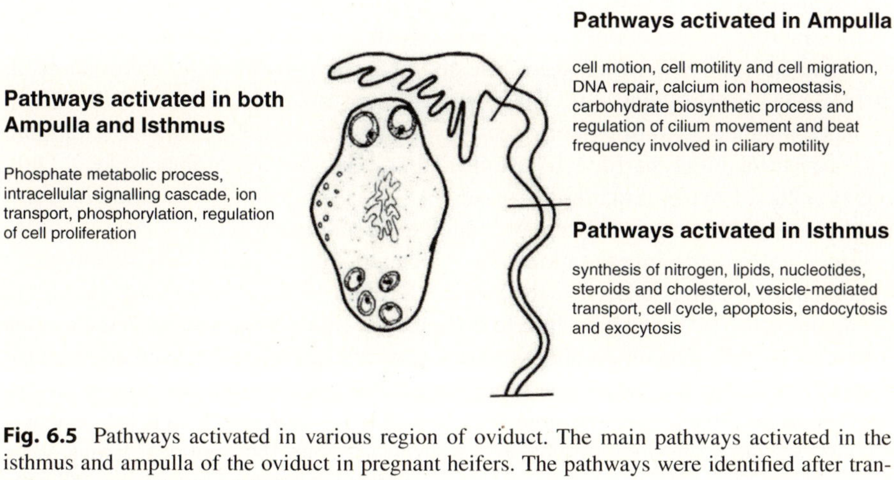

Fig. 6.5 Pathways activated in various region of oviduct. The main pathways activated in the isthmus and ampulla of the oviduct in pregnant heifers. The pathways were identified after transcriptome determination (Maillo et al. 2016)

quantitative PCR, which confirmed *TRA1*, *ERP70*, *GRP76/HSPA5*, *AGR2*, and *OVGP1* as upregulated during estrus, and *C3*, *MS4A8B*, and *18sRNA* as upregulated during diestrus. A bovine oviduct and endometrium was also developed (Bauersachs et al. 2004). The gene expression profile can be modulated by early bovine embryos. A recent investigation (Schmaltz-Panneau et al. 2014) demonstrated that oviduct epithelial cells may differentially regulate 34 genes after co-incubation with early embryos, inducing the upregulation of several genes involved in immune defenses. The oviduct–embryo interaction was further confirmed by a study on the effect of the presence of embryos, either single or multiples, on oviduct transcriptome (Maillo et al. 2015). The transcriptome of both embryos and oviduct was carried out by means of microarray hybridization and demonstrated that the presence of a single embryo in the oviductal hystmus failed to induce any difference in the transcriptome. On the contrary, multiple embryos induced the upregulation of 123 genes, and the downregulation of 155 genes, in oviduct epithelial cells. These results are somehow in contradiction with other previously reported (Schmaltz-Panneau et al. 2014), since most of the genes that were found to be differentially regulated were not related to immune functions.

6.3.2 The Uterus and Development of Endometritis

A thorough review summarizing the results of transcriptome studies of bovine, porcine, and equine endometrium has been recently published (Bauersachs and Wolf 2015). Research on uterus proteome and transcriptome has focused on two components: the uterus fluid and the endometrium. An adequate balance of nutrients and growth factors is the prerequisite to maintain growth and development during embryo implantation. Uterine fluid provides a balanced environment. Both plasma-derived and uterine epithelium-secreted proteins contribute to proteome. The proteome of the uterine flushes collected from beef heifers on day 7 after insemination was determined by LC-MS. The study compared uterine flushes from animals with viable embryos, at morula/blastocysts stage, with those collected from animals with degenerated embryos, which were arrested at 2- to 16-cell stage (Beltman et al. 2014). The proteins that were identified as more abundant in the viable group included platelet-activating factor, acetylhydrolase 1b—catalytic subunit 3, tubulin, b 4A class IVa, tubulin a 1d, cytochrome c-1, and dihydropyrimidinase-like 2. Only one protein, S100 calcium binding protein A4, was significantly increased in histotroph from the degenerated group but failed to be validated.

Two preliminary investigations, one focusing on pregnant and nonpregnant endometrium, using 2-DIGE techniques (Berendt et al. 2005), and the other (Ledgard et al. 2009) using 2-DGE to compare the proteins expressed in the uterine luminal fluid of pregnant and nonpregnant cows, described the bovine uterine proteome. The proteomic profiles between uterine fluid and plasma was afterward carried out following a gel free approach and iTRAQ labeling of tryptic peptides. Cation exchange chromatography was used to separate the peptides, which were finally identified by means of LC-MS/MS. Results were validated by Western blotting, when antibodies were available (Faulkner et al. 2012). A total of 35 proteins were found to be overexpressed in uterine fluid as compared to plasma, the top three proteins exhibiting the highest concentration being triosephosphate isomerase, protein S100-A12, and macrophage migration inhibitory factor (MIF), with a fold change of 11.3, 11.1, and 8.4, respectively. A number of 18 proteins were found to be overexpressed in plasma, as compared to uterine fluid, the three proteins present with the highest concentration being alpha-2-macroglobulin, apolipoprotein A-I, and fibrinogen γ-B chain with fold change of 6.1, 8.6, and 10.9, respectively. Besides the obvious findings of proteins related to embryo growth and development, steroid, vitamins and mineral transport proteins and proteins related to protection from stress, one of the most remarkable finding was the expression of MIF in uterine fluid. Macrophage migration inhibitory factor is involved in dampening of the immune response, and might sustain the immune tolerance level required to prevent maternal rejection of the embryos.

A very comprehensive overview highlighting the transcriptomes pathways activated during the preimplantation phase of farm animals, including also cows, identified the main gene families associated with embryo–maternal interaction (Bauersachs and Wolf 2012). The review is mostly focused on physiological status, and provide information on gene expression profiles of specific stages of estrus cycles and early pregnancy (Bauersachs et al. 2005; Bauersachs et al. 2008; Mitko et al. 2008; Salilew-Wondim et al. 2010), molecular markers of uterine receptivity and early pregnancy (Bauersachs et al. 2006, 2012), gene differences between caruncular and intercaruncular endometrium at implantation (Mansouri-Attia et al. 2009), and the impact of estrogen and progesterone on transcriptome profile (Shimizu et al. 2010).

Gene expression profiling obtained by means of NGS was recently published, reporting that a total of 216 genes where differentially expressed between uterus of pregnant and nonpregnant cows (Van Hoeck et al. 2015). Gene expression studies on endometritis from HF heifers during mid-luteal phase of the estrus cycle were compared to that of LF heifers by means of microarray hybridization gene differential expression (Killeen et al. 2014). The study focused on intercaruncular endometrial tissue. Out of 419 genes that were found to be differentially expressed between LF and HF cows, a number of 171 genes were found to be upregulated in the LF cow endometrium, and 248 downregulated, as compared to HF heifers. Key genes positively contributing to endometrial related fertility included NPPC and GJA1, involved in cellular growth and proliferation; MMP19 and HMGB1, involved in angiogenesis; FASN and PPARA, involved in lipid metabolism; FST and TGFB1, involved in cellular and tissue morphology and development; SLC1A3 and SLC25A24, involved in metabolic exchange, and IL-33, involved in inflammation.

The studies about the molecular differences between retained and normal placenta are still very limited. Endometritis-associated marker proteins were identified by means of a 2-DGE analysis followed by MALDI-TOF on endometritis affected endometrium (Choe et al. 2010). Several proteins were found to be upregulated in bovine endometritis, namely desmin, α-actin-2, heat-shock protein (HSP) 27, peroxiredoxin-6, luteinizing hormone receptor isoform 1, collectin-43 precursor, deoxyribonuclease-I (DNase-I), and MHC class I heavy chain (MHC-Ih). On the contrary, transferrin, interleukin-2 precursor, hemoglobin β subunit, and potassium channel tetramerization domain containing 11 (KCTD11) were found to be downregulated as compared to normal endometrium. Both desmin and α-actin-2 were identified as playing important roles in endometritis, although their possible use as endometritis biomarkers is debated, on the background that these two proteins are widely diffused among mammalian cells, and their expression is very often associated to cell activation events and cancer (a quick Medline search crossing the keywords desmin and biomarker resulted in 2223 citations, on June 2016).

Reduced fertility in dairy cattle is often caused by uterine disorders, the most frequent of which is endometritis (Knutti et al. 2000). Endometrial gene expression changes were detected between dairy cows affected with subclinical endometritis and those with a healthy endometrium (Hoelker et al. 2012). Two sample groups were collected: at time 0 and at time 7, from healthy and subclinical endometritis affected animals. Endometrial transcriptomics was assessed by microarray hybridization. A total of 3185 genes was found to be differentially expressed in the four groups. Ten genes were differentially expressed between subclinical endometritis affected and healthy cows. Protein kinase inhibitor b, calcium-activated chloride channel-2, lysozyme, S100 calcium binding protein, and transcribed locus were upregulated in subclinical endometritis affected cows, whereas PDZ domain containing 1, peroxidasin homologue, DDHD domain containing 2, glycosylphosphatidylinositol specific phospholipase D1, and sulfotransferase family 1B were downregulated. At T7, in what was defined as clinical metritis, eleven transcripts were differentially expressed, of which proline-serine-threonine phosphatase interacting protein 2, transcribed locus, and hypothetical LOC509393 were upregulated in subclinical endometritis, and zinc finger protein Helios, N-acetylglucosamine-1-phosphatase transferase gamma subunit, major histocompatibility complex class II DQ a 5, chromodomain helicase DNA binding protein 2, vascular cell adhesion molecule 1, and rho/rac guanine nucleotide exchange factor GEF 2 were found to be downregulated.

A very recent study (Salilew-Wondim et al. 2016) investigated both the transcriptome and miRNome profiles of endometrium from healthy Holstein-Friesian cows at 40–60 days postpartum as compared with animals with subclinical and clinical metritis. The transcriptome profile identified the upregulation of 92 genes in clinical endometritis as compared with healthy animals, and 111 downregulated genes. Subclinical endometritis induced a significant dysregulation of 28 genes, as compared to healthy animals, 26 of which were also upregulated during clinical endometritis. The functional classification of dysregulated genes pointed toward the alteration of immune system pathways, including chemotaxis, cell adhesion, and G-protein coupled receptor signaling pathways. Remarkably, most of differentially expressed genes involved in immunity pathways were also downregulated in cows affected by subclinical endometritis. The results were validated by quantitative PCR, which was carried out on 13 differentially expressed genes. The impact of lipopolysaccharide (LPS) on gene expression profile was also studied on an in vitro model that included endometrial stromal and epithelial cells. Cell cultures were incubated with different concentration of LPS, simulating subclinical and clinical endometritis. Three genes, namely *MLLlT11*, *INHBA,* and *PTHLH* were upregulated in both endometrial and stromal cells co-incubated with LPS. Three other genes, namely *JUN*, *PTGDS*, and *EMID2* were found to be downregulated in LPS-stimulated cell cultures. These results were consistent with those reported in transcriptome analysis.

A microRNA expression profile was also carried out in clinical and subclinical endometritis affected animals, and results were compared with that of healthy animals. As compared with healthy animals, a number of 10 miRNAs, namely miR-608, miR-625*, miR-218-1*, miR-888*, miR-1184, and miR-1264 were detected only in endometritis affected animals, whereas five microRNAs, namely miR-938, miR-519c-3p, miR-1265, miR-498, and miR-488 were exclusively detected in healthy animals. The expression level of seven microRNA was significantly increased in clinical endometritis as compared with healthy animals, whereas the expression of 28 microRNA was downregulated. Two microRNA, namely has-miR-608 and hsa-miR-526b were upregulated of 1978.7 and 1238.7 fold, respectively. Conversely, has-miR196b and has-miR1265 were found to be downregulated of 107.2 and 3147.5 fold, respectively, in clinically affected animals as compared to healthy ones.

The expression of microRNA during subclinical metritis was also investigated (Hailemariam et al. 2014). The study was carried out starting from cytobrush samples collected from animals with or without subclinical endometritis. Total RNA was extracted, and corresponding cDNA was used to hybridize microRNA PCR array consisting of 352 probes. A number of 23 microRNAs were found to be differentially expressed, of which 15 were upregulated and 8 were downregulated in animals with subclinical metritis. Remarkably, miR-423-3p was upregulated with a change of 1341-fold. Of the 23 microRNA differentially regulated from ex vivo samples with subclinical metritis, 11 were found to be differentially regulated also in an in vitro experiment using LPS-challenged endometrial cells, which was used to confirm and validate expression microarray results. The pathways and biological functions regulated by differentially expressed microRNA included cellular growth and proliferation, cell death and cellular motility, although the most significant gene network modulated by microRNA, which were differentially expressed during subclinical metritis, included nfk-b pathway, which plays a crucial role in immune and inflammatory response.

6.3.3 The Conceptus and the Placenta

The conceptus fluid is the fluid that fills the embryonic membranes, the amnion and the allantois. Conceptus fluid-derived proteins provide an important source of biomarkers for diagnosis and fetal viability prognosis. In bovine species, high levels of fetal losses occur early in pregnancy, ranging between 2 and 4% of naturally conceived bovine pregnancies (Forar et al. 1995), 7.7% in artificial insemination derived pregnancy (Lòpez-Gatius et al. 2004), and from 52.5 to 63.2% in assisted reproduction technologies (Taverne et al. 2002). Proteomic analysis of bovine conceptus fluid during early pregnancy was carried out, proteins were separated by 2-DGE and identified by MALDI-TOF or by LC-MS after digestion with trypsin. A total of 139 proteins were identified. As expected, the role of many proteins was associated with transport, metabolism, and development, but many others, such as ser/cys protease inhibitors, were associated with defensive/immune functions (Riding et al. 2008a).

The differences in protein expression between naturally conceived, in vitro fecundation, and somatic cell nuclear transfer were then determined by means of 2-DIGE (Riding et al. 2008b). Specific antibacterial proteins belonging to the cathelicidin family were identified in the concepts derived from in vitro fecundation.

The formation of placenta plays a crucial role during pregnancy. Moreover, retention of placenta, e.g., the failure to expel the placenta within 24 h from calving, has negative effects on the general health of the cow and her subsequent reproductive performance (Attuparam et al. 2016). A proteomic study, aimed to assess whether placental dysfunction in somatic cell nuclear transfer was responsible of fetal losses, investigated the differential protein patterns of placentae collected from cases of postnatal death of Korean Native calves after somatic cell nuclear transfer fecundation (Kim et al. 2005). 2-DGE was carried out, and proteins were identified by means of mass spectrometry. Results were compared with those from samples obtained from normal placentae. A total of 33 proteins were upregulated in somatic cell nuclear transfer placentae, whereas 27 proteins were downregulated. The results were validated by Western blotting, and demonstrated that one of the upregulated proteins in somatic cell nuclear transfer was TIMP-2 protein, which plays an important role in remodeling of extracellular matrix during pregnancy.

A 2-DGE reference map for bovine placenta during late pregnancy was published, allowing the identification of 273 proteins and providing the background for studies on molecular mechanism of placenta diseases during late pregnancy (Kim et al. 2010).

A proteomic comparison between proteins extracted from placentae that were properly released and those that were retained also was conducted (Kankofer et al. 2014). Samples were collected from carunculae and fetal villi, and proteins were separated by 1D and 2D electrophoresis, but no identification of differentially expressed proteins was performed. Modification of protein expression profiles between the two groups was assessed by means of computer-aided analysis, and presented the evidence of a qualitative and quantitative modification of the protein profile. A further study provided insight into the protein pattern differences between normal placentae and those that were retained for more than 12 h after parturition (Kankofer et al. 2015). The protein profiles were determined by 2-DIGE, and differentially expressed proteins were individually identified by means of MALDI-TOF analysis. Maternal and fetal parts were separated manually, and fetal proteins were identified as being differentially expressed between fetal healthy/retained and maternal healthy/retained. Although the number of proteins identified was limited (five proteins), and the results have to be considered as preliminary, yet results highlighted several differences. Ras-related protein Rab-7b was overexpressed only in healthy maternal placenta. Short transient receptor potential channel 5 was overexpressed in the maternal part of both retained and not retained placenta, whereas Rab GDP dissociation inhibitor beta was overexpressed in the fetal part of both placentae. Finally, transforming growth factor 2 was highly expressed in both maternal and fetal part of retained placenta.

The molecular mechanism behind the detachment of fetal membrane after birth in bovine species remains still unclear. The gene expression profiling of peripartal placentome was carried out by means of microarray hybridization (Streyl et al.

2012). The study focused only on healthy animals, and aimed to define the molecular pathways involved in the release of fetal membranes. Samples of placentomes were collected before the delivery (ante partum, AP) and during the delivery (Intrapartum, IP). Fetal parts of the placenta were not separated from the maternal part, the aim being to get insights into the transcriptome of the whole placental unit. Transcriptome analysis was carried out following microarray hybridization technique using an Affimetrix GeneChip Bovine Genome Array, and thereafter validated by means of quantitative PCR. A number of 514 genes were found to be upregulated in IP placentomes, whereas 59 genes were found to be upregulated in AP placentomes. Validation was also carried out by means of immunohistochemistry, which confirmed the localization of most of the expressed proteins in the placentomes. This study demonstrated the change of gene expression from AP to IP. Before delivery the genes with higher expression rate were mostly related to mitosis and tissue differentiation. During parturition, a shift occurs toward genes involved in placental related pathways, namely degradation of extracellular matrices, such as metalloproteases for example, innate immune response and apoptosis.

6.4 Future Perspectives: Metabolomics Studies and piRNA

To the best of the authors' knowledge, few metabolomics studies focused on fertility have been performed in bovine species. Metabolomics workflow is presented in Fig. 6.6. The metabolome is defined as the complete set of metabolites within a cell, tissue, or biological fluid. Metabolomics provides a powerful approach because metabolites and their concentrations directly reflect the underlying biochemical activity and state of cells and tissues. Unlike other methods, the metabolites detected are not predefined, allowing analysis of previously undescribed biomarkers. Thus metabolomics best mirror the molecular phenotype in that physiological and pathological situation. Metabolomic techniques are of growing application in human infertility (Xia et al. 2014; Krisher et al. 2015; Cordeiro et al. 2015; RoyChoudhury

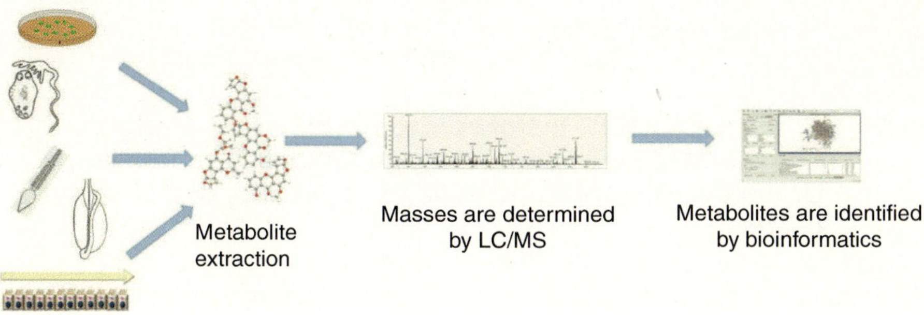

Fig. 6.6 The metabolomics workflow. Metabolites are first isolated from biological samples and their mass determined by LC/MS. The raw data are processed by means of bioinformatics to perform nonlinear retention time alignment and identify peaks. The *m/z* values for the peaks of interest are compared to the values in metabolite databases to obtain putative identifications, which are then validated by comparing tandem mass spectrometry (MS/MS) data to that of standard compounds

et al. 2016; Zhang et al. 2015; Jafarzadeh et al. 2015; Hu et al. 2016; Minai-Tehrani et al. 2015; Zhou et al. 2016). In bovine species, a pioneering investigation on metabolite composition of the follicular fluid was carried out by means of gas-chromatography metabolomics to unravel the differences in fertility between heifers and lactating cows (Bender et al. 2010). The aim was to determine the metabolome of follicular fluids of cows and heifers over different phases of follicle development, including newly selected dominant follicles, preovulatory follicles prior to estrus, and post-LH surge follicles. Among 24 fatty acids and 9 aqueous metabolites that were found to be significantly different between cows and heifers, saturated fatty acids, including palmitic and stearic acid, were found to be increased in follicular fluid from cows. Remarkably, levels of polyunsaturated fatty acids, such as docosahexaenoic acid, were found to be increased in follicular fluid from heifers. The importance of polyunsaturated fatty acid was confirmed by a recent study on follicular fluid metabolome, which investigated the differences in the follicular fluid and serum metabolite profile between dairy cows with different fertility index (Moore et al. 2015). Gas-chromatography metabolome results demonstrated that 9 fatty acids, among the other several PUFA, were affected by genotype.

The metabolome of uterine fluid 15 days after synchronized ovulation and artificial insemination (AI) was determined on anovular Holstein cows, Holstein cows that were estrous cyclic and Jersey/Holstein crossbred cows that were estrous cyclic was carried out by LC/MS, finding at least 28 features which were different between the three groups (Ribeiro et al. 2016).

PiRNA are a germ-cell-specific class of small RNA interacting with Argonaute proteins, the so-called Piwi proteins, both of them being essential for germ cell development and function (Ketting 2011; Luteijn and Ketting 2013). Piwi proteins and piRNAs are expressed in bovine oocytes (Rosenkranz et al. 2015; Roovers et al. 2015; Russell et al. 2016). These pioneering studies are poised to shed new light into the complex relationship between oocyte, spermatozoa, and uterine environment.

On the background of the advancing in transcriptomics and proteomics techniques, metabolomics and piwi investigation suggest that we are on the verge of description of entirely new pathways, which may provide new information contributing to solve the many questions that remain in this field and possibly provide new biomarkers for bovine subfertility.

6.5 Conclusions: New Insights into the New Knowledge Contributed by Systems Biology Approach to Better Understanding of Subfertility and Fertility of Dairy Cattle

Postgenomic applications in veterinary medicine, including transcriptomics, proteomics, and metabolomics, are increasing exponentially. The number of OMICS studies carried out in bovine reproduction pales if compared to those carried out in humans and a true system biology approach to understand subfertility issues in dairy cow is far to be implemented. Until the "OMICS" revolution, as it is named today, the knowledge of the physiological bases of subfertility in dairy cows was pursued through independent analysis of single transcript, proteins, and metabolite. The systems biology approach provides an integrated network of the single elements, the knowledge of which provides greater information than the sum of individual parts. Transcriptomics and proteomics provided an evident leap forward the understanding of basic biology of reproduction. The differences between X and Y spermatozoa as identified by proteomics techniques rank between the most easily implementable in the field, allowing selection of females for dairy production. The very recent definition of preovulatory follicles of less fertile cows also provide some clues to understand the molecular basis of subfertility.

The massive information gathered though OMICS already made available a huge amount of potential biomarkers. Not of them are easily implementable: for the proteins, the availability of enzymatic assays or antibodies is the prerequisite for the detection and quantification in biological fluids. For what concerns mRNA, the main issue here is that mRNA is labile, and it can hardly be detected in biological fluid. MicroRNA represent probably the most promising biomarkers. They are cheap to be measured, being detectable with a Real-Time PCR, they are very much resistant in the environment, making them identifiable in biological fluids such as saliva, uterine fluid, and also hairs. Remarkably, modification of their expression has been already found to be related to endometritis, or to assess male fertility, their expression being found to be different from high-fertility to low-fertility bulls.

For what concerns the metabolome, the most interesting knowledge obtained by systems biology was the witch of the fatty acids' asset from heifers (enriched with PUFA) to cows (with more Saturated fatty acids), and its relationship with fertility index. This information might be readily implemented into the field, on the background that fatty acid content may be modified through diet. Several issues remain to be addressed, and most of the molecular pathways and complex gene expression patterns driving subfertility in cows remain undisclosed.

OMICS technologies remain still quite expensive, proteomics in particular. Yet, technology moves rapidly forward and the costs for omics application is constantly dropping. The goal of a $1000 genome has been almost reached, and it is expected that further drops in omics experiment costs will result in an exponential increase of transcriptomics and proteomic studies, which will likely be extended to other omics disciplines such as glycomics and metabolomics in veterinary medicine.

References

Attupuram NM, Kumaresan A, Narayanan K, Kumar H (2016) Cellular and molecular mechanisms involved in placental separation in the bovine: A review. Mol Reprod Dev 83(4):287–97. doi:10.1002/mrd.22635. Epub 2016 Mar 18. Review. PubMed PMID: 26970238

Bauersachs S, Wolf E (2012) Transcriptome analyses of bovine, porcine and equine endometrium during the pre-implantation phase. Anim Reprod Sci 134(1–2):84–94

Bauersachs S, Wolf E (2015) Uterine responses to the preattachment embryo in domestic ungulates: recognition of pregnancy and preparation for implantation. Annu Rev Anim Biosci 3:489–511

Bauersachs S, Rehfeld S, Ulbrich SE, Mallok S, Prelle K, Wenigerkind H, Einspanier R, Blum H, Wolf E (2004) Monitoring gene expression changes in bovine oviduct epithelial cells during the oestrous cycle. J Mol Endocrinol 32(2):449–466

Bauersachs S, Ulbrich SE, Gross K, Schmidt SE, Meyer HH, Einspanier R, Wenigerkind H, Vermehren M, Blum H, Sinowatz F, Wolf E (2005) Gene expression profiling of bovine endometrium during the oestrous cycle: detection of molecular pathways involved in functional changes. J Mol Endocrinol 34(3):889–908

Bauersachs S, Ulbrich SE, Gross K, Schmidt SE, Meyer HH, Wenigerkind H, Vermehren M, Sinowatz F, Blum H, Wolf E (2006) Embryo-induced transcriptome changes in bovine endometrium reveal species-specific and common molecular markers of uterine receptivity. Reproduction 132(2):319–331

Bauersachs S, Mitko K, Ulbrich SE, Blum H, Wolf E (2008) Transcriptome studies of bovine endometrium reveal molecular profiles characteristic for specific stages of estrous cycle and early pregnancy. Exp Clin Endocrinol Diabetes 116(7):371–384

Bauersachs S, Ulbrich SE, Reichenbach HD, Reichenbach M, Büttner M, Meyer HH, Spencer TE, Minten M, Sax G, Winter G, Wolf E (2012) Comparison of the effects of early pregnancy with human interferon, alpha 2 (IFNA2), on gene expression in bovine endometrium. Biol Reprod 86(2):46

Belleannée C, Labas V, Teixeira-Gomes AP, Gatti JL, Dacheux JL, Dacheux F (2011) Identification of luminal and secreted proteins in bull epididymis. J Proteome 74(1):59–78

Belleannée C, Calvo É, Caballero J, Sullivan R (2013) Epididymosomes convey different repertoires of microRNAs throughout the bovine epididymis. Biol Reprod 89(2):30

Beltman ME, Mullen MP, Elia G, Hilliard M, Diskin MG, Evans AC, Crowe MA (2014) Global proteomic characterization of uterine histotroph recovered from beef heifers yielding good quality and degenerate day 7 embryos. Domest Anim Endocrinol 46:49–57

Bender K, Walsh S, Evans AC, Fair T, Brennan L (2010) Metabolite concentrations in follicular fluid may explain differences in fertility between heifers and lactating cows. Reproduction 139(6):1047–1055

Berendt FJ, Fröhlich T, Schmidt SE, Reichenbach HD, Wolf E, Arnold GJ (2005) Holistic differential analysis of embryo-induced alterations in the proteome of bovine endometrium in the preattachment period. Proteomics 5(10):2551–2560

Besenfelder U, Havlicek V, Brem G (2012) Role of the oviduct in early embryo development. Reprod Domest Anim 47(Suppl 4):156–163

Bhojwani M, Rudolph E, Kanitz W, Zuehlke H, Schneider F, Tomek W (2006) Molecular analysis of maturation processes by protein and phosphoprotein profiling during in vitro maturation of bovine oocytes: a proteomic approach. Cloning Stem Cells 8(4):259–274

Byrne K, Leahy T, McCulloch R, Colgrave ML, Holland MK (2012) Comprehensive mapping of the bull sperm surface proteome. Proteomics 12(23–24):3559–3579

Chen X, Zhu H, Wu C, Han W, Hao H, Zhao X, Du W, Qin T, Liu Y, Wang D (2012) Identification of differentially expressed proteins between bull X and Y spermatozoa. J Proteome 77:59–67

Choe C, Park JW, Kim ES, Lee SG, Park SY, Lee JS, Cho MJ, Kang KR, Han J, Kang D (2010) Proteomic analysis of differentially expressed proteins in bovine endometrium with endometritis. Korean J Physiol Pharmacol 14(4):205–212

Cordeiro FB, Cataldi TR, Perkel KJ, do Vale Teixeira da Costa L, Rochetti RC, Stevanato J, Eberlin MN, Zylbersztejn DS, Cedenho AP, Turco EG (2015) Lipidomics analysis of follicular fluid by ESI-MS reveals potential biomarkers for ovarian endometriosis. J Assist Reprod Genet 32(12):1817–1825

D'Amours O, Frenette G, Fortier M, Leclerc P, Sullivan R (2010) Proteomic comparison of detergent-extracted sperm proteins from bulls with different fertility indexes. Reproduction 139(3):545–556

Demant M, Deutsch DR, Fröhlich T, Wolf E, Arnold GJ (2015) Proteome analysis of early lineage specification in bovine embryos. Proteomics 15(4):688–701

Deutsch DR, Fröhlich T, Otte KA, Beck A, Habermann FA, Wolf E, Arnold GJ (2014) Stage-specific proteome signatures in early bovine embryo development. J Proteome Res 13(10):4363–4376

Druart X, Rickard JP, Mactier S, Kohnke PL, Kershaw-Young CM, Bathgate R, Gibb Z, Crossett B, Tsikis G, Labas V, Harichaux G, Grupen CG, de Graaf SP (2013) Proteomic characterization and cross species comparison of mammalian seminal plasma. J Proteome 91:13–22

Du Y, Wang X, Wang B, Chen W, He R, Zhang L, Xing X, Su J, Wang Y, Zhang Y (2014) Deep sequencing analysis of microRNAs in bovine sperm. Mol Reprod Dev 81(11):1042–1052

Fagerlind M, Stålhammar H, Olsson B, Klinga-Levan K (2015) Expression of miRNAs in bull spermatozoa correlates with fertility rates. Reprod Domest Anim 50(4):587–594

Faulkner S, Elia G, Mullen MP, O'Boyle P, Dunn MJ, Morris D (2012) A comparison of the bovine uterine and plasma proteome using iTRAQ proteomics. Proteomics 12(12):2014–2023

Feugang JM, Rodriguez-Osorio N, Kaya A, Wang H, Page G, Ostermeier GC, Topper EK, Memili E (2010) Transcriptome analysis of bull spermatozoa: implications for male fertility. Reprod Biomed Online 21(3):312–324

Forar AL, Gay JM, Hancock DD (1995) The frequency of endemic fetal loss in dairy cattle: a review. Theriogenology 43(6):989–1000

Forde N, McGettigan PA, Mehta JP, O'Hara L, Mamo S, Bazer FW, Spencer TE, Lonergan P (2014) Proteomic analysis of uterine fluid during the pre-implantation period of pregnancy in cattle. Reproduction 147(5):575–587

Frenette G, Girouard J, Sullivan R (2006) Comparison between epididymosomes collected in the intraluminal compartment of the bovine caput and cauda epididymidis. Biol Reprod 75(6):885–890

Galligan DT (1999) The economics of optimal health and productivity in the commercial dairy. Rev Sci Tech 18(2):512–519

Gatti JL, Castella S, Dacheux F, Ecroyd H, Métayer S, Thimon V, Dacheux JL (2004) Post-testicular sperm environment and fertility. Anim Reprod Sci 82–83:321–339

Gilbert I, Bissonnette N, Boissonneault G, Vallée M, Robert C (2007) A molecular analysis of the population of mRNA in bovine spermatozoa. Reproduction 133(6):1073–1086

Gonella-Diaza AM, Andrade SC, Sponchiado M, Pugliesi G, Mesquita FS, Van Hoeck V, Strefezzi Rde F, Gasparin GR, Coutinho LL, Binelli M (2015) Size of the ovulatory follicle dictates spatial differences in the oviductal transcriptome in cattle. PLoS One 10(12):e0145321

Govindaraju A, Uzun A, Robertson L, Atli MO, Kaya A, Topper E, Crate EA, Padbury J, Perkins A, Memili E (2012) Dynamics of microRNAs in bull spermatozoa. Reprod Biol Endocrinol 10:82

Graf A, Krebs S, Zakhartchenko V, Schwalb B, Blum H, Wolf E (2014) Fine mapping of genome activation in bovine embryos by RNA sequencing. Proc Natl Acad Sci U S A 111(11): 4139–4144

Hailemariam D, Ibrahim S, Hoelker M, Drillich M, Heuwieser W, Looft C, Cinar MU, Tholen E, Schellander K, Tesfaye D (2014) MicroRNA-regulated molecular mechanism underlying bovine subclinical endometritis. Reprod Fertil Dev 26(6):898–913

He Z, Kokkinaki M, Pant D, Gallicano GI, Dym M (2009) Small RNA molecules in the regulation of spermatogenesis. Reproduction 137(6):901–911

Hermo L, Jacks D (2002) Nature's ingenuity: bypassing the classical secretory route via apocrine secretion. Mol Reprod Dev 63(3):394–410

Hoelker M, Salilew-Wondim D, Drillich M, Christine GB, Ghanem N, Goetze L, Tesfaye D, Schellander K, Heuwieser W (2012) Transcriptional response of the bovine endometrium and embryo to endometrial polymorphonuclear neutrophil infiltration as an indicator of subclinical inflammation of the uterine environment. Reprod Fertil Dev 24(6):778–793

Hossain MM, Ghanem N, Hoelker M, Rings F, Phatsara C, Tholen E, Schellander K, Tesfaye D (2009) Identification and characterization of miRNAs expressed in the bovine ovary. BMC Genomics 10:443

Hu W, Chen M, Wu W, Lu J, Zhao D, Pan F, Lu C, Xia Y, Hu L, Chen D, Sha J, Wang X (2016) Gene-gene and gene-environment interactions on risk of male infertility: focus on the metabolites. Environ Int 91:188–195

Ioannidis J, Donadeu FX (2016) Circulating miRNA signatures of early pregnancy in cattle. BMC Genomics 17(1):184

Jafarzadeh N, Mani-Varnosfaderani A, Minai-Tehrani A, Savadi-Shiraz E, Sadeghi MR, Gilany K (2015) Metabolomics fingerprinting of seminal plasma from unexplained infertile men: a need for novel diagnostic biomarkers. Mol Reprod Dev 82(3):150

Jensen PL, Beck HC, Petersen J, Hreinsson J, Wånggren K, Laursen SB, Sørensen PD, Christensen ST, Andersen CY (2013) Proteomic analysis of human blastocoel fluid and blastocyst cells. Stem Cells Dev 22(7):1126–1135

Jensen PL, Grøndahl ML, Beck HC, Petersen J, Stroebech L, Christensen ST, Yding Andersen C (2014a) Proteomic analysis of bovine blastocoel fluid and blastocyst cells. Syst Biol Reprod Med 60(3):127–135

Jensen PL, Beck HC, Petersen TS, Stroebech L, Schmidt M, Rasmussen LM, Hyttel P (2014b) Proteomic analysis of the early bovine yolk sac fluid and cells from the day 13 ovoid and elongated preimplantation embryos. Theriogenology 82(5):657–667

Kankofer M, Wawrzykowski J, Hoedemaker M (2014) Profile of bovine proteins in retained and normally expelled placenta in dairy cows. Reprod Domest Anim 49(2):270–274

Kankofer M, Wawrzykowski J, Miller I, Hoedemaker M (2015) Usefulness of DIGE for the detection of protein profile in retained and released bovine placental tissues. Placenta 36(2):246–249

Kasvandik S, Sillaste G, Velthut-Meikas A, Mikelsaar AV, Hallap T, Padrik P, Tenson T, Jaakma Ü, Kõks S, Salumets A (2015) Bovine sperm plasma membrane proteomics through biotinylation and subcellular enrichment. Proteomics 15(11):1906–1920

Ketting RF (2011) The many faces of RNAi. Dev Cell 20(2):148–161

Killeen AP, Morris DG, Kenny DA, Mullen MP, Diskin MG, Waters SM (2014) Global gene expression in endometrium of high and low fertility heifers during the mid-luteal phase of the estrous cycle. BMC Genomics 15:234

Kim HR, Kang JK, Yoon JT, Seong HH, Jung JK, Lee HM, Sik Park C, Jin DI (2005) Protein profiles of bovine placenta derived from somatic cell nuclear transfer. Proteomics 5(16):4264–4273

Kim HR, Han RX, Yoon JT, Park CS, Jin DI (2010) A two-dimensional electrophoresis reference map for the bovine placenta during late pregnancy. Proteomics 10(3):564–573

Knutti B, Küpfer U, Busato A (2000) Reproductive efficiency of cows with endometritis after treatment with intrauterine infusions or prostaglandin injections, or no treatment. J Vet Med A Physiol Pathol Clin Med 47(10):609–615

Krisher RL, Schoolcraft WB, Katz-Jaffe MG (2015) Omics as a window to view embryo viability. Fertil Steril 103(2):333–341

Kropp J, Khatib H (2015 Sep) Characterization of microRNA in bovine in vitro culture media associated with embryo quality and development. J Dairy Sci 98(9):6552–6563

Lalancette C, Thibault C, Bachand I, Caron N, Bissonnette N (2008) Transcriptome analysis of bull semen with extreme nonreturn rate: use of suppression-subtractive hybridization to identify functional markers for fertility. Biol Reprod 78(4):618–635

Ledgard AM, Lee RS, Peterson AJ (2009) Bovine endometrial legumain and TIMP-2 regulation in response to presence of a conceptus. Mol Reprod Dev 76(1):65–74

Ledgard AM, Berg MC, McMillan WH, Smolenski G, Peterson AJ (2012) Effect of asynchronous transfer on bovine embryonic development and relationship with early cycle uterine proteome profiles. Reprod Fertil Dev 24(7):962–972

Lòpez-Gatius F, Murugavel K, Santolaria P, López-Béjar M, Yániz JL (2004) Pregnancy rate after timed artificial insemination in early post-partum dairy cows after Ovsynch or specific synchronization protocols. J Vet Med A Physiol Pathol Clin Med 51(1):33–38

Lucy MC (2001) Reproductive loss in high-producing dairy cattle: where will it end? J Dairy Sci 84(6):1277–1293

Luteijn MJ, Ketting RF (2013) PIWI-interacting RNAs: from generation to transgenerational epigenetics. Nat Rev Genet 14(8):523–534

Magalhães MJ Jr, Martins LF, Senra RL, Santos TF, Okano DS, Pereira PR, Faria-Campos A, Campos SV, Guimarães JD, Baracat-Pereira MC (2016) Differential abundances of four forms of Binder of SPerm 1 in the seminal plasma of Bos Taurus indicus bulls with different patterns of semen freezability. Theriogenology 86(3):766–777.e2

Maillo V, Gaora PÓ, Forde N, Besenfelder U, Havlicek V, Burns GW, Spencer TE, Gutierrez-Adan A, Lonergan P, Rizos D (2015) Oviduct-embryo interactions in cattle: two-way traffic or a one-way street? Biol Reprod 92(6):144

Maillo V, de Frutos C, O'Gaora P, Forde N, Burns GW, Spencer TE, Gutierrez-Adan A, Lonergan P, Rizos D (2016) Spatial differences in gene expression in the bovine oviduct. Reproduction 152(1):37–46

Maniwa J, Izumi S, Isobe N, Terada T (2005) Studies on substantially increased proteins in follicular fluid of bovine ovarian follicular cysts using 2-D PAGE and MALDI-TOF MS. Reprod Biol Endocrinol 3:23

Mansouri-Attia N, Aubert J, Reinaud P, Giraud-Delville C, Taghouti G, Galio L, Everts RE, Degrelle S, Richard C, Hue I, Yang X, Tian XC, Lewin HA, Renard JP, Sandra O (2009) Gene expression profiles of bovine caruncular and intercaruncular endometrium at implantation. Physiol Genomics 39(1):14–27

Massicotte L, Coenen K, Mourot M, Sirard MA (2006) Maternal housekeeping proteins translated during bovine oocyte maturation and early embryo development. Proteomics 6(13):3811–3820

Memili E, Peddinti D, Shack LA, Nanduri B, McCarthy F, Sagirkaya H, Burgess SC (2007) Bovine germinal vesicle oocyte and cumulus cell proteomics. Reproduction 133(6):1107–1120

Minai-Tehrani A, Jafarzadeh N, Gilany K (2015) Metabolomics: a state-of-the-art technology for better understanding of male infertility. Andrologia 48(6):609–616

Mondéjar I, Acuña OS, Izquierdo-Rico MJ, Coy P, Avilés M (2012) The oviduct: functional genomic and proteomic approach. Reprod Domest Anim 47(Suppl 3):22–29

Mitko K, Ulbrich SE, Wenigerkind H, Sinowatz F, Blum H, Wolf E, Bauersachs S (2008) Dynamic changes in messenger RNA profiles of bovine endometrium during the oestrous cycle. Reproduction 135(2):225–40. doi:10.1530/REP-07-0415. PubMed PMID:18239051

Michot VP, Eozenou P, Lefebvre C, Bonnet R, Capitan A,Uzbekova A, Sellem S, Ponsart E, Schibler CL (2015) Dairy cattle reproduction is a tightly regulated genetic process: Highlights on genes, pathways, and biological processes Animal Frontiers, American Society of Animal Science 5(1):32–41. doi:10.2527/af.2015-0006

Moore SG, O'Gorman A, Brennan L, Fair T, Butler ST (2015) Follicular fluid and serum metabolites in Holstein cows are predictive of genetic merit for fertility. Reprod Fertil Dev. doi:10.1071/RD15182

Mortarino M, Tedeschi G, Negri A, Ceciliani F, Gottardi L, Maffeo G, Ronchi S (1998) Two-dimensional polyacrylamide gel electrophoresis map of bull seminal plasma proteins. Electrophoresis 19(5):797–801

Mortarino M, Vigo D, Maffeo G, Ronchi S (1999) Two-dimensional polyacrylamide gel electrophoresis map of bovine ovarian fluid proteins. Electrophoresis 20(4–5):866–869

Moura AA, Chapman DA, Koc H, Killian GJ (2007) A comprehensive proteomic analysis of the accessory sex gland fluid from mature Holstein bulls. Anim Reprod Sci 98(3–4):169–188

Moura AA, Souza CE, Stanley BA, Chapman DA, Killian GJ (2010) Proteomics of cauda epididymal fluid from mature Holstein bulls. J Proteome 73(10):2006–2020

Muhammad Aslam MK, Kumaresan A, Rajak SK, Tajmul M, Datta TK, Mohanty TK, Srinivasan A, Yadav S (2015) Comparative proteomic analysis of Taurine, Indicine, and crossbred (Bos taurus × Bos indicus) bull spermatozoa for identification of proteins related to sperm malfunctions and subfertility in crossbred bulls. Theriogenology 84(4):624–633

Mullen MP, Elia G, Hilliard M, Parr MH, Diskin MG, Evans AC, Crowe MA (2012) Proteomic characterization of histotroph during the preimplantation phase of the estrous cycle in cattle. J Proteome Res 11(5):3004–3018

Munoz M, Corrales FJ, Caamaño JN, Díez C, Trigal B, Mora MI, Martín D, Carrocera S, Gómez E (2011) Proteome of the early embryo-maternal dialogue in the cattle uterus. J Proteome Res 11(2):751–766

Ostermeier GC, Miller D, Huntriss JD, Diamond MP, Krawetz SA (2004) Reproductive biology: delivering spermatozoan RNA to the oocyte. Nature 429(6988):154

Park YJ, Kwon WS, Oh SA, Pang MG (2012) Fertility-related proteomic profiling bull spermatozoa separated by percoll. J Proteome Res 11(8):4162–4168

Peddinti D, Nanduri B, Kaya A, Feugang JM, Burgess SC, Memili E (2008) Comprehensive proteomic analysis of bovine spermatozoa of varying fertility rates and identification of biomarkers associated with fertility. BMC Syst Biol 2:19

Ribeiro ES, Monteiro AP, Bisinotto RS, Lima FS, Greco LF, Ealy AD, Thatcher WW, Santos JE (2016) Conceptus development and transcriptome at preimplantation stages in lactating dairy cows of distinct genetic groups and estrous cyclic statuses. J Dairy Sci 99(6):4761–4777

Riding GA, Jones A, Holland MK, Hill JR, Lehnert SA (2008a) Proteomic analysis of bovine conceptus fluids during early pregnancy. Proteomics 8(1):160–177

Riding GA, Hill JR, Jones A, Holland MK, Josh PF, Lehnert SA (2008b) Differential proteomic analysis of bovine conceptus fluid proteins in pregnancies generated by assisted reproductive technologies. Proteomics 8(14):2967–2982

Roncoletta M, Morani Eda S, Esper CR, Barnabe VH, Franceschini PH (2006) Fertility-associated proteins in Nelore bull sperm membranes. Anim Reprod Sci 91(1–2):77–87

Roovers EF, Rosenkranz D, Mahdipour M, Han CT, He N, Chuva de Sousa Lopes SM, van der Westerlaken LA, Zischler H, Butter F, Roelen BA, Ketting RF (2015) Piwi proteins and piRNAs in mammalian oocytes and early embryos. Cell Rep 10(12):2069–2082

Rosenkranz D, Han CT, Roovers EF, Zischler H, Ketting RF (2015) Piwi proteins and piRNAs in mammalian oocytes and early embryos: From sample to sequence. Genom Data 5:309–313

Rousso D, Kourtis A, Mavromatidis G, Gkoutzioulis F, Makedos G, Panidis D (2002) Pyriform head: a frequent but little-studied morphological abnormality of sperm. Arch Androl 48(4):267–272

Royal MD, Darwash AO, Flint APF, Webb R, Woolliams JA, Lamming GE (2000) Declining fertility in dairy cattle: changes in traditional and endocrine parameters of fertility. Anim Sci 70: 487–401

RoyChoudhury S, Singh A, Gupta NJ, Srivastava S, Joshi MV, Chakravarty B, Chaudhury K (2016) Repeated implantation failure versus repeated implantation success: discrimination at a metabolomic level. Hum Reprod 31(6):1265–1274

Russell SJ, Stalker L, Gilchrist G, Backx A, Molledo G, Foster RA, LaMarre J (2016) Identification of PIWIL1 isoforms and their expression in bovine testes, oocytes, and early embryos. Biol Reprod 94(4):75

Salilew-Wondim D, Hölker M, Rings F, Ghanem N, Ulas-Cinar M, Peippo J, Tholen E, Looft C, Schellander K, Tesfaye D (2010) Bovine pretransfer endometrium and embryo transcriptome fingerprints as predictors of pregnancy success after embryo transfer. Physiol Genomics 42(2):201–218

Salilew-Wondim D, Ibrahim S, Gebremedhn S, Tesfaye D, Heppelmann M, Bollwein H, Pfarrer C, Tholen E, Neuhoff C, Schellander K, Hoelker M (2016) Clinical and subclinical endometritis induced alterations in bovine endometrial transcriptome and miRNome profile. BMC Genomics 17(1):218

Sayed D, Abdellatif M (2011) MicroRNAs in development and disease. Physiol Rev 91(3):827–887

Schmaltz-Panneau B, Cordova A, Dhorne-Pollet S, Hennequet-Antier C, Uzbekova S, Martinot E, Doret S, Martin P, Mermillod P, Locatelli Y (2014) Early bovine embryos regulate oviduct epithelial cell gene expression during in vitro co-culture. Anim Reprod Sci 149(3–4):103–116

Shimizu T, Krebs S, Bauersachs S, Blum H, Wolf E, Miyamoto A (2010) Actions and interactions of progesterone and estrogen on transcriptome profiles of the bovine endometrium. Physiol Genomics 42A(4):290–300

Shojaei Saadi HA, van Riemsdijk E, Dance AL, Rajamanickam GD, Kastelic JP, Thundathil JC (2013) Proteins associated with critical sperm functions and sperm head shape are differentially expressed in morphologically abnormal bovine sperm induced by scrotal insulation. J Proteome 82:64–80

Soggiu A, Piras C, Hussein HA, De Canio M, Gaviraghi A, Galli A, Urbani A, Bonizzi L, Roncada P (2013) Unravelling the bull fertility proteome. Mol BioSyst 9(6):1188–1195

Stowe HM, Calcatera SM, Dimmick MA, Andrae JG, Duckett SK, Pratt SL (2014) The bull sperm microRNAome and the effect of fescue toxicosis on sperm microRNA expression. PLoS One 9(12):e113163

Streyl D, Kenngott R, Herbach N, Wanke R, Blum H, Sinowatz F, Wolf E, Zerbe H, Bauersachs S (2012) Gene expression profiling of bovine peripartal placentomes: detection of molecular pathways potentially involved in the release of foetal membranes. Reproduction 143(1):85–105

Talbot NC, Powell AM, Caperna TJ, Garrett WM (2010) Proteomic analysis of the major cellular proteins of bovine trophectoderm cell lines derived from IVP, parthenogenetic and nuclear transfer embryos: reduced expression of annexins I and II in nuclear transfer-derived cell lines. Anim Reprod Sci 120(1–4):187–202

Taverne MA, Breukelman SP, Perényi Z, Dieleman SJ, Vosa PL, Jonker HH, de Ruigh L, Van Wagtendonk-de Leeuw JM, Beckers JF (2002) The monitoring of bovine pregnancies derived from transfer of in vitro produced embryos. Reprod Nutr Dev 42(6):613–624

Thepparat T, Katawatin S, Vongpralub T, Duangjinda M, Thammasirirak S, Utha A (2012) Separation of bovine spermatozoa proteins using 2D-PAGE revealed the relationship between tektin-4 expression patterns and spermatozoa motility. Theriogenology 77(9):1816–1821

Van Hoeck V, Scolari SC, Pugliesi G, Gonella-Diaza AM, Andrade SC, Gasparin GR, Coutinho LL, Binelli M (2015) Gene expression profiling by high throughput sequencing to determine signatures for the bovine receptive uterus at early gestation. Genom Data 5:94–96

Virant-Klun I, Krijgsveld J (2014) Proteomes of animal oocytes: what can we learn for human oocytes in the in vitro fertilization programme? Biomed Res Int 2014:856907

Xia L, Zhao X, Sun Y, Hong Y, Gao Y, Hu S (2014) Metabolomic profiling of human follicular fluid from patients with repeated failure of in vitro fertilization using gas chromatography/mass spectrometry. Int J Clin Exp Pathol 7(10):7220–7229

Zachut M, Sood P, Levin Y, Moallem U (2016) Proteomic analysis of preovulatory follicular fluid reveals differentially abundant proteins in less fertile dairy cows. J Proteome 139:122–129

Zhang X, Diao R, Zhu X, Li Z, Cai Z (2015) Metabolic characterization of asthenozoospermia using nontargeted seminal plasma metabolomics. Clin Chim Acta 450:254–261

Zhou X, Wang Y, Yun Y, Xia Z, Lu H, Luo J, Liang Y (2016) A potential tool for diagnosis of male infertility: plasma metabolomics based on GC-MS. Talanta 147:82–89

Retained Placenta: A Systems Veterinary Approach

7

Elda Dervishi and Burim N. Ametaj

Abstract

Retained placenta (RP) or retention of fetal membranes (RFM) is a common condition with high incidence in cattle that is typically defined as failure to expel the placenta within 24 h after parturition. Retained placenta is financially important to dairy industry because of treatment related costs, reduced milk production, milk withdrawal, labor, and veterinary services. Although RP has been a permanent issue of dairy industry, the real reason for failure to expel the fetal membranes and the pathogenesis still remain unknown. Mounting information suggest that retained placenta is a multifactorial health issue which involves cellular aspects of immune system, gene expression as well as protein and metabolite alterations. Recent pioneering research work utilizing omics sciences is bringing forth new findings that deserve to be discussed further. Utilization of systems biology in researching the pathobiology of the disease and causative agents promises to throw light on the etiopathology of RP.

7.1 Introduction

Retained placenta (RP) or retention of fetal membranes (RFM) is a common health issue of dairy cows with high incidence and typically defined as failure to expel placental membranes within 24 h after calving. The average incidence rate of RP in dairy cows varies between 4 and 16%; however, it is not uncommon to observe greater incidence rates in some problematic herds (Eiler 1997).

E. Dervishi, Ph.D. (✉) • B.N. Ametaj, D.V.M., Ph.D., Ph.D.
Department of Agricultural, Food and Nutritional Science, University of Alberta, Edmonton, AL, Canada, T6G 2P5
e-mail: dervishi@ualberta.ca; burim.ametaj@ualberta.ca

B.N. Ametaj (ed.), *Periparturient Diseases of Dairy Cows*,
DOI 10.1007/978-3-319-43033-1_7

Originally RP was hypothesized to be related to deficiencies of selenium, vitamin E, or calcium; however, multiple investigations involving supplementation of those nutrients to dairy cows showed a modest influence on the incidence rate of RP. There is growing support for the idea that RP is related more to an immune dysfunction that results in failure to degrade the placentome, which connects maternal and placental tissues (LeBlanc 2008).

Retained placenta is associated with economic losses related to costs of treatment, lowered milk production, milk withdrawal, and greater risk for other periparturient diseases. In two early publications it was estimated that a single case of RP costs around US $285 or UK £239 (Laven and Peters 1996; Guard 1999). Financial losses to the dairy industry are increased more because cows affected by RP are at greater risk of other diseases like uterine infections, displaced abomasum, and udder infections. Up to 6% of RP cases cows are culled due to very low milk production and veterinary examinations.

Because of the high prevalence and economic importance to dairy industry, it would be of interest that we develop an understanding of the etiopathology of the RP in dairy cows and its causes and predisposing factors.

7.2 Physiology of Parturition, and Expulsion of Fetal Membranes

Placenta plays a major role in the communication between the mother and the fetus, providing all the necessary nutrients and oxygen and removing waste materials from the growing fetus and protecting the fetus. Morphologically the ruminant placenta is classified as cotyledonary (consisting of abundant blood vessels and connective tissue) (Lemley et al. 2015). In dairy cows, fetus has no direct contact with the maternal blood and such type of placentas are known as noninvasive or syndesmochorial (i.e., there is a complete intact layer of epithelium in both maternal and fetal components). In noninvasive placentas the nutrients are transported from the mother to the growing fetus through anatomical structures known as placentomes (King 1993). The fetal side of the placenta is known as cotyledon and it is contributed by the chorion, whereas the maternal side is called a caruncle and originates from the caruncular regions of the uterus, whereas the placentome is the point of interface between the two. In 1961, Bjorkman and Sollen proposed that the tissue turgor and pressure is one of the mechanisms that keep the caruncle and the cotyledon together. In cows, the total number of placentomes ranges between 70 and 120 and they are dispersed throughout the whole placenta (Senger 2003). Placentomes contain very highly vascularized villi that derive from the trophoblast and the corresponding endometrial crypts into which the villi fit. Nutrients are transported from the maternal to the fetal side of blood circulation across six layers of cells shared equally between mother's endometrium and the fetal trophoblast villi. Placentomes increase their radius during pregnancy between 2.5 and 3 cm near parturition (Senger 2003).

Initiation of parturition starts with activation of the fetal hypothalamus-pituitary-adrenal (HPA) axis (Fig. 7.1). However, the stimulus that leads to fetal HPA axis activation is not yet identified. The activation of fetal HPA axis is associated with the release of adrenocorticotropic hormone (ACTH) by pituitary gland, production of cortisol from the fetus, and synthesis of estrogen from the placenta (Flint et al. 1979). Fetal cortisol triggers the conversion of placental progesterone to estrogen and also induces synthesis of endometrial prostaglandins in both estrogen-dependent and estrogen-independent ways (Whittle et al. 2000). In fact, one of the major effects of estrogen on uterine myometrium is to stimulate upregulation of oxytocin receptors and secretion of prostaglandin F2α (PGF2α) (Fuchs et al. 1999). Prostaglandin F2α triggers lysis of the corpus luteum, production of relaxin, and contraction of uterine smooth muscles, which pushes the fetus toward the cervix (Janszen et al. 1993). Given that corpus luteum (CL) is an important source of progesterone synthesis, its lysis decreases concentrations of progesterone in the blood. During pregnancy progesterone exerts an inhibitory effect on the activity of collagenase, however, at the end of parturition concentrations of estrogen are elevated stimulating the activity of collagenase. Lysis of the CL also is accompanied with secretion of relaxin. Relaxin is involved in stimulation of collagenase production, an enzyme that degrades collagen and induces cervical dilation, pelvic relaxation, and separation of interpubic ligaments. In addition, increased estrogen is associated with increased levels of proteins like connexins, which form gap junctions and

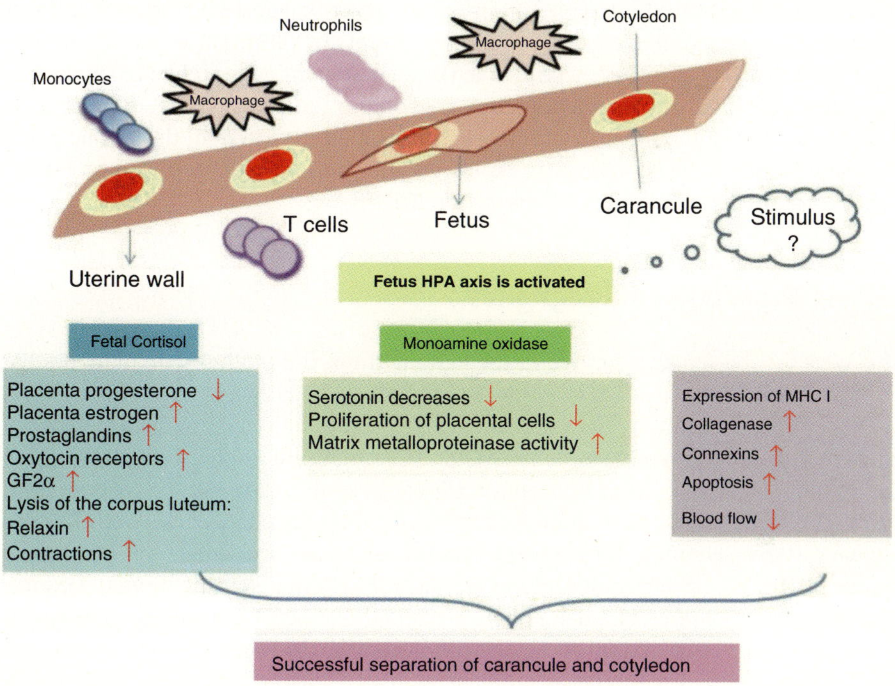

Fig. 7.1 Summary of the physical and hormonal changes for a successful expulsion of placenta

facilitate movements of ions and electrical impulses that help in coordination of myometrial contractions (Fig. 7.1).

Another compound that is related to calving is serotonin, which has been shown to play a significant role in regulating the attachment of placental membranes in dairy cows (Fecteau and Eiler 2001). Several studies have reported that high concentrations of serotonin in both fetal and maternal membranes, during pregnancy, help maintain placental attachment by stimulating proliferation of placental cells and inhibiting the activity of the matrix metalloproteinase (Eiler and Hopkins 1993; Fecteau and Eiler 2001). On the other hand, the release of monamine oxidase, from the fetus, close to parturition, results in degradation of serotonin, which promotes placental separation and parturition (Lee et al. 1989).

Detachment of the caruncle–cotyledon connection is a must for the placenta to be expelled from the uterus. The process of detachment is facilitated by relaxin, which promotes breakdown of collagen at the fetal cotyledon–maternal caruncle interface (Musah et al. 1987). Moreover, during labor, contractions of the myometrium result in pressure changes in the cotyledonary villi leading to physical separation of the fetal–maternal unit. A study by Eiler and Hopkins (1992) demonstrated that collagenase is involved in the breakdown of caruncle–cotyledon attachment allowing for the expulsion of the placenta following parturition. Matrix degradation enzymes including matrix metallopeptidase 2, 3, and 9 also play a role in expulsion of fetal membranes (Walter and Boos 2001; Takagi et al. 2007; Streyl et al. 2012). An increased number of apoptotic cells in the maternal crypt epithelium, maternal stroma, and fetal chorionic epithelium occur, which suggest an important role of apoptosis in expulsion of fetal membranes (Boos et al. 2003). These results are supported by Streyl et al. (2012) who showed upregulation of genes related to apoptosis and suggested their involvement in the expulsion of the placenta. In addition, once parturition has begun there is a decrease in the blood flow to both the caruncle and cotyledon, which shrinks the vessels, lowers the capillary pressure, and facilitates separation of the membranes (Manspeaker 2010).

7.3 Protection of Pregnancy

In order for the fetus to survive in the endometrium the mother's immune system behaves in such a way that tolerates its implantation, although the fetus is made partly of foreign proteins from the sire. This is achieved by an interaction between maternal and fetal immune systems. Maternal recognition of pregnancy takes place at around 16–17 days after conception (Spencer et al. 2004). As the embryo goes through cleavage stages, maternal uterus goes through changes as well. Corpus luteum starts secreting progesterone which decreases muscular activity of the uterus, stimulates proliferation of the uterine epithelium and growth of cotyledons. Progesterone increases blood supply, growth and coiling of the uterine glands, and leukocyte infiltration. Thirty-three days after mating, the fetal chorionic membrane has already formed a fragile attachment with two to four of the cotyledons surrounding the fetus.

An important cell surface molecule that contributes in the immune tolerance of the mother is major histocompatibility complex (MHC I). The main function of the MHC is to bind to peptide fragments derived from pathogens or foreign organisms and display them on the cell surface to be recognized by a certain set of leukocytes. There are classical and nonclassical MHC I. The classical MHC I is able to present multiple antigens including foreign antigens and are known as polymorphic (Tilburgs et al. 2010). If host cells express antigens that are deemed foreign, they are attacked and killed by cytotoxic T lymphocytes (CTL) (Baker et al. 1999). Nonclassical MHC I present another type of antigen known as the "zero" antigen (they are not polymorphic). The "zero" antigen is positioned within a groove in the MHC I and is recognized by maternal white blood cells as a "self" antigen, although it does not originate from the host itself. All the cells that express the "zero" antigen are protected by the host immune cells, otherwise the cells that do not express this antigen are killed by Natural Killer cells (Lash et al. 2010; Lash and Bulmer 2011).

It should be pointed out that in cattle both classical MHC I with paternal antigens and nonclassical MHC I with "zero" antigens are expressed during pregnancy (O'Gorman et al. 2010). The paternal antigens are expressed on specific cells known as binuclear cells, which play a significant role in supporting the pregnancy. The origin of binuclear cells is fetal trophoblast, although their precise origin remains undetermined. Binuclear cells are able to migrate from the trophoblast and spread through the endometrium, to merge with endometrial cells and to form trinuclear cells. Trinuclear cells lose the antigens deriving from the father (i.e., sire) and do not express MHC I (Davies et al. 2000; Bainbridge et al. 2001). Trinuclear cells also secrete a compound known as lactogen, which helps to stabilize pregnancy by stimulating production of steroid hormones in the ovaries and placenta and also by influencing maternal metabolism to support fetal growth and development (Patel et al. 1996).

7.3.1 Parturition and Release of Fetal Membranes

Research in dairy cows indicates that both the maternal and fetal immune systems contribute to initiation of the calving process. Even though the details of this interaction are not fully understood, research suggests that the initial step for ejection of fetal membranes starts when the immune system of the mother recognizes in the fetal membranes the antigenic compounds belonging to the father (Davies et al. 2004).

Interestingly, classical MHC I with paternal antigens have been reported to be expressed by cows at the time of parturition (Newman and Hines 1979; Hines and Newman 1981). Expression of MHC I is essential for maturation of the placenta. The villi of the trophoblast are the contact area in the placentomes, and up to 1 month before parturition, there is a thinning process of the endometrial epithelium that ends up in complete disappearance of that layer. This histological change leads to slackening of the contact area, so that the trophoblast (i.e., fetus) epithelium establishes direct contact with the connective tissue of the endometrium (Grünert 1986).

Finally, when paternal antigens are presented by classical MHC I protein on the surface of chorion cells, the antigens are recognized by T lymphocytes (CD8+ cells) (Adams et al. 2007), leading to elevated migration of T cells to the surface of the placenta. Once the immunological recognition of the trophoblast classical MHC I occurs, then, the destruction of the placentome (a physiological process) and parturition starts. This results in the breakdown of the maternal and fetal attachment and the expelling of the placenta.

7.4 Predisposing Factors

Multiple factors have been suggested to influence the incidence of RP (Fig. 7.1) including dystocia, twin birth, stillbirth, hypocalcemia, high environmental temperatures, age, premature birth, placentitis, several nutritional factors, various stressors, certain diseases, bacterial endotoxins, and neutrophil inactivation (Maas 2008; McNaughton and Murray 2009a, b; Ametaj et al. 2010). We will discuss some of the most important factors of RP below. A summary of predisposing factors and consequences of RP are summarized in Fig. 7.2.

7.4.1 Inflammation

The placentome, as previously described, consists of the fetal cotyledon and maternal caruncle. As pregnancy advances the chorionic villi present in the cotyledons, which associate with the caruncle, begin to invade the caruncular crypts. The

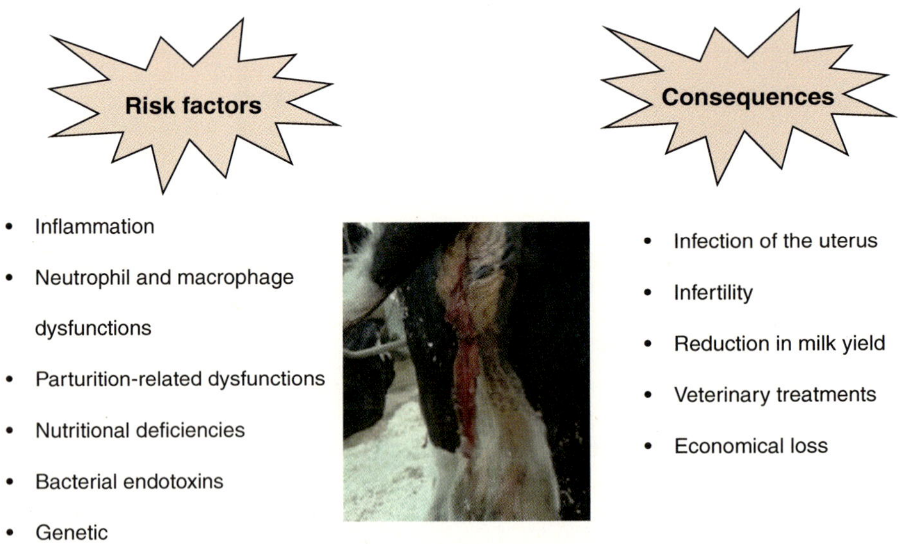

Fig. 7.2 Risk factors and consequences of retained placenta in dairy cows

invasion brings closer together the attachment of the caruncle and cotyledon into a lock-and-key manner. Swelling or edema of the chorionic villi, which occurs following bacteria-related inflammation, results in retained placenta due to the continued pressure that holds the placentome together (McNaughton and Murray 2009a, b). Under normal conditions the blood flow to both the caruncle and cotyledon decreases to facilitate separation (Manspeaker 2010). However, during inflammation of both the caruncle and cotyledon the secretory activity of the endometrium and contractile activity of the myometrium are impaired.

7.4.2 Nutritional Deficiencies

Several nutritional deficiencies have been reported to impair the expulsion of fetal membranes. For example, vitamin E and selenium (Se) have been shown to impair neutrophil functions and are risk factors for the development of RP. Likewise, dietary supplementation with vitamin E and Se prior to calving has been demonstrated to lower oxidative stress and the incidence rate of RP as well as mastitis (Goff 2006). Periparturient hypocalcemia or milk fever is another metabolic risk factor for RP as lower calcium impairs the ability of the uterus to contract (i.e., uterine atony) (McNaughton and Murray 2009a, b).

7.4.3 Parturition-Related Dysfunctions

Abnormal deliveries, including twins, dystocia, abortions, or premature calving, and induced labor increase the incidence rate of RP (Maas 2008; Manspeaker 2010). Abortion during late pregnancy (after 120 days of gestation) increases the incidence rate of RP noticeably to nearly 60%. Pharmacological induction of parturition after 120–150 days of gestation greatly increases the incidence as well, although the placenta is often delivered spontaneously around 8–10 days later. Induction of parturition via dexamethasone and prostaglandins, within 30 days of the expected parturition date, is associated with an increased risk of RP up to 85% (McNaughton and Murray 2009a, b). It should be noted, however, that while abnormal delivery is a significant risk factor for RP, it only contributes to around a third of cases and there is still a 4.1% incidence rate of RP with normal pregnancy and delivery (Davies et al. 2004).

7.4.4 Neutrophil and Macrophage Dysfunctions

Host immune functions are essential to normal separation of placental membranes (Davies et al. 2004). Early research has evidenced impairment of neutrophil chemotactic functions that has been seen prior to calving in cows that subsequently develop RP. For instance, Gunnink (1984) reported lower migration capabilities of neutrophils from cows with RP. Furthermore, Kimura et al. (2002) reported reduced migration, phagocytic, and oxidative activities in cows that developed RP. Neutrophils also had lowered myeloperoxidase prior to calving, suggesting lowered killing capabilities. Moreover, cows that developed RP had lower concentrations of IL-8 in the plasma, a chemokine that attracts neutrophils to cotyledons and increases secretion of collagenase, which helps to detach maternal and placental tissues (Kimura et al. 2002). In another study, LeBlanc (2008) also reported lowered concentrations of IL-8 in the plasma at 2 weeks prepartum, reduced neutrophil migration, and decreased oxidative burst activity of neutrophils in cows that later developed RP. Immediately after calving, the immune system is responsible for degradation of the placentome and separation of placental membranes from the maternal tissue. Therefore, impaired immune function is the major contributing factor to RP (LeBlanc 2008).

Numerous factors have been reported to negatively affect the activity of neutrophils. Some of those factors include hormones like cortisol, progesterone, estrogen, and glucagon (Watson et al. 1987; Preisler et al. 2000; Burton et al. 2005; Galvão et al. 2010; Chaveiro and Moreira da Silva 2010). In addition, combination of negative energy balance around parturition, insulin resistance, and a decrease in concentrations of glucose and glycogen in the blood as well as an increase in the concentrations of NEFA and BHBA has been shown to alter neutrophil migration, phagocytosis, and oxidative burst capacity (Zerbe et al. 2000; Hammon et al. 2006; Kim et al. 2005; Vazquez-Añon et al. 1994).

Interestingly Miyoshi et al. (2002) showed that cows with RP had differences in the distribution of macrophages and T cells in the placentomes as compared to normal cows. They demonstrated that there is a physical difference between placentomes from RP cows when compared with normal placentas. They also reported that the number of macrophages that reside in the stroma of caruncles was estimated to be one third to one half of the total number of cellular components of this tissue. T cells usually are presented in lower numbers in the stroma of the caruncles, and they are not present in the deep layer of the lamina propia of caruncles. No neutrophils or eosinophils were observed in the stroma of the caruncles. Another important finding of this study was that the intensity and number of macrophages showing acid phosphatase activity in caruncles from RP cows was decreased. Low activity of acid phosphatase suggests lower phagocytic activity of the macrophages (Miyoshi et al. 2002). The authors proposed that macrophages may act as scavengers in the caruncle tissue during postpartum and control normal termination of the pregnancy by tissue attenuation and detachment of the fetal membranes.

7.4.5 Role of Bacterial Endotoxins

Lochia of cows with RP has been reported to contain large concentrations of lipopolysaccharides (LPS). Retained placenta delays elimination of lochia from the uterus, which become an excellent medium for the growth of bacteria, impairing uterine involution and aiding the development of uterine infections (Dohmen et al. 2000; Sheldon et al. 2009). A study conducted by our lab (Zebeli et al. 2011) showed that cows administered parenterally with increasing doses of LPS over three consecutive weeks around parturition had greater incidence of RP compared to controls, suggesting a possible role of LPS in the pathobiology of RP.

In the presence of LPS a whole molecular cascade is activated in order to clear bacterial infection. Bovine endometrial cells express TLR4, which binds to LPS deriving from *E. coli* and other Gram-negative bacteria. However, although a whole variety of chemical compounds are released to attract neutrophils and macrophages to the site of infection including nitric oxide, tumor necrosis factor, IL-6, and IL-8, LPS lowers expression of L-selectin and impairs neutrophil migration to the infected tissue (Diez-Fraile et al. 2003; Takeda and Akira 2004).

7.4.6 Genetic Factors

It has been suggested that RP is a heritable trait in cattle (Joosten et al. 1991). Heritability of RP in cows varies between and within breeds. Reported values of heritability oscillate between 0.004 and 0.22 (Lin et al. 1989; Lyons et al. 1991, Wassmuth et al. 2000; Heringstad et al. 2005, 2009; Heringstad, 2010; Benedictus et al. 2013). For example, Van Dorp et al. (1998) reported a heritability of 0.01 in Holstein cows in the first lactation. Meanwhile greater values have been reported for Simmental cattle (0.14) in the first lactation (Schnitzenlehner et al. 1998). These authors found a genetic correlation of 0.79 between RP in the first and second lactations. Using a sire-maternal grandsire model the heritability of RP in Meuse-Rhine-Yssel cattle was estimated to be 0.22 (SEM = 0.07). The authors suggested that heritability of RP may be greater in dual purpose breeds (Benedictus et al. 2013).

Moreover, cows that have RP are at greater risk of other diseases as genetic correlations between dystocia, RP, metritis, and mastitis were moderate in size and positive (Lin et al. 1989). Heringstad et al. (2005) reported low to medium genetic correlations for RP. In dairy cows breeding objectives have been focused mostly on increasing milk yield; however, there have been little efforts to select cows for resistance to periparturient diseases. Presence of an underlying genetic component of RP suggests the potential for improvement in disease resistance through selection of resistant cows.

7.5 Consequences of Retained Placenta

7.5.1 Infection of the Uterus

Retained placenta increases the risk of uterine infections (Fig. 7.2). Given that placental membranes hang out of the vulva bacteria or other infectious agents can enter the vulva and uterus more readily. Indeed, the placenta moves back and forth through the external opening of the vulva and vagina, increasing the contact of the reproductive tract with various infectious agents. Retention of the placenta slows down the uterine involution process and interferes with the timely expulsion of the lochia (Maas 2008). Delays in the expulsion of lochia create conditions within the reproductive tract that support the growth of bacteria, which are commonly associated with uterine disease, like *Escherichia coli* and *Arcanobacterium pyogenes* (Azawi 2008).

7.5.2 Infertility

There is a strong association of RP with uterine infections. Retained placenta–metritis complex has been reported to be related to lower fertility by increasing the calving to first service interval (i.e., days open), decreasing pregnancy rate to first service, increasing the number of services per conception, and extending the calving interval (Sandals et al. 1979). A study by Halpern et al. (1985) including 1,111 heifers and 2,493 cows showed that duration of retained placenta was associated with decreases in conception rate, and delays of 18 or 57 days to conception when duration of RP was 5 or 7 days, respectively. Moreover, Fourichon et al. (2000) estimated that pregnancy rate in affected cows is lowered by around 15% relative to unaffected cows.

7.5.3 Reduction in Milk Yield

The decrease in milk sold caused by RP occurs not only from possible decreases in yield but also because of mandatory milk withdrawal (Guard 1999). Retained placenta has been found to have a significant negative effect on milk yield for several weeks following calving with the estimated reduction in 305-day production around 7% (Rajala and Gröhn 1998). It should be noted that although cows with RP might not always have lower milk yield, it is still discarded as milk from cows with RP is considered unfit for human consumption (Guard 1999; Laven and Peters 1996). Moreover, in a recent study by our lab cows with RP had lower milk production when compared with healthy cows. Cows with RP produced 9.79 L of milk/day less compared to their healthy counterparts, which implies a loss of $2.74/day/cow, without taking into consideration the cost of medications and veterinary services (Dervishi et al. 2016).

7.6 System Biology Approach to Retained Placenta

As the omics technology advances, the etiopathology and the underlying pathophysiological mechanisms of RP still remain incomplete. The systems veterinary approach offers a whole new methodology to integrate the information generated by multiple disciplines including genomics, transcriptomics, proteomics, and metabolomics to better understand the pathobiology of RP and other metabolic diseases in dairy cattle. New omics approaches offer a great possibility to explore new quantitative trait locus (QTL), genes, transcripts, proteins, and metabolites involved in the etiopathology of RP.

Genome-Wide Association Studies (GWAS) utilize information on genetic markers or single nucleotide polymorphisms (SNP) spread across the genome to determine associations with a trait of interest (Goddard and Hayes 2009). Genome-Wide Association Studies have been useful for identifying SNP markers and genes associated with a particular disease. Application of these methodologies have identified a genomic region such as QTL on BTA5 and BTA9 that is associated with RP and with fertility treatments in dairy cows (Schulman et al. 2004; Olsen et al. 2011). Those studies suggest the potential to include immune response traits in genomic selection indices to decrease occurrence of disease and improve animal health in the dairy industry (Thompson-Crispi et al. 2014).

In addition, transcriptomics has been used to study the mechanisms underlying detachment of fetal membranes after birth in cows (Streyl et al. 2012). The latter authors reported that 12–15 days before parturition overexpressed genes were related to mitotic cell cycle and tissue differentiation, microtubule cytoskeleton, transmembrane transporters, and regulation of signal transduction. Meanwhile after parturition (after the end of the expulsion phase), the genes with upregulated mRNA levels were almost all related to three different physiological processes: innate immune response (20 genes), apoptosis (15 genes), and degradation of extra cellular matrix (11 genes), which play a fundamental role in placental detachment. Among the immune response genes upregulated after parturition were *CD14*, *CD36*, chemokine C-X-C motif ligand 2 (*CXCL2*), *CXCL14*, chemokine C-C motif receptor 4 (*CCR4*), macrophage expressed 1 (*MPEG1*), and macrophage scavenger receptor 1 (*MSR1*). Upregulation of *MPEG1* and *MSR1* indicates increased numbers of monocytes and/or macrophages (Streyl et al. 2012). Activation of immune response is required for the release of fetal membranes and the authors suggested an important role of inflammatory cells, such as macrophages and neutrophils, in the process of detachment of fetal membranes (Streyl et al. 2012). However, it should be noted that a prolonged immune response might have detrimental effects on the host. Recent data from our lab showed that cows that retained their fetal membranes had activation of innate immunity starting at −8 weeks before parturition (Dervishi et al. 2016) and up to +8 weeks after parturition.

In several recent studies proteomics techniques were used to investigate the pathobiology of RP in cows and screen for early clinical diagnostic markers of disease (Kankofer et al. 2014, 2015; Fu et al. 2016). Kankofer et al. (2015) conducted a 1D and 2D gel proteomic approach to compare the protein profile of caruncle (maternal side) and fetal villi of bovine placenta that was either properly expelled or retained. They reported differences in the number of fractions and the intensity of staining between maternal side of the released and retained tissues as well as respective fetal membranes. In addition, the same research group used 2D-DIGE to evaluate protein fingerprints of bovine placentas that were properly released and to compare this profile with proteins in retained fetal membranes (Kankofer et al. 2015). Out of 1,174 spots, they selected only 5 spots for identification. Specifically, short transient receptor potential channel 5 (TrpC5) showed high expression in maternal side of retained and not retained placenta. Rab GDP dissociation inhibitor beta also was highly expressed in fetal side of retained and not retained placentas. Furthermore, transforming growth factor b2 (TGF-β2) was highly expressed in maternal and fetal sides of retained placenta while Ras-related protein Rab-7b was increased only in healthy maternal caruncles. Proline dehydrogenase 2 (PRODH2) was similarly expressed in all examined samples. The authors suggested that those proteins may be considered as having an impact on the process of proper release or the retention of fetal membranes (Kankofer et al. 2014). Moreover, Fu et al. (2016) showed that proteins bovine serum albumin (BSA), alpha enolase (ENO1), apolipoprotein A-I (APOA1), annexin A8-like 1 (ANXA8L1), serine proteinase inhibitor, glutathione transferase, and transketolase were differently expressed in cows with RP. They concluded that the potential causes and influencing processes involved in RP might be related to fibrinolysis, pyruvate metabolism, inflammatory response, and oxidative stress.

Metabolomic investigations for evaluation of RP have not been reported yet. Most of the studies published have been focused on a limited number of blood indicators or metabolites, for example, non-esterified fatty acids (NEFA), beta-hydroxybutyrate (BHBA), cholesterol, lactate, and glucose as well as a few proteins of innate immunity like interleukins, haptoglobin (Hp), and serum amyloid alpha (SAA) (Seifi et al. 2007; Quiroz-Rocha et al. 2009; Ospina et al. 2010). From these studies valuable insight has been acquired and few metabolites have been identified to be associated with retained placenta. For example, elevated concentrations of several serum innate immunity variables IL-1, IL-6 TNF, SAA, and lactate reflected pathophysiological events occurring in RP cows at −8 weeks prior to occurrence of the disease (Dervishi et al. 2016). Enhanced concentrations of several serum innate immunity variables and lactate reflect pathophysiological events occurring prior to occurrence of disease. In another study, Pohl et al. (2015) concluded that RP in multiparous cows is associated with high Hp concentration at 5 days in milk.

In a recent study, Fadden and Bobe (2016) reported that serum visfatin may serve as a chronic disease indicator and assist in early detection of cows at increased risk to develop RP and other diseases. However, the studies conducted to date report a limited number of blood indicators or metabolites. Metabolomics has been successfully used in dairy cows affected with ketosis, milk fever, subclinical mastitis and also to identify biomarkers of disease state in dairy cows (Zhang et al. 2013; Sun et al. 2014; Hailemariam et al. 2014a, b; Dervishi et al. 2016). In one of these pioneering studies, alterations in blood plasma metabolites start prior to the onset of the clinical signs of transition diseases (Hailemariam et al. 2014b; Dervishi et al. 2016). Metabolomics is fairly new in animal and veterinary sciences, and together with the other omics sciences it is the building block of systems veterinary. Metabolomic studies for prediction of RP have not been previously reported; therefore, it would be of great interest to conduct such studies with the final goal to develop screening biomarkers of RP and to better understand the causal agents and the pathobiology of the disease. These biomarkers will greatly benefit the dairy industry as they can help

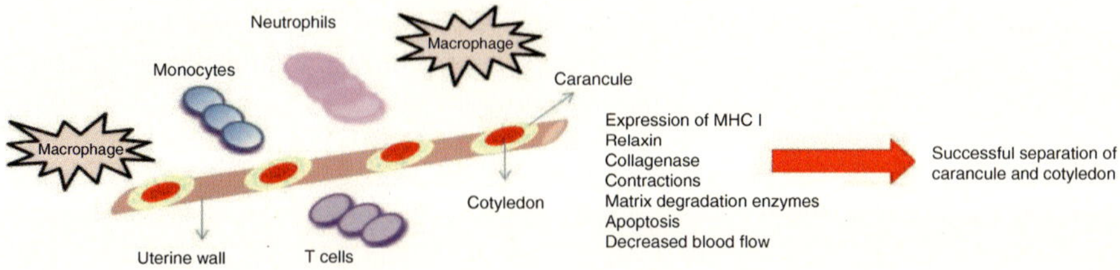

Genomic	Transcriptomic	Proteomic	Metabolites
QTLs on BTA5 and BTA9	*CD14, CD80, CD36, CXCL5*	TrpC5	NEFA
	CXCL2, CXCL14,CCL27, CCR1,	Rab GDP dissociation inhibitor β	BHBA
	CCR4, MPEG1,	(TGF-β2)	Cholesterol
	MSR1, MMP14, AREGB,	Rab-7b	Lactate
	SERPINE1, ADAMTS4, FBLN5,	PRODH2	Glucose
	LRP2, MAN1C1, MMP1,	BSA	IL-1, IL-6
	MMP3, MMP9	ENO1	TNF
	DNASE1L3, IFI6, IFI16, IFIH1,	APOA1	Haptoglobin
	GADD45B, NFKBIA, NR4A1,	ANXA8L1	Serum amyloid alpha (SAA)
	CTSB, RHOB, IGF1R, EIF2AK2,	Serine proteinase inhibitor	Visfatin
	TNFRSF1B, CCAR1, BIRC5,	Glutathione transferase	
	XIAP, CFLAR, CTNNAL1	Glutathione transketolase	

Fig. 7.3 Contribution of the omics sciences in the understanding of the etiopathology of retained placenta

identifying animals with greater risk to develop RP prior to occurrence of disease. The information and contribution generated thus far from omics sciences is summarized in Fig. 7.3.

Conclusions

Retained placenta is a condition that affects dairy cows immediately after calving. It is associated with financial costs related to milk withdrawal, decreased milk yield, and treatment expenses. The key factor leading to RP is immune dysfunction preventing separation of maternal–fetal membranes, and although a large risk factor for RP is abnormal delivery, this contributes to only one third of the cases with an overall incidence of roughly 4.1% (Davies et al. 2004).

Currently, preventive measures mostly aim at prevention of nutritional deficiencies, which have been associated with RP while there are few, if any, methods to prevent immune dysfunctions that are the key factor for development of the disease. There is much promise to the new research implicating the role of endotoxins in the etiology of RP in providing the opportunity to develop preventive interventions and avoid economic losses.

Omics sciences including genomics, transcriptomics, proteomics, and metabolomics are in their infancy and have been applied separately. It would be desirable to utilize all of those sciences altogether in order to have a multi-angle approach. Those sciences have the potential to provide the molecular basis of RP but also will help to identify genes, proteins, metabolites, and metabolic pathways, which may help in better understanding the pathobiology of the disease, identify causal agent(s), and serve as screening or diagnostic biomarkers that would identify cows at risk of developing RP prior to disease occurence. Those biomarkers have the potential to greatly benefit the dairy industry if preventive strategies will be developed in the near future.

References

Adams AP, Oriol JG, Campbell RE et al (2007) The effect of skin allografting on the equine endometrial cup reaction. Theriogenology 68:237–247

Ametaj BN, Zebeli Q, Iqbal S (2010) Nutrition, microbiota, and endotoxin-related diseases in dairy cows. Rev Bras Zootec 39:433–444

Azawi OI (2008) Postpartum uterine infection in cattle. Anim Reprod Sci 105:187–208

Baker JM, Bamford AI, Antczak DF (1999) Modulation of allospecific CTL responses during pregnancy in equids: an immunological barrier to interspecies matings? J Immunol 162:4496–4501

Bainbridge D, Sargent I, Ellis S (2001) Increased expression of major histocompatibility complex (MHC) class I transplantation antigens in bovine trophoblast cells before fusion with maternal cells. Reproduction 122:907–913

Benedictus L, Koetsa AP, Kuijperse FHJ, Joostenb I, van Eldika P, Heuven HCM (2013) Heritable and non-heritable genetic effects on retained placenta in Meuse-Rhine-Yssel cattle. Anim Reprod Sci 137:1–7

Bjorkman NH, Sollen P (1961) A morphological study on retention secundinarum in cattle. Acta Vet Scand 2:347–362

Boos A, Janssen V, Mulling C (2003) Proliferation and apoptosis in bovine placentomes during pregnancy and around induced and spontaneous parturition as well as in cows retaining the fetal membranes. Reproduction 126:469–480

Burton JL, Madsen SA, Ling-Chu CH et al (2005) Gene expression signatures in neutrophils exposed to glucocorticoids: a new paradigm to help explain "neutrophil dysfunction" in parturient dairy cows. Vet Immunol Immunopathol 105:197–219

Chaveiro A, Moreira da Silva F (2010) Invitro effect of the reproductive hormones on the oxidative burst activity of polymorphonuclear leucocytes from cows: a flow cytometric study. Reprod Domest Anim 45:40–45

Davies CJ, Fisher PJ, Schlafer DH (2000) Temporal and regional regulation of major histocompatibility complex class I expression at the bovine uterine/placental interface. Placenta 21:194–202

Davies CJ, Hill JR, Edwards JL et al (2004) Major histocompatibility antigen expression on the bovine placenta: its relationship to abnormal pregnancies and retained placenta. Anim Reprod Sci 82:267–280

Dervishi E, Zhang G, Hailemariam D et al (2016) Occurrence of retained placenta is preceded by an inflammatory state and alterations of energy metabolism in transition dairy cows. J Anim Sci Biotechnol 7:26

Diez-Fraile A, Meyer E, Duchateau L et al (2003) L-selectin and β2-integrin expression on circulating bovine polymorphonuclear leukocytes during endotoxin mastitis. J Dairy Sci 86:2334–2342

Dohmen MJW, Joop K, Sturk A et al (2000) Relationship between intra-uterine bacterial contamination, endotoxin levels and the development of endometritis in postpartum cows with dystocia or retained placenta. Theriogenology 54:1019–1032

Eiler H, Hopkins FM (1992) Bovine retained placenta: effects of collagenase and hyaluronidase on detachment of placenta. Biol Reprod 46:580–585

Eiler H, Hopkins FM (1993) Successful treatment of retained placenta with umbilical cord injections of collagenase in cows. J Am Vet Med Assoc 203:436–443

Eiler H (1997) Retained placenta. In: Youngquist RS (ed) Current therapy in large animal. Theriogenology. W.B. Saunders Company, Philadelphia, pp 340–348

Fadden AN, Bobe G (2016) Serum visfatin is a predictive indicator of retained placenta and other diseases in dairy cows. J Vet Sci Med Diagn 5:1

Fecteau KA, Eiler H (2001) Placenta detachment: unexpected high concentrations of 5-hydroxytryptamine (serotonin) in fetal blood and its mitogenic effect on placental cells in the bovine. Placenta 22:103–110

Flint APF, Ricketts AP, Craig VA (1979) The control of placental steroid synthesis at parturition in domestic animals. Anim Reprod Sci 2:239–251

Fourichon C, Seegers H, Malher X (2000) Effect of disease on reproduction in the dairy cow: a meta-analysis. Theriogenology 53:1729–1759

Fu S, Liu Y, Nie T et al (2016) Comparative proteomic analysis of cow placentas with retained foetal membranes. Thai J Vet Med 46:261–270

Fuchs AR, Rust W, Fields MJ (1999) Accumulation of cyclooxygenase-2 gene transcripts in uterine tissues of pregnant and parturient cows: stimulation by oxytocin. Biol Reprod 60:341–348

Galvão KN, Flaminio MJBF, Brittin SB (2010) Association between uterine disease and indicators of neutrophil and systemic energy status in lactating Holstein cows. J Dairy Sci 93:2926–2937

Goff JP (2006) Major advances in our understanding of nutritional influences on bovine health. J Dairy Sci 89:1292–1301

Goddard ME, Hayes BJ (2009) Mapping genes for complex traits in domestic animals and their use in breeding programmes. Nat Rev Genet 10:381–391

Grünert E (1986) In: Current therapy in theriogenology: etiology and pathogenesis of retained bovine placenta. Saunders, Philadelphia

Guard C (1999) Set up fresh and milking cows for successful A. I Hoard's Dairyman: 8–9

Gunnink JW (1984) Prepartum leukocytic activity and retained placenta. Vet Q 6:52–54

Halpern E, Erb H, Smith RD (1985) Duration of retained fetal membranes and subsequent fertility in dairy cows. Theriogenology 23:5807–5813

Hailemariam D, Mandal R, Saleem F et al (2014a) Identification of predictive biomarkers of disease state in transition dairy cows. J Dairy Sci 97:2680–2693

Hailemariam D, Mandal R, Saleem F et al (2014b) Metabolomics approach reveals altered plasma amino acid and sphingolipid profiles associated with pathological states in transition dairy cows. Curr Metab 3:184–195

Hammon DS, Evjen IM, Dhiman TR et al (2006) Neutrophil function and energy status in Holstein cows with uterine health disorders. Vet Immunol Immunopathol 113:21–29

Heringstad B, Chang YM, Gianola D et al (2005) Genetic analysis of clinical mastitis, milk fever, ketosis, and retained placenta in three lactations of Norwegian red cows. J Dairy Sci 88(2005):3273–3281

Heringstad B, Wu XL, Gianola D (2009) Inferring relationships between health and fertility in Norwegian red cows using recursive models. J Dairy Sci 92:1778–1784

Heringstad B (2010) Genetic analysis of fertility-related diseases and disorders in Norwegian red cows. J Dairy Sci 93:2751–2756

Hines HC, Newman MJ (1981) Production of foetally stimulated lymphocytotoxic antibodies by multiparous cows. Anim Blood Groups Biochem Genet 12:201–206

Janszen BMP, Bevers MM, Ravenshorst MM et al (1993) Relationship between prostaglandin-induced luteolysis and temporary inhibition of myometrial activity in late pregnant cows with ear implants containing progestagen. Reproduction 97:457–461

Joosten I, Van Eldik P, Elving L et al (1991) Factors affecting occurrence of retained placenta in cattle. Effect of sire on incidence. Anim Reprod Sci 25:11–22

Kankofer M, Wawrzykowski J, Hoedemake M (2014) Profile of bovine proteins in retained and normally expelled placenta in dairy cows. Reprod Domest Anim 49:270–274

Kankofer M, Wawrzykowski J, Miller I et al (2015) Usefulness of DIGE for the detection of protein profile in retained and released bovine placental tissues. Placenta 36:246–249

Kim IH, Na KJ, Yang MP (2005) Immune responses during the peripartum period in dairy cows with postpartum endometritis. J Reprod Dev 51:757–764

King GJ (1993) Reproduction in domesticated animals. Elsevier, Amsterdam

Kimura K, Goff JP, Kehrli ME et al (2002) Decreased neutrophil function as a cause of retained placenta in dairy cattle. J Dairy Sci 85:544–550

Lash GE, Robson SC, Bulmer JN (2010) Review: functional role of uterine natural killer (uNK) cells in human early pregnancy decidua. Placenta 31:S87–S92

Lash GE, Bulmer JN (2011) Do uterine natural killer (uNK) cells contribute to female reproductive disorders? J Reprod Immunol 88:156–164

Laven RA, Peters AR (1996) Bovine retained placenta: aetiology, pathogenesis and economic loss. Vet Rec 139:465–471

LeBlanc SJ (2008) Postpartum uterine disease and dairy herd reproductive performance. A review. Vet J 176:102–114

Lee SL, Dunn J, FS Y et al (1989) Serotonin uptake and configurational change of bovine pulmonary artery smooth muscle cells in culture. J Cell Physiol 138:145–153

Lemley CO, Camacho LE, Vonnahme KA (2015) In Bovine Reproduction: Maternal Recognition and Physiology of Pregnancy, John Wiley & Sons, Ames, Iowa, IA, USA

Lin HK, Oltenacu PA, Dale van Vleck L et al (1989) Heritabilities of and genetic correlations among six health problems in Holstein cows. J Dairy Sci 72:180–186

Lyons DT, Freeman AE, Kuck AL (1991) Genetics of health traits I Holstein cattle. J Dairy Sci 74:1092–1100

McNaughton AP, Murray RD (2009a) Structure and function of the bovine fetomaternal unit in relation to the causes of retained fetal membranes. Vet Rec 165:615–622

Maas J (2008) Treating and preventing retained placenta in beef cattle. UCD Vet Views California Cattlemen's Magazine

Manspeaker JE (2010) Retained placentas. IRM-21. Dairy integrated reproductive management. University of Maryland

McNaughton AP, Murray RD (2009b) Structure and function of the bovine fetomaternal unit in relation to the causes of retained fetal membranes. Vet Rec 165:615–622

Miyoshi M, Sawamukai Y, Iwanaga T (2002) Reduced phagocytotic activity of macrophages in the bovine retain placenta. Reprod Domest Anim 37:53–56

Musah AI, Schwabe C, Willham RL et al (1987) Induction of parturition, progesterone secretion, and delivery of placenta in beef heifers given relaxin with cloprostenol or dexamethasone1. Biol Reprod 37:797–803

Newman MJ, Hines HC (1979) Production of foetally stimulated lymphocytotoxic antibodies by primiparous cows. Anim Blood Groups Biochem Genet 10:87–92

O'Gorman GM, Al Naib A, Ellis SA et al (2010) Regulation of a bovine nonclassical major histocompatibility complex class I gene promoter. Biol Reprod 83:296–306

Olsen HG, Hayes BJ, Kent MP (2011) Genome-wide association mapping in Norwegian Red cattle identifies quantitative trait loci for fertility and milk production on BTA12. Anim Genet 42:466–474

Ospina PA, Nydam DV, Stokol T (2010) Evaluation of nonesterified fatty acids and β-hydroxybutyrate in transition dairy cattle in the northeastern United States: Critical thresholds for prediction of clinical diseases. J Dairy Sci 93:546–554

Patel OV, Hirako M, Takahashi T et al (1996) Plasma bovine placental lactogen concentration throughout pregnancy in the cow; relationship to stage of pregnancy, fetal mass, number and postpartum milk yield. Domest Anim Endocrinol 13:351–359

Pohl A, Burfeind O, Heuwieser W (2015) The associations between postpartum serum haptoglobin concentration and metabolic status, calving difficulties, retained fetal membranes, and metritis. J Dairy Sci 98(7):4544–4551

Preisler MT, Weber PSD, Tempelman RJ et al (2000) Glucocorticoid receptor expression profiles in mononuclear leukocytes of periparturient Holstein cows. J Dairy Sci 83:38–47

Quiroz-Rocha GF, LeBlanc S, Duffield T et al (2009) Evaluation of prepartum serum cholesterol and fatty acids concentrations as predictors of postpartum retention of the placenta in dairy cows. J Am Vet Med Assoc 234:790–793

Rajala PJ, Gröhn YT (1998) Effects of dystocia, retained placenta, and metritis on milk yield in dairy cows. J Dairy Sci 81:3172–3181

Sandals WCD, Curtis RA, Cote JF et al (1979) The effect of retained placenta and metritis complex on reproductive performance in dairy cattle—a case study. Can Vet J 20:131–135

Senger PL (2003) Pathways to pregnancy and parturition, 2nd edn. Current Conceptions, Inc., Pullman

Schnitzenlehner S, Essl A, Sölkner J (1998) Retained placenta: estimation of nongenetic effects, heritability and correlations to important traits in cattle. J Anim Breed Genet 115:467–478

Schulman NF, Viitala SM, de Koning DJ (2004) Quantitative trait loci for health traits in finnish ayrshire cattle. J Dairy Sci 87:443–449

Seifi HA, Dalir-Naghadeh B, Farzaneh N (2007) Metabolic changes in cows with or without retained fetal membranes in transition period. J Vet Med 54:92–97

Sheldon IM, Cronin J, Goetze L et al (2009) Defining postpartum uterine disease and the mechanisms of infection and immunity in the female reproductive tract in cattle. Biol Reprod 81:1025–1032

Spencer TE, Burghardt RC, Johnson GA et al (2004) Conceptus signals for establishment and maintenance of pregnancy. Anim Reprod Sci 82–83:537–550

Streyl D, Kenngott R, Herbach N et al (2012) Gene expression profiling of bovine peripartal placentomes: detection of molecular pathways potentially involved in the release of foetal membranes. Reproduction 143:85–105

Sun Y, Xu C, Li C et al (2014) Characterization of the serum metabolic profile of dairy cows with milk fever using H-NMR spectroscopy. Vet Q 34(3):159–163

Takeda K, Akira S (2004) TLR signaling pathways. Semin Immunol 16:3–9

Takagi M, Yamamoto D, Ohtani M et al (2007) Quantitative analysis of messenger RNA expression of matrix metalloproteinases (MMP-2 and MMP-9), tissue inhibitor-2 of matrix metalloproteinases (TIMP-2), and steroidogenic enzymes in bovine placentomes during gestation and postpartum. Mol Reprod Dev 74:801–807

Thompson-Crispi KA, Sargolzaei M, Ventura R et al (2014) A genome-wide association study of immune response traits in Canadian Holstein cattle. BMC Genomics 15:559

Tilburgs T, Scherjon SA, Claas FHJ (2010) Major histocompatibility complex (MHC)-mediated immune regulation of decidual leukocytes at the fetal-maternal interface. J Reprod Immunol 85:58–62

Van Dorp TE, Dekkers JC, Martin SW et al (1998) Genetic parameters of health disorders, and relationships with 305-day milk yield and conformation traits of registered Holstein cows. J Dairy Sci 81:2264–2270

Vazquez-Añon M, Bertics S, Luck M et al (1994) Peripartum liver triglyceride and plasma metabolites in dairy cows. J Dairy Sci 77:1521–1528

Walter I, Boos A (2001) Matrix metalloproteinases (MMP-2 and MMP-9) and tissue inhibitor-2 of matrix metalloproteinases (TIMP-2) in the placenta and interplacental uterine wall in normal cows and in cattle with retention of fetal membranes. Placenta 22:473–483

Wassmuth R, Boelling D, Madsen P et al (2000) Genetic parameters of disease incidence, fertility and milk yield of first parity cows and the relation to feed intake of growing bulls. Acta Agric Scand Sect A Anim Sci 50:93–102

Watson ED, Stokes CR, Bourne FJ (1987) Influence of administration of ovarian steroids on the function of neutrophils isolated from the blood and uterus of ovariectomized mares. J Endocrinol 112:443–448

Whittle WL, Holloway AC, Lye SJ et al (2000) Prostaglandin production at the onset of ovine parturition is regulated by both estrogen-independent and estrogen-dependent pathways. Endocrinology 141:3783–3791

Zebeli Q, Sivaraman S, Dunn SM (2011) Intermittent parenteral administration of endotoxin triggers metabolic and immunological alterations typically associated with displaced abomasum and retained placenta in periparturient dairy cows. J Dairy Sci 94:4968–4983

Zerbe H, Schnider N, Leibold W et al (2000) Altered functional and immunophenotypical properties of neutrophilic granulocytes in postpartum cows associated with fatty liver. Theriogenology 54:771–786

Zhang H, Wu L, Xu C et al (2013) Plasma metabolomic profiling of dairy cows affected with ketosis using gas chromatography/mass spectrometry. BMC Vet Res 9(1):186

8 Omic Approaches to a Better Understanding of Mastitis in Dairy Cows

Manikhandan Mudaliar, Funmilola Clara Thomas, and Peter David Eckersall

Abstract

Mastitis, which is caused by infection of the mammary gland is the most important disease problem facing dairy farmers. While the disease has been studied for decades in order to determine better diagnosis and treatment, it is only recently that the full panoply of advanced biotechnological methodologies has been applied that are needed to bring a systems biology approach to investigations. Molecular investigations using analyte-specific immunoassays, such as for the acute phase proteins haptoglobin and mammary-associated serum amyloid A3 in milk have introduced possibilities for monitoring the host inflammatory response. The omics revolution in biology, with genomics being harnessed especially for identification of the causative pathogens of mastitis, has enhanced dissection of the mammary microbiome. The application of proteomics, peptidomics and metabolomics to the diagnosis and pathophysiology of mastitis, in contrast, is in its infancy though the potential of these advanced tools of biological research is clear as they are applied in a systems biology analysis of this major health problem of dairy cows.

M. Mudaliar, BVSc & AH, MSc (✉)
Glasgow Molecular Pathology Node, College of Medical, Veterinary and Life Science, University of Glasgow, Glasgow G61 1QH, UK
e-mail: Manikhandan.Mudaliar@glasgow.ac.uk

F.C. Thomas, DVM, PhD
Biochemistry Unit, Department of Veterinary Physiology and Pharmacology, College of Veterinary Medicine, Federal University of Agriculture, Abeokuta, Nigeria
e-mail: funmithomas2007@gmail.com

P.D. Eckersall, BSc, PhD, MRCPath, MAE
Institute of Biodiversity Animal Health and Comparative Medicine, College of Medical, Veterinary and Life Science, University of Glasgow, Glasgow, G61 1QH, UK
e-mail: david.eckersall@glasgow.ac.uk

B.N. Ametaj (ed.), *Periparturient Diseases of Dairy Cows*,
DOI 10.1007/978-3-319-43033-1_8

8.1 Introduction

Mastitis refers to the inflammation of the udder or mammary gland. This usually follows invasion by microorganisms, although other physical or chemical causes such as trauma or harmful toxins/chemicals can also lead to mastitis. There are two main forms of mastitis, clinical (CM) and subclinical mastitis (SCM) both of which are common in the peri-parturient period. For a long time, bovine mastitis has remained prominent as one of the most costly and prevalent diseases in the dairy industry (Hillerton and Berry 2005; Halasa et al. 2007; Hettinga et al. 2008b; Akers and Nickerson 2011). Losses accrued due to mastitis are related to cessation or reduction of milk production (accounting for up to two-thirds of total losses (Akers and Nickerson 2011)), costs of treatment, culling, extra labour, wasted time and discarded milk as well as veterinary charges. It is often difficult to estimate the total costs of mastitis due to the myriad of factors that can contribute into losses during mastitis episodes (Heikkila et al. 2012). Moreover, drug residues in milk, as a result of treatment, pose the danger of inducing antibiotic resistance in pathogens and can constitute other public health hazards when milk and other dairy products are consumed by humans.

8.1.1 Aetiology

Microorganisms including bacteria, fungi, algae and viruses are a variety of the pathogens capable of invading the mammary gland leading to mastitis (Nicholas 2011; Tomasinsig et al. 2012; Wellenberg et al. 2002; Green and Bradley 2013; Pachauri et al. 2013; Reyher et al. 2012a, b; Schukken et al. 2012; Zadoks and Fitzpatrick 2009). More than 200 different pathogens have been reported to be able to cause mastitis in the bovine species (Zadoks et al. 2011). Bacteria are, however, the most prevalent cause of mastitis. From an epidemiological viewpoint, pathogens causing mastitis can be generally classified as environmental or contagious pathogens. Amongst the former, *Escherichia coli* usually causes severe clinical mastitis that elicits massive increases in inflammatory and immune indices, usually resulting in disease which may either be rapidly eliminated or may become systemic and consequently fatal (Baeker et al. 2002; Pyorala et al. 2011). The environmental pathogens are present in the cow's environment such as bedding and transmitted to the teat by direct contact. On the other hand, contagious pathogens are generally considered host adapted to cause mastitis, and are transmitted from one cow, udder or quarter to the other in a herd, and include *Staphylococcus aureus, Streptococcus dysgalactiae, Streptococcus agalactiae* and some strains of *Streptococcus uberis*, amongst others. Subclinical or chronic forms of mastitis are usually associated with contagious pathogens, because these organisms are adapted to survive for long periods in the mammary gland; thus, such infections are difficult to eliminate and often lead to a rise in somatic cells in milk (Bradley 2002). Besides the epidemiological classification, the pathogens causing bovine mastitis can be classified as major pathogens and minor pathogens based on their virulence, the impact on milk production and the severity of damage they cause to the udder (Table 8.1). The major pathogens are more virulent and generally cause clinical mastitis (although some of them like *Streptococci and Staphylococcus aureus* mostly cause subclinical mastitis) while the minor pathogens are less damaging to the udder and generally cause subclinical mastitis.

8.1.2 Pathogenesis

Infection of the mammary gland usually occurs via the teat canal irrespective of the source of origin whether from the environment or from a contagious source. Upon transmission to the outer edge/skin of the teat, the pathogens invade to the

Table 8.1 Pathogens causing mastitis in dairy cows

Major pathogens	Minor pathogens	Uncommon pathogens
Gram-Positive Bacteria	• Coagulase-Negative *Staphylococcus* spp.	**Bacteria**
• *Streptococcus* spp.	*S. hyicus*	• *Shigella* spp.
S. agalactiae	*S. chromogenes*	• *Proteus* spp.
S. dysgalactiae	*S. xylosus*	• *Citrobacter* spp.
S. uberis, etc.	*S. sciuri*	• *Yersinia* spp.
	S. warneri	• *Raoultella* spp.
• *Staphylococcus aureus*	*S. simulans*	• *Leptospira* spp.
	S. epidermidis	• *Mycobacterium* spp.
Gram-Negative Bacteria	*S. intermedius*, etc.	
• *Escherichia coli*	• *Serratia* spp.	**Fungi** (Pachauri et al. 2013)
• *Klebsiella* spp.	*S. marcescens*	• *Aspergillus jumigatus*
K. pneumoniae	*S. liquefaciens*	• *Aspergillus nidulans*
K. oxytoca, etc.	• *Corynebacterium bovis*	• *Trichosporon* spp.
	• *Arcanobacterium pyogenes*	• *Pichia* spp.
• *Enterobacter* spp.	• *Nocardia* spp.	• *Candida* spp.
• *Pseudomonas aeruginosa*		• *Saccharomyces* spp.
		• *Cryptococcus* spp.
• *Mycoplasma* spp.		• *Torulopsis* spp.
M. bovis		
M. californicum etc.		**Algae**
		• *Prototheca* spp. (Costa et al. 1996; Janosi et al. 2001; Tomasinsig et al. 2012)
		Virus (Wellenberg et al. 2002)
		• Bovine herpesvirus (BHV1 and BHV4)
		• Parainfluenza virus

milk inside the teat cistern, adhere to or invade mammary epithelium to avoid elimination by neutrophils or milk flow, and may multiply rapidly. Depending on the nature and ability of the pathogen they may further invade the mammary tissue. Once the microbes penetrate the physical barrier of the teat canal, the host innate immune system detects the microbes through the pattern-recognition receptors (PRRs), particularly via the toll-like receptors (TLRs) (Ezzat Alnakip et al. 2014). Binding of microbial components with TLRs activates the TLR signalling pathway that mediates several intracellular signal transduction cascades triggering the production of pro-inflammatory cytokines leading to inflammation and eventually elimination of the microbes by leukocytes (Akira et al. 2006). Migration of immune cells, particularly neutrophils, and desquamation of mammary epithelium accompanied with reduced milk production result in a several-fold increase in SCC per unit volume of milk. Bovine neutrophils migrate to the mammary epithelium by diapedesis, and they constitute more than 90% of the total leukocytes in mammary gland during inflammation. At the site of infection, the neutrophils engulf, phagocytose and destroy the invading microbes via an oxygen-dependent respiratory burst system producing hydroxyl and oxygen radicals, and an oxygen-independent system using peroxidases, lysozymes, hydrolytic enzymes and lactoferrin (Ezzat Alnakip et al. 2014). However, this mechanism does not work well with all pathogens. It works well in the case of *Escherichia coli, but Staphylococcus aureus* survives inside the phagolysosome and *Streptococcus* uberis inactivates neutrophils so that they don't even engulf the bacteria. If the microbes are eliminated rapidly resulting in the removal of the inflammatory stimuli, the neutrophil recruitment ceases and the SCC returns to normal levels. However, if the microbes survive the immediate host defence response, then the infection and inflammation continue to spread to the adjacent mammary tissues.

Following pathogen invasion and establishment in the gland, either of the two major forms of mastitis may result, namely CM or SCM. CM occurs showing the signs of inflammation—swelling, redness, pain and heat (or generalised fever) of the udder or quarter, as well as physical and chemical changes in milk such as the presence of flakes, clots or blood, increased proteolysis of milk caseins, increase in sodium and chloride ions, a decrease in lactose and release of intracellular enzymes into milk. SCM occurs with no noticeable physical signs of inflammation, but is commonly indicated by an increase in somatic cell counts (SCC) in milk produced from affected quarters due to the migration of leukocytes from blood into milk. Either of these two forms of mastitis may occur as a peracute, acute or chronic infection. Clinical mastitis is usually peracute or acute in duration while SCM is often chronic. When chronic mastitis occurs, it is usually characterised by high SCC and reduction in milk production and can persist for long periods from lactation to lactation. *S. aureus* is one of the most common cause of chronic mastitis. All forms of mastitis have a negative impact on the quality and quantity of milk produced from affected animals; however, it is believed that SCM is more costly overall than CM (Zhao and Lacasse 2008).

8.2 Mastitis: Molecular and Diagnostic Investigations

Clinical signs of inflammation of the udder, namely painful swelling, heat, hyperaemia and in some cases generalised fever are indicative of CM. In milk, the presence of blood clots, flakes and change in colour towards a bloody or serum-like appearance, also points to the presence of mastitis. A definite diagnosis is usually made by bacteriological culture and isolation of causative organisms from milk in combination with SCC.

8.2.1 Analytes of Mastitis Detection and Monitoring

Bacteriological culture is generally accepted as the most reliable means of detecting intramammary infections (Dohoo et al. 2011); however, major limitations of being time consuming, expensive and not practically adaptable to cow side or on line use are associated with this method of diagnosis. Recently, polymerase chain reaction-single strand conformation polymorphism (PCR-SSCP) has been introduced as a useful alternative for the identification of pathogenic causes of mastitis with Gurjar et al. (2012) presenting several case reports of diagnosis of IMI caused by *S. aureus*, *Mycoplasma bovis*, *S. uberis* and *Enterobacter* spp. in farms using the DNA-based molecular technique. Commercial PCR-based mastitis diagnosis kits have also been introduced (Spittel and Hoedemaker 2012). However, even in cases of CM, there have been reports of no pathogens being detected in milk samples even with the PCR method (Kalmus et al. 2013).

Rapid migration of leukocytes from blood into the udder in order to help combat invading pathogens occurs in mastitis (Leitner et al. 2000), and damage to mammary epithelial cells also occurs. These leukocytes and damaged mammary epithelial cells are the components of the SCC measurable in milk during IMI. Neutrophils are the predominant type of leukocytes found in milk in acute mastitis and macrophages to a lesser extent. The SCC is the estimated number of somatic cells in milk and has been used as a gold standard for confirming mastitis in cases of SCM (Pyorala 2003) and is also used to assess milk quality (Larsen et al. 2010b). Methods for the determination of SCC are direct microscopy (also called the Breed method) and Fossomatic counter methods which are based on fluoro-optical properties and the Coulter milk counter by counting cells as they flow through an electric field. The SCC levels have been shown to remain high in milk long after the resolution of a mammary infection/inflammation (Pyorala 2003) thus further compromising its specificity in mastitis diagnosis. In addition, SCC has also been questioned in relation to its correlation to milk protein quality (Åkerstedt et al. 2008).

The California mastitis test (CMT) is an indirect estimation of SCC designed by Schalm and Noorlander in 1957 (Sargeant et al. 2001), and is based on the formation of a gelatinous precipitate in milk mixed with a detergent reagent as a result of the interaction of DNA released from cells with the detergent. The CMT is inexpensive, fast and can easily be used as a cow side test. Although some variability in detecting abnormal milk using CMT has been reported (Kawai et al. 2013), CMT has been found to be sufficiently sensitive in the mid to late lactation stage and specific in early lactation (Sargeant et al. 2001). Good correlations have been found between CMT and other indicators of mammary inflammation such as SCC and electrical conductivity (EC) (Kaşikçi et al. 2012; Kamal et al. 2014), and a significant correlation between CMT scores and ultrasonographic teat measurements of different breeds of cows have been found (Seker et al. 2009).

There have been other approaches to the detection and monitoring of mastitis. Measuring EC of milk samples has been shown to indicate SCM and CM (Milner et al. 1996) based on leakage of extracellular ions such as Na^+ and Cl^- into milk and the subsequent loss of predominantly lactose and K^+ (Pyorala 2003). EC was found to show similar sensitivity in detecting SCM as did SCC and CMT in the study of Kaşikçi et al. (2012). Infra-red thermography (IRT) is a non-invasive method that utilises heat absorbed following emission of infra-red radiation to generate images and can be used as an indicator of inflammation (Metzner et al. 2014). However, studies by Honiven et al. and Pezeshki et al. found the use of IRT in SCM and early mastitis not to be reliable, the latter observing that the changes in udder skin temperature (UST) occurred hours after the appearance of local signs of inflammation (Pezeshki et al. 2011; Hovinen et al. 2008).

Certain enzyme levels increase in milk during IMI, originating from phagocytes, ruptured epithelial cells and from serum contribute to the change in the physical and chemical properties of milk seen during mastitis. *N*-Acetyl-β-D-glucosaminidase (NAGase), β-glucuronidase (Nagahata et al. 1987; Larsen and Aulrich 2012) and catalase (Kitchen 1976) are lysosomal enzymes the activities of which increase in milk as they are released from neutrophils to facilitate the phagocytic process on pathogens. Assays of these enzymes are an important diagnostic test for mastitis (Polat et al. 2010). NAGase is also present in lysosomes of mammary epithelial cells and following cell lysis, NAGase is released into the milk (Zhao and Lacasse 2008). The concentration of adenosine triphosphate (ATP), the energy metabolite of living cells, has been shown to correlate with SCC (Olsson et al. 1986) following its release by these SCC into milk. ATP level has been successfully used to group milk samples by health status, and was also found to correlate with acute phase proteins (APP) (Gronlund et al. 2005). Lactose is the predominant form of milk sugar and is synthesised in the mammary gland secretory cells (Golgi apparatus). Since the synthetic ability of the mammary cells is affected in mastitis due to cell damage, lactose concentrations are known to fall (Pyorala 2003). Several studies have also demonstrated the correlation of a reduction in lactose concentration with mastitis or SCC (Sharma and Misra 1966; Malek dos Reis et al. 2013). Berning and Shook, however found that the change in lactose concentration does not correlate well with SCC and is not very indicative of IMI (Berning and Shook 1992).

There have therefore been many alternative approaches explored for improving the diagnosis of mastitis, but it is clear that none have demonstrated sufficient superiority in value to the established methods of CMT, SCC and bacteriology. Before examining the potential of omic technologies to make an impact in this regard an additional relatively new technology should be examined: the use of acute phase proteins (APPs) in milk as biomarkers of mastitis.

8.2.2 Acute Phase Proteins as Biomarkers of Mastitis

At the start of the present century it was discovered that APPs, which are established biomarkers of infection and inflammation in bovine serum or plasma (Ceciliani et al. 2012), also have dramatic increases in milk during mastitis (Eckersall et al. 2001). Since then the potential of APPs in diagnostic assessment for the disease has been extensively examined for their potential as additional mastitis biomarkers.

Following the release of cytokines and other pro-inflammatory mediators, predominantly interleukin-1 (IL-1), interleukine-6 (IL-6) and tumour necrosis factor-alpha (TNFα), by macrophages in the mammary gland upon pathogen invasion (Tassi et al. 2013), several local and systemic innate immune responses are elicited. This includes the acute phase response (APR), comprising the release of APPs from the liver into the blood (Jensen and Whitehead 1998) and from the mammary glands into milk (Ceciliani et al. 2012) (Fig. 8.1). APPs are a group of proteins predominantly produced in the liver, that are changed (increased or decreased in quantity) by over 25% during inflammation, infection or stress and released into blood (Lomborg et al. 2008; McDonald et al. 2001; Ceron et al. 2005). The APPs occur not only in serum, but also in other body fluids such as milk, colostrum, nasal

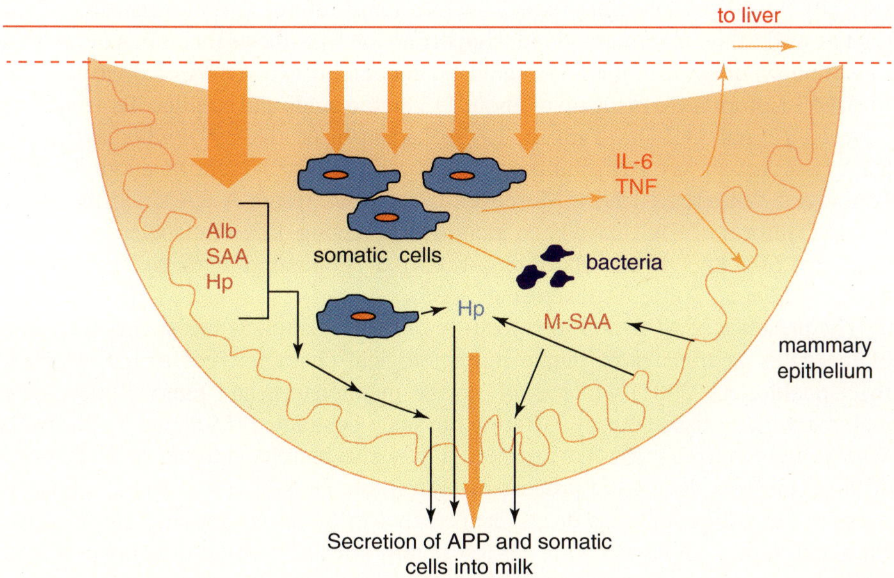

Fig. 8.1 The mechanism of acute phase protein production in the bovine mammary gland and secretion into milk during mastitis. The presence of pathogenic bacteria causes the influx of somatic cells (neutrophils) into the mammary gland which are stimulated to produce pro-inflammatory cytokines and these stimulate the mammary epithelium to produce M-SAA3 and Hp. The somatic cells also secrete Hp while serum proteins such as albumin as well as serum Hp and SAA cross the blood-mammary barrier

secretion, saliva and synovial fluid (Eckersall et al. 2001; McDonald et al. 2001; Molenaar et al. 2009).

Following the first report of the correlation of major APPs mammary-associated serum amyloid A3 (M-SAA3) and haptoglobin (Hp) in milk from mammary glands with mastitis (Eckersall et al. 2001), many other workers have described similar findings. The level of APPs have been related to SCC (Nielsen et al. 2004; O'Mahony et al. 2006; Åkerstedt et al. 2008; Viguier et al. 2009; Pyorala et al. 2011; Thomas et al. 2016b), milk composition and protein quality (Åkerstedt et al. 2008) and with severity of the IMI (Pyorala et al. 2011). Only a small variation in their levels were observed in healthy cow's milk over 42 consecutive milkings, showing that these APPs are stable and able to discriminate between healthy and inflamed tissues reliably (Åkerstedt et al. 2008). It is now established that the major bovine APPs (Hp and MSAA3) are synthesised in the mammary gland (Hiss et al. 2004; Eckersall et al. 2006; Larson et al. 2006; Thielen et al. 2007; Lai et al. 2009; Molenaar et al. 2009). However, it is evident that Hp is also synthesised by the neutrophils that can migrate into the gland during inflammation (Cooray et al. 2007).

8.2.2.1 Mammary-Associated Serum Amyloid A3 (M-SAA3)

Mammary-associated SAA3, the isoform in milk, is expressed in all mammalian species that have so far been studied. Its expression is highly conserved, suggesting its importance and functional integrity in mammals. It is also synthesised in histologically normal extrahepatic tissues such as lungs, uterus and the gastrointestinal tract as well as the mammary gland, supporting the hypothesis that this APP plays a role in innate defence of the body against pathogen invasion (Berg et al. 2011). The M-SAA3 isoforms are made up of about 113 amino acids and a molecular weight of between 12 and 14 kDa (Yamada 1999; Takahashi et al. 2009), whereas human SAA (isoform 1) has 104 amino acids and a molecular weight of about 12 kDa (Uhlar and Whitehead 1999). Moreover, the various serum A-SAA and the milk-specific isoform of SAA have been identified as having different isoelectric points, with the M-SAA3 being alkaline (pI 9.6) as compared with other isoforms (McDonald et al. 2001; Jacobsen et al. 2005).

The presence of M-SAA3 in colostrum of healthy animals has suggested a role of the APPs in transfer of innate immunological characteristics to the newborn (McDonald et al. 2001) while a role in the maintenance and remodelling of the mammary gland tissue during lactation has been postulated (Molenaar et al. 2009; Watson and Kreuzaler 2011). Moreover, it may also be related to the stress induced APR on the cow during the process of parturition, as APP levels remain elevated only for a few days and then drop. The presence of M-SAA3 and also Hp in colostrum and milk from cows over a 10-day period after parturition has shown that in healthy cows their concentrations normalise in 4 days post-parturition while in animals with mastitis the concentrations remain elevated (Thomas et al. 2016d).

The value of M-SAA3 in diagnosis and prognosis of mastitis, particularly SCM has also been investigated by various groups in bovine (Suojala et al. 2008; Gerardi et al. 2009; Pyorala et al. 2011; Kovacevic-Filipovic et al. 2012; Szczubial et al.

2012; Thomas et al. 2016b) and other species including ovine (Winter et al. 2006; Miglio et al. 2013). Using both experimental models (Eckersall et al. 2001; Nielsen et al. 2004; Gronlund et al. 2005; Jacobsen et al. 2005; Eckersall et al. 2006) and farm collected samples (Pyorala et al. 2011; Kovacevic-Filipovic et al. 2012; Kalmus et al. 2013) the potential of M-SAA3 in mastitis detection, assessment of severity of infection and prognostic evaluation of cases has been examined. The milk concentration of M-SAA3 has been correlated to other established inflammatory indices of intramammary inflammation (Jacobsen et al. 2005; O'Mahony et al. 2006; Gerardi et al. 2009; Kovacevic-Filipovic et al. 2012) and has been found to be more sensitive to changes in the inflammatory condition of the mammary gland than its serum counterpart (Eckersall et al. 2001; Jacobsen et al. 2005; O'Mahony et al. 2006).

8.2.2.2 Haptoglobin

Increased milk Hp in IMI was initially thought to occur only as a result of increased permeability of blood milk barrier during udder infections (Eckersall et al. 2001); however a number of studies have demonstrated the synthesis of Hp in mammary epithelial cells as well as in milk somatic cells, which are predominantly neutrophils (Hiss et al. 2004; Thielen et al. 2007; Lai et al. 2009). Several isotypes of Hp were demonstrated in the granules of granulocytes of healthy cattle (Cooray et al. 2007) indicating it is constitutively expressed with the different isotypes possibly being due to post-translational modifications. Isoelectric points (p*I*) ranging from 8 to 9.5 for the 40 kDa (β) subunit and ranging from 6 to 8 for the 20 kDa (α) of Hp, were also shown.

Measuring milk Hp has been shown to be a more sensitive indicator of mastitis than serum Hp, as Hp in milk increases several more folds during IMI than serum (Eckersall et al. 2001; Gronlund et al. 2005), and the increase appears earlier than in serum, for example, a rise by just 3 h post challenge in milk as against 9 h in serum, following an intramammary LPS challenge (Hiss et al. 2004; Eckersall et al. 2006). Milk Hp has also been evaluated by several studies for its ability as a diagnostic marker for subclinical mastitis (Gultiken et al. 2012), although the authors did not find Hp analysis useful in diagnosing and evaluating treatment of SCM and CM (Wenz et al. 2010) as well as other conditions such as metabolic status during early lactation (Hiss et al. 2009). Milk Hp has been found to be elevated in the early prepartum period (Crawford et al. 2005; Hiss et al. 2009). In the study of Kalmus et al., milk Hp performed better as an indicator of IMI than M-SAA3, due to the fact that the latter was not able to correctly identify inflammation by *Arcanobacterium pyogenes*, the cause of a purulent form of acute mastitis (Kalmus et al. 2013).

8.2.2.3 Measuring APP in Milk

Milk Hp and also M-SAA3 are measured by immunoassays, such as ELISAs, although an immunodiffusion assay was used in its initial identification in milk (Eckersall et al. 2001). Pedersen et al. utilised a sandwich ELISA to measure milk Hp (Pedersen et al. 2003) while Hiss et al. also described an ELISA using purified bovine Hp and polyclonal antibody raised against the bovine Hp (Hiss et al. 2004). The same type of immunoassay has been used to measure milk M-SAA3 (Eckersall et al. 2001, 2006), but Hp has been shown to be more stable in milk than M-SAA3 (Thomas et al. 2016c). As ELISA, in its usual format, has a limitation of being time consuming and requiring laboratory facility, the availability of a rapid on line biosensor system, applicable to milking machinery for the diagnosis of udder infections would be of great value in the control of mastitis. A biosensor format assay to determine Hp concentration in milk was described by Åkerstedt et al. (2008); the assay was based on the high affinity of Hp for haemoglobin and was a competitive (indirect) assay. Although not as sensitive as the ELISA, some level of success was achieved with this biosensor assay. However, this assay requires expensive machinery and well-trained personnel to use. A rapid and economical on-farm test for either or both APPs could be of great value in mastitis evaluation.

8.3 Omic Investigations of Mastitis

The major advances which have been made in the last decades in the omic technologies have been applied to research in bovine mastitis providing the largest boost to our understanding. Thus genomic, proteomic, and to a lesser extent metabolomics and microbiome investigations have provided resources for the systems biology approach with an increasing insight into the molecular interactions of host and pathogen.

8.3.1 Genomic Investigations of Mastitis

Investigation of the genomics of mastitis was greatly enhanced by the first draft of the *Bos taurus* genome sequence, which was completed in October 2004 (www.genome.gov/12512874; http://may2009.archive.ensembl.org/Bos_taurus/Info/Index 2009). In the Year of the Ox (2009), there was the release of a new assembly (UMD2) of the *Bos taurus* genome (Zimin et al. 2009), and reports published of improved assembly (Btau 4.0) and annotations of the *Bos taurus* genome by the 'Bovine Genome Sequencing and Analysis Consortium' (BGSAC) and the 'Bovine HapMap Consortium' involving over 300 scientists from 25 different countries (Tellam et al. 2009; Liu et al. 2009; Zimin et al. 2009; Reese et al. 2010). Currently, there are two assemblies of *Bos taurus* genome available. They are (1) the *Bos taurus* genome 'reference' assembly—the University of Maryland assembly release 3.1.1 (UMD3.1.1) and (2) the *Bos taurus* genome 'alternate' assembly—the Baylor College of Medicine Human Genome Sequencing Center assembly version Btau_5.0.1. In addition, the genome sequence of *Bos indicus* (Nellore bull, from Brazil) is also available (Canavez et al. 2012). Availability of the reference sequence, improvements in sequencing technologies and the lessening cost of sequencing have all enhanced the pace of application of genomics in mastitis research.

However even before the recent major advances in genomic high-throughput technology, Rupp and Boichard argued that although host genetic variability for resistance to mastitis had a low heritability, it was as an important factor underlying mastitis resistance even in the presence of other confounding factors such as infections (Rupp and Boichard 2003). The genetics of the immune response in mastitis and its role in disease resistance has subsequently been the topic of review (Rainard and Riollet 2006; Thompson-Crispi et al. 2014). Of particular interest was the association of chemokine CXCR1 gene polymorphism CXCR1 +735 (previously reported as CXCR2 +777, but later revised to CXCR1 +735 in the improved gene annotations) with subclinical mastitis in Holsteins (Youngerman et al. 2004; Galvao et al. 2011). The CXCR1 gene codes for the interleukin-8 receptor, which is present on the surface of the neutrophil and mediates migration of neutrophils to sites of inflammation, and is hence regarded as a potential candidate for modifying mastitis susceptibility. There are tens of single nucleotide polymorphisms in this gene (Pighetti et al. 2012; Zhou et al. 2013), and CXCR1c.−1768T>A (rs41255711) was reported to be associated with mastitis resistance due to its location in the transcription binding site of the gene (Leyva-Baca et al. 2008). However, there are contradictory reports from subsequent association studies between the CXCR1 gene polymorphisms and susceptibility to mastitis. For instance, no statistical significance between two of the CXCR1 gene polymorphisms and somatic cell score was found in the German Holstein-Friesian population, although a large variance was caused by the loci (Goertz et al. 2009). On the contrary, a recent study in Polish Holsteins found statistically significant associations between CXCR1 +472 SNP and test day SCC (Pawlik et al. 2015), even though this study was underpowered to observe associations with *S. aureus* mastitis.

Toll-like receptors (TLR) play an important role in detecting invading pathogens and the induction of host defence responses (Takeda and Akira 2005; Mogensen 2009). There is considerable evidence, at both transcript and protein levels, of increased expression of TLR2 and TLR4 in the udder during mastitis (Goldammer et al. 2004; Reinhardt and Lippolis 2006). Using single gene PCR amplification and sequencing method, Russell and co-workers identified associations between the SNPs in the bovine Toll-like receptor 1 (TLR1) gene and the occurrences of clinical mastitis in a British Holstein-Friesian herd (Russell et al. 2012), although a previous study did not detect significant association between clinical mastitis and SNPs in TLR2 and TLR4 genes (Opsal et al. 2008).

With the improved *Bos taurus* genome assembly and the developments in genomics technologies, genome-wide association studies (GWAS) for mastitis susceptibility in cows have become possible, and a number of SNPs associated with mastitis or somatic cell score (SCS) have been reported (Wang et al. 2015; Sharma et al. 2015; Abdel-Shafy et al. 2014; Waldmann et al. 2013; Meredith et al. 2013; Tiezzi et al. 2015; Sahana et al. 2014; Ibeagha-Awemu et al. 2016). Currently the Cattle Quantitative Trait Locus Database (Cattle QTLdb), which archives the curated data from published associations and Quantitative Trait Loci (QTLs) has 81,652 QTLs representing 519 different traits including 163 QTLs for clinical mastitis, 1,070 QTLs for somatic cell score and 77 QTLs for SCC (Hu et al. 2016;

www.animalgenome.org 2016). Meredith and colleagues conducted a genome-wide association study (GWAS) for many production traits including SCC on two large cohorts of Holstein-Friesian cattle in Ireland and detected significant association of 9 SNPs with somatic cell score in the sires using a single SNP regression method (Meredith et al. 2012). Similarly, Wijga and colleagues performed a GWAS using phenotypic and genotypic data of first-lactation Holstein cows (n = 1484) from four European research herds (from different countries—Ireland, the Netherlands, Scotland and Sweden) and identified associations of two loci (SNPs ARS-BFGL-NGS-101491 and BTB-02087354) with the standard deviation of test-day SCC (Wijga et al. 2012). Recently, Ibeagha-Awemu and colleagues used genotyping-by-sequencing method on an Illumina platform to identify SNPs in 1,246 Canadian Holstein cows, and performed GWAS for milk traits (Ibeagha-Awemu et al. 2016). This study identified associations of 52 SNPs with SCC located in the genomic regions of 48 genes, most of them with immunity or inflammatory functions.

In parallel with the host-centric genomics studies on mastitis, there have been many developments focusing on the pathogen involved in host–pathogen interactions. In the UK, *Streptococcus uberis,* an environmental bacterium, has emerged as the top pathogen responsible for CM and SCM, with the frequency of 23.5% for CM in culture positive samples (Bradley et al. 2007; Zadoks and Fitzpatrick 2009). While there are multiple strains of *S. uberis* isolated from dairy cows, as this organism being an environmental bacterium, only few virulent strains produce mastitis. Multiple genomics-based approaches have been used to study the mechanisms that impart pathogenicity to the virulent strains of *S. uberis*. With the development of techniques to generate random mutations in *S. uberis* (Ward et al. 2001), the phenotypic and genotypic characteristics of a large number of randomly generated genome mutations in *S. uberis* could be studied (Leigh et al. 2004). This was followed by the identification of a gene sequence encoding for a protein called 'adhesion molecule' (*sua* gene) in *S. uberis,* which was hypothesized to be a virulence factor in *S. uberis* pathogenesis (Luther et al. 2008). The *sua* gene was reported to be conserved in 12 strains of *S. uberis* (Luther et al. 2008). Later, sequencing and assembly of the whole genome of *S. uberis* (strain 0140J) and detailed comparative genomics analysis showed niche adaptations in the genome for utilising nutritional flexibility derived from a diversity of metabolic options that would enable this pathogen to live in challenging and changing environmental conditions (Ward et al. 2009). With the availability of reference genome sequence and the advent of the next-generation sequencing technology, allelic profiles of many ovine and bovine isolates of *S. uberis* could be generated (Davies et al. 2016; Gilchrist et al. 2013). Comparisons of the allelic profiles of the host-specific populations identified distinct host-specific allelic profiles including well-defined allelic profiles for virulence genes (Gilchrist et al. 2013). Similarly, comparisons of the allelic profiles of 494 isolates of *S. uberis* showed a small subset of sequence types causing most infections in the study cohort (Davies et al. 2016).

Recently, Tassi and colleagues examined the pathogenicity of two *S. uberis* strains (host-adapted FSL Z1–048 and non-adapted FSL Z1–124) isolated from mastitis in the same herd and during the same time period, and found that the non-adapted FSL Z1-124 was avirulent whereas the host-adapted strain caused clinical mastitis (Tassi et al. 2013) in experimentally challenged cows. Concurring with this result, a recent study comparing four different strains of *S. uberis* has also shown similar strain-specific variation in pathogenicity (Notcovich et al. 2016). In vitro study of the strain-dependent differences in virulence showed that the virulent strain had both increased adhesion to mammary epithelial cells and better abilities to evade killing by bovine monocyte derived macrophages (Tassi et al. 2015). As with *S. uberis*, other bacterial species such as *E. coli*, *S. aureus* and *S. epidermidis* that produce bovine mastitis have been studied using genomic approaches for subtyping, strain-specific pathogenicity and for identification of novel virulence genes (Blum et al. 2015; Boss et al. 2016; Kempf et al. 2016; Le Marechal et al. 2011; Lindsay 2014; Savijoki et al. 2014; Goldstone et al. 2016).

8.3.2 Transcriptomic Investigations of Mastitis

The transcriptome is dynamic and sensitive to numerous physical, biological, environmental and temporal changes. Understanding transcriptomic changes can provide much valuable insight into the molecular mechanisms underlying biological processes. Microarrays and RNA sequencing are the two main technologies currently used in global transcriptome profiling studies, although quantitative reverse transcription polymerase chain reaction (qPCR) technology has been widely used in mastitis research for candidate gene expression studies. Using these technologies, differential gene expression studies have been undertaken to compare expression of genes in mammary epithelial cells and milk SCC during the course of experimental infections (Moyes et al. 2009, 2016; Younis et al. 2016; Lawless et al. 2013; Wang et al. 2016; Swanson et al. 2009). The immune response mounted against invading pathogens in mastitis is a complex process, and involves resident and recruited immune cells, mammary epithelial and endothelial cells. In both acute and chronic mastitis, there is a manyfold increase in the number of SCC and changes in the composition of cell types that constitute SCC. The predominant cell type (66–88%) present in SCC of a healthy cow is macrophage, however, during IMI, the proportion changes in favour of neutrophils, which would go as high as 90% of the total SCC during mastitis (Pyorala 2003), and accordingly the transcriptome profile of SCC changes during IMI.

Linking with the immunoassay analysis of APPs in milk during experimental *S. aureus* mastitis (Eckersall et al. 2006) as mentioned in section 8.2.1, it was apparent (Whelehan et al. 2011) that in mammary tissue there was overexpression of genes for the APPs, Hp and SAA3 as well as for toll-like receptors, antimicrobial peptides and cytokines. The greatest up-regulation at 48 h after challenge was found to be for Hp and M-SAA3 which is consistent with the concentrations found in milk from the cows with mastitis. The expression of host response genes in either teat cistern or mammary parenchyma over the first 3 h after challenge has been examined with either *S. aureus* or *E. coli* sequential intramammary challenge (Petzl et al. 2016). The early responses were seen in teat cistern in the first hour and subsequently in mammary parenchyma, and transcripts encoding for chemokines, cytokines and antimicrobial molecules were over 25 times greater with *E. coli* than with *S. aureus* and a number of the immune mediators were only expressed in response to *E. coli*. Similarly, a previous study that compared transcript expression in SCC during IMI with *E. coli* or *S. aureus* showed increased expression of pro-inflammatory cytokines such as IL-6, IL-8, IL-12, granulocyte macrophage-colony stimulating factor (CSF2) and TNF-α during IMI with both bacterial species; however, the magnitude of gene expression was greater with *E. coli* (Lee et al. 2006). The differences in gene expression pattern between *E. coli* and *S. aureus* are also supported by a recent meta-analysis study (Younis et al. 2016), which showed *S. aureus* inducing innate immunity in mammary epithelia via Toll-like and NOD-like receptors, while suppressing acquired immune responses through suppression of cell motility and antigen presentation. More importantly, genes necessary for milk production including the genes encoding for lipid biosynthesis—Farnesyl-Diphosphate Farnesyltransferase 1 (FDFT1) and 1-acylglycerol-3-phosphate O-acyltransferase (AGPAT6) were down-regulated during IMI with *E. coli*. Down-regulation of genes coding for lipid biosynthesis and metabolism such as Farnesyl diphosphate synthase (FDPS) and 3-Hydroxy-3-methylglutaryl-coenzyme A synthase 1 (HMGCS1) was also observed in mammary alveolar tissue experimentally infected with *S. uberis* (Swanson et al. 2009). In this study, one of the fore or hind udder quarters in each cow was infected with *S. uberis* while the non-infected quarter served as control. The same study showed up-regulation of genes linked to acute phase response signalling, e.g. serum amyloid A3 (SAA3), Hp and LPS-binding protein (LBP), oxidative stress, e.g. superoxide dismutase 2 (SOD2) and selenoprotein P (SEPP1), and immune response, e.g. complement component 3 (C3), IL-6, IL-8, IL-10, toll-like receptor-2 (TLR-2) and TNF-α.

Inhibition of lipid biosynthesis and activation of acute phase signalling and immune response genes were further confirmed by a large study that compared gene expression profiles of mammary gland biopsies between non-infected and infected cows with *S. uberis* (Moyes et al. 2009). This study also showed enrichment of 20 canonical pathways including APR signalling, liver X receptor (LXR)/retinoid X receptor (RXR) activation, peroxisome proliferator-activated receptor (PPAR) activation, IL-10 signalling and IL-6 signalling pathways in the differentially expressed genes. The pattern of expression of parenchymal genes for antimicrobial peptides, in response to coagulase-negative or -positive *Staphylococci* differs between the peptide class. The beta-defensins were up-regulated in response to the bacteria but the cathelicidins present in healthy tissue were not affected by the infection (Kosciuczuk et al. 2014). This confirms earlier reports that the cathelicidins were up-regulated in mastitis in neutrophils (mammary somatic cells) but not in the epithelium of the mammary gland (Tomasinsig et al. 2010). MicroRNAs (miRNAs) are post-transcriptional regulators of gene expression, and a Next-Generation Sequencing (NGS)-based miRNA expression profiling in primary bovine mammary epithelial cells challenged with *S. uberis* 0140J showed 21 miRNAs were differentially expressed during infection with *S. uberis* suggesting regulatory role of miRNAs in IMI (Lawless et al. 2013). Similar NGS-based miRNA expression profiling in milk exosomes in *S. aureus* infection showed miRNAs bta-miR-142-5p and bta-miR-223 could potentially be used as biomarkers for early detection of *S. aureus* infections (Sun et al. 2015).

8.3.3 Proteomic Investigation of Mastitis

As a major function of milk is to provide protein for the nutrition of the neonate, understanding the changes that occur to this important component of milk is fundamental to examination of the host response to mastitis. Recent advances in proteomics with application to farm animal science (Almeida et al. 2015) have allowed unprecedented depth of investigation of the changes in proteins and peptides during the disease. For proteomic investigation in dairy cows, the Bovine PeptideAtlas, a database of bovine proteome from different tissues including milk, was recently created within the PeptideAtlas framework (Bislev et al. 2012). Applications of proteomics technologies in bovine milk related research has recently been reviewed (Roncada et al. 2012; Bendixen et al. 2011). Eckersall and colleagues have recently emphasised the importance of proteomics in farm animal science (Eckersall et al. 2012). Proteomics methodologies in recent years have been adopted for the discovery of biomarkers of bovine mastitis in order to identify, validate and screen biomarker candidates for bovine mastitis (Viguier et al. 2009; Boehmer et al. 2010a; Lippolis and Reinhardt 2010; Bendixen et al. 2011; Eckersall et al. 2012; Bassols et al. 2014; Mudaliar et al. 2016). These studies have used both field cases and experimental models, from mastitis caused by different pathogens and representing diverse clinical phases, in order to validate results leading to diagnostic applications. A variant of the use of proteomics in dairy health assessment has been established by identification of mastitis-causing bacteria in milk using a Matrix-assisted Laser Desorption Ionisation–Mass Spectrometry (MALDI-MS) (Barreiro et al. 2012). This method employs bacterial ribosomal proteins as fingerprinting markers to identify specific microorganisms, from a dedicated MALDI biotype reference library after a pre-concentration step. However, a high bacteria count is required for accuracy and only a few species of bacteria have been evaluated in milk using this method.

Milk protein is composed of two major groups of high-abundance proteins, the insoluble caseins and soluble whey proteins. There are several types of caseins including α-caseins (α-CN), β-caseins (β-CN) and κ-caseins (κ-CN); all these constitute about 80% of the total milk proteins. The remaining 20%, whey proteins are made up of β-lactoglobulin, α-lactalbumin, immunoglobulins, bovine serum albumin, bovine lactoferrin, lactoperoxidase as well as cytokines and other immune proteins (Pepe et al. 2013). There are additional low-abundance milk proteins presenting a wide repertoire of functions recently identified by proteomics from which likely biomarkers of disease conditions of the mammary gland may be found.

In order to overcome the effect of the high-abundance proteins in milk, fractionation steps are often carried out on samples and include centrifugation, acidification, filtration, the use of peptide ligand libraries and various precipitation methods to rid the samples of the high-abundance proteins (D'Amato et al. 2009; Nissen et al. 2013). These recent advances in fractionation techniques have helped to resolve the limitation posed by the presence of high-abundance proteins (Boehmer et al. 2010a). Thus the different fractions of milk proteins, especially whey and the milk fat globule membrane (MFGM), have been examined and new proteins not previously known to be in milk have been identified (Reinhardt and Lippolis 2006; Reinhardt et al. 2013).

Mass exclusion filters, one dimension electrophoresis and commercial depletion kits have been used for the purpose of fractionating milk proteins prior to proteomic analysis (Boehmer 2011). In the study by Nissen et al. (2013), ultracentrifugation at a very high speed, before carrying out a proteomics experiment, was found to be the most reproducible and robust method of obtaining the milk proteome compared to other milk protein fractionation techniques such as acidification or filtration. The use of combinatorial peptide ligand libraries has been developed and has been successfully employed for fractionation of peptides and identification of new proteins in milk (D'Amato et al. 2009). Enrichment, for example, by cysteine tagging has also been used to enhance the identification of low-abundance caseins in milk containing cysteine as against the abundant α-s_1 CN and β-CN that do not (Holland et al. 2006).

Investigations of bovine milk proteome in the early 2000s were hampered by non-availability of reference genome/proteome. Even though attempts were made to identify differential protein expression in milk between normal and mastitis health conditions. In one of the early works, caseins from bovine milk were depleted using ammonium sulphate salt precipitation and protein expression in whey between healthy and mastitic conditions was compared (Hogarth et al. 2004). This study used a two-dimensional gel electrophoresis (2-DE) method to separate and quantify whey proteins and a MALDI-MS to identify proteins. Although 2-DE is a semi-quantitative method and the study was constrained by a limited availability of protein reference sequences, increased expression of bovine serum albumin (BSA) and serotransferrin and decreased expression of caseins, β-lactoglobulin and α-lactalbumin were observed in clinical mastitis (Hogarth et al. 2004). The rapid growth of protein databases following sequencing of the *Bos taurus* genome in 2004 and improvements in chromatography and mass spectrometry enabled identification and quantification of less abundant proteins in milk. Multiple variants of the semi-quantitative 2-DE methods along with several mass spectrometry techniques to identify proteins have been used to study bovine mastitis (Turk et al. 2012; Bian et al. 2014; Pongthaisong et al. 2016). Combined use of a Liquid Chromatography-tandem Mass Spectrometry (LC-MS/MS), 2-DE and MALDI-MS identified 95 proteins (gene products) including 15 host defence related proteins such as cathelicidin, SAA and lactoferrin in bovine milk during colostrum and peak production stages and in IMI with *S. uberis* (Smolenski et al. 2007), suggesting a complex nature for the milk proteome and the role of milk proteins in defence against IMI. Similarly, differential expression of whey proteins before and 18 h after infection with *E. coli* was studied using 2-DE and MALDI-MS after depleting caseins using ultracentrifugation (Boehmer et al. 2008). This study showed decreased amounts of caseins and increased levels of serum albumin, α-1-acid glycoprotein, transthyretin, serotransferrin, complements C3 and C4, cathelicidins and apolipoproteins in milk collected 18 h after infection with *E. coli* (Boehmer et al. 2008). The decrease in caseins was attributed to proteolysis by indigenous bovine milk proteases plasmin, elastase and cathepsin D, and this was confirmed by a study that induced mastitis using LPS infusion (Hinz et al. 2012). As with proteomics analysis using milk, mammary tissues and blood serum during mastitis have been analysed using 2-DE methods (Yang et al. 2009; Alonso-Fauste et al. 2012). Particularly, differential protein expression between healthy and mastitic conditions using either bovine serum or whey showed greater proteomic changes in whey than in serum (Alonso-Fauste et al. 2012), suggesting milk rather than blood could be the better body fluid to look for biomarkers for mastitis.

In recent years, several quantitative proteomics methods have been developed and applied to the investigation of bovine mastitis to study the pathophysiology of bovine mastitis and to identify biomarkers. Using a 4-plex isobaric tag for relative and absolute quantitation (iTRAQ) method, differential expression of whey proteins was compared before LPS infusion with 4 h or 7 h after LPS stimulation (Danielsen et al. 2010). In response to LPS stimulation, there were over threefold increase in peptidoglycan recognition protein, cathelicidins, SAA, Annexin A1, and over twofold increase in Hp, ceruloplasmin, serotransferrin, fibrinogen, plasminogen, apolipoproteins A-1, A-2 and A-4, and complement C3 and C4 at 7 h post-stimulation (Danielsen et al. 2010). Likewise, an iTRAQ proteomics method has been used to compare protein expression in whey, MFGM and exosomes from milk obtained from healthy and *S. aureus* infected cows, and identified a total of 2971 proteins including 94 proteins that were significantly differentially expressed between the healthy and infected milk (Reinhardt et al. 2013). This is by far the largest number of proteins quantified from milk, and this could be attributed to the 2-dimensional chromatography (an offline first dimension and an online second dimension) and fractioning of milk into whey, MFGM and exosomes. Comparably, using an iTRAQ method, the proteomes of mammary tissues from cows with IMI due to methicillin resistant *S. aureus* and healthy cows were analysed to identify mechanisms associated with mammary tissue damage, and up-regulation of collagens and fibrinogens, and down-regulation of caseins and apolipoprotein A-4 were observed (Huang et al. 2014). Down-regulation of caseins in mammary tissues during mastitis could suggest either lower production of caseins in mammary tissues or proteolysis of caseins as noted elsewhere in this chapter. Interestingly, differences in bacterial proteome between strains of *E. coli* from persistent and transient mastitis analysed using an 8-plex iTRAQ showed increased expression of proteins involved in bacterial mobility in strains causing persistent infections (Lippolis et al. 2014).

A label-free quantitative proteomics method was used to analyse temporal changes in whey proteome during *E. coli* mastitis (Boehmer et al. 2008, 2010a, b; Boehmer 2011). Ibeagha-Awemu and colleagues analysed the proteome of mastitis milk from naturally occurring *E. coli* and *S. aureus* infections and compared them with normal milk proteome using LC-MS/MS method (Ibeagha-Awemu et al. 2010). They also performed an in vitro challenge study using inactivated *E. coli* strain P4 or *S. aureus* strain Smith CP and mammary alveolar cells (MAC-T cells), an immortalized mammary epithelial cell line, to compare their proteomics results. Their study concluded that the differences in the proteomics profiles could be attributed to pathogens, rather than the host, and identified significant enrichment of acute phase response signalling, coagulation system and complement system pathways in the differentially expressed proteins. Kim and colleagues challenged healthy cows with three different strains of *S. aureus* bacteria, the SCV Heba 3231 strain that causes chronic SCM, the 3231 parent strain and the Newbould 305 strain that causes acute CM and compared the host immune responses over a period up to 21 days post infection by cytokine assays and differential milk proteome analysis using the LC-MS/MS method, and found marked differences in the temporal expression of cytokines (Kim et al. 2011). The progression of the use of proteomic methods of increasing sensitivity can be seen from Table 8.2 which shows the proteins identified by studies of milk or mammary gland during mastitis. High-abundance protein such as albumin or lactoferrin have been identified as increasing in most studies, while lower abundance proteins such as Hp or cathelicidin have been identified in a lower number of more recent investigations.

8.3.4 Peptidomic Investigation of Mastitis

The peptidome, which is a subset of the proteome, is the collection of all peptides within a biological system at a given time and is one of the newly emerged 'omics' technologies. It is the detection, identification and quantification of all peptides with their post-translational modifications within a cell, tissue organism or biological sample. In clinical research, peptidomics technology has proven particularly useful in the areas such as urinary markers of disease (Albalat et al. 2011) and neuroendocrine research for biomarker and drug discovery (Menschaert et al. 2010). Although peptidomics primarily focuses on the simultaneous identification of endogenously derived peptides within a biological fluid/system, it can encompass peptide products of protein degradation (Dallas et al. 2015). It can be used to elucidate proteolytic regulation of bioactive peptides as a key to understanding the physiology and identifying possible drug targets (Kim et al. 2013).

A number of investigations of the peptides in milk have been made possible as a result of peptidomics. Antimicrobial peptides amongst other peptides exhibiting diverse properties such as immunomodulation have been identified following endogenous proteolysis of the major milk proteins (caseins and lactalbumin) in human milk (Dallas et al. 2013). Furthermore, several antimutagenic properties have also been associated with peptides obtained after hydrolysis of milk protein constituents such as caseins and lactalbumin (Larsen et al. 2010b). Peptides in milk increase during episodes of mastitis, mostly as a result of the action of proteases such as plasmin, elastase, cathepsins A and B (Guerrero et al. 2015). Aminopeptidases, in addition to these proteolytic enzymes may leak into milk from blood through a disrupted blood milk barrier, or be secreted into milk by somatic cells or mammary epithelial cells as a tool for killing bacteria, or arise from microorganisms' metabolism. Proteases originating from leucocytes that increase in the mammary gland

Table 8.2 Proteins recognised as either increasing or decreasing by proteomic investigations during mastitis

Sr. number	Milk protein	Reference reporting increase in mastitis	Reference reporting decrease in mastitis
1	α-S2-casein precursor	14	
2	β-Casein B	14	
3	β-Casein	14	
4	κ-Casein precursor	14	
5	14-3-3 Protein zeta chain, epsilon	2, 10, 15	
6	15 kDa selenoprotein		13
7	6-Phosphogluconate dehydrogenase, decarboxylating (EC 1.1.1.44)	15	
8	Acidic leucine-rich nuclear phosphoprotein 32	15	
9	Acidic ribosomal protein 60S P2	15	
10	Actin, cytoplasmic-1	2, 3, 7, 13, 15	
11	Actin-related, ARP3	9	
12	Adenosylhomocysteinase (AdoHcyase) (EC 3.3.1.1)	15	
13	Adenyl cyclase-associated protein 1	7, 15	
14	Adenylate kinase 2	15	
15	Adipose differentiation-related protein	5	
16	Albumin	1, 2, 3, 4, 5, 6, 7, 10, 11	
17	Aldehyde dehydrogenase (NAD) 2 precursor	2	
18	Alkaline phosphatase	15	
19	Alpha-1-acid glycoprotein	3, 7, 10, 12, 15	
20	Alpha-1-antitrypsin	3, 5, 6, 7	
21	Alpha-1-B glycoprotein	6, 15	
22	Alpha-2-antiplasmin (Alpha-2-AP))	15	
23	Alpha-2-glycoprotein 1, zinc-binding	10	6, 9, 13, 15
24	Alpha-2-macroglobulin	5, 6, 7, 10, 15	
25	Alpha-actinin-1	6, 15	
26	Alpha-actinin-4	6, 15	
27	Alpha-fetoprotein	7	
28	Alpha-lactalbumin		1, 3, 4, 6, 11, 13, 15
29	Ankyrin 3, node of Ranvier	2	
30	Annexin A1	2, 5, 10, 15	
31	Annexin A11	10	
32	Annexin A2	12	
33	Annexin A3, A6 and A7	10	
34	Antitestosterone antibody	5	
35	Antithrombin-III	5, 7	

(continued)

Table 8.2 (continued)

Sr. number	Milk protein	Reference reporting increase in mastitis	Reference reporting decrease in mastitis
36	Apolipoprotein A-1	9	
37	Apolipoprotein E (Apo-E)	13	15
38	Apolipoproteins	2, 3, 4, 5, 6, 7, 10, 15	
39	Apoptosis-associated speck-like protein	15	
40	ARP3 (actin-related protein 3, yeast) homolog	2	
41	Aspartate aminotransferase, cytoplasmic	15	
42	Bactenecin 5 (cathelicidin-2)	3, 5, 7, 9, 10 , 12, 13, 15	3
43	Bactenecin 7 (cathelicidin-3)	3, 5	3
44	Beta-1,4 Galactosyltransferase-1	7, 13	
45	Beta-2-glycoprotein-1	7	
46	Beta-2-microglobulin	7	2, 3
47	Beta-actin (ACTB)	9	
48	Beta-casein precursor	8	
49	Beta-defensin	10	
50	Beta-lacatoglobulin		1, 3, 4, 5, 6, 8, 11, 13, 14, 15
51	BPI fold-containing family B member 1	15	
52	Brain acid soluble protein 1	15	
53	Butyrophilin subfamily 1 member A1		6, 13, 15
54	Calcium uniporter channel component	12	
55	Calcium-binding protein 45 kDa (Cab45)		15
56	Calgranulin B (Protein S100-A9)	2, 5, 7, 9, 15	
57	Calmodulin (CaM)	15	
58	Calpain-1 catalytic subunit (EC 3.4.22.52)	15	
59	Calumenin	15	
60	Capping protein (actin filament) muscle Z-line, alpha 1	2, 10	
61	CapZ-interacting protein (Protein kinase substrate CapZIP)	15	
62	Caseins		1, 3, 11, 12
63	Cathelicidin (general)	3, 5, 6, 9, 12, 10	3, 6
64	Cathelicidin-4	9, 12, 13	
65	Cathelicidin-5	10, 15	
66	Cathelicidin-6	10	
67	Cathelicidin-7	9, 10, 15	
68	Cathepsin B (EC 3.4.22.1) (BCSB)	15	
69	Cathepsin S (EC 3.4.22.27)	15	

Table 8.2 (continued)

Sr. number	Milk protein	Reference reporting increase in mastitis	Reference reporting decrease in mastitis
70	CCR4-NOT transcription complex subunit 10	15	
71	CD9 antigen	5	13
72	Ceruloplasmin	5, 6	
73	Chitinase-3-like protein-1	7, 6, 9, 15	
74	Clusterin	5, 7, 13	
75	Coactosin-like protein	15	
76	Coagulation factor XIII A chain	15	
77	Cofilin 1 (nonmuscle)	6, 7, 15	12
78	Collagen, type VI, alpha 3-like isoform 1, 2, 3, 4	12	
79	Complement C1s subcomponent	15	
80	Complement C2	10, 15	
81	Complement C3	3, 5, 7, 9, 10, 13	14
82	Complement C4	3, 5, 7, 15	
83	Complement C5a anaphylatoxin	15	
84	Complement component 3	5, 6, 9, 10	
85	Complement component C7		13
86	Complement factor B	6, 7, 10, 15	
87	Complement factor H	6, 10, 15	
88	Copper amine oxidase	5	
89	Coronin, actin binding protein, 1A	2, 6, 10, 15	
90	Corticoliberin	7	
91	Cyclic dodecapeptides (cathelicidin-1)	2, 3, 4, 5, 7, 9, 10, 12, 13, 14, 15	
92	Cystatin-B (Stefin-B)	15	
93	Cystatin-C (Colostrum thiol proteinase inhibitor) (Cystatin-3)		15
94	Cysteine-rich PDZ-binding protein (Cysteine-rich interactor of PDZ three) (Cysteine-rich interactor of PDZ3)		15
95	Cytochrome b-245 heavy chain	10	
96	Diacylglycerol kinase (DAG kinase) (EC 2.7.1.107)	15	
97	Dipeptidyl peptidase 1 (EC 3.4.14.1) (Cathepsin C)	15	
98	Disintegrin and metalloproteinase domain-containing protein 10	15	
99	DnaJ homolog subfamily B member 12	15	
100	Dystroglycan (Dystrophin-associated glycoprotein 1)		15
101	ECM1 protein	15	

(continued)

Table 8.2 (continued)

Sr. number	Milk protein	Reference reporting increase in mastitis	Reference reporting decrease in mastitis
102	EF-hand domain-containing protein D2 (Swiprosin-1)	15	
103	ELAV-like protein	15	
104	Elongation factor 1-alpha 1 (EF-1-alpha-1) (Elongation factor Tu)	15	
105	Elongation factor 2 (EF-2)	15	
106	Endopin 2B	5, 9	
107	Endopin-1	5, 6, 9, 12	
108	Enolase 1 (alpha)	2, 6, 7, 9, 12, 16	
109	Epididymal secretory protein E1 precursor		3, 6
110	Eukaryotic translation initiation factor 5A-1	15	
111	Ezrin (Cytovillin) (Villin-2) (p81)	15	
112	F-actin-capping protein subunit alpha-1 (CapZ alpha-1)	15	
113	F-actin-capping protein subunit beta (CapZ beta)	15	
114	Factor XIIa inhibitor (XIIaINH)		15
115	Fat storage-inducing transmembrane protein 2	7	
116	Fatty acid-binding protein		3, 6, 12, 13, 14, 15
117	Fetuin (α-2-HS-glycoprotein)	2, 3, 5, 6, 7, 15	
118	FGG protein	12	
119	Fibrinogen	2, 3, 4, 5, 6, 7, 9, 10, 12, 15	3
120	Fibroblast growth factor-binding	5	13
121	Fibronectin (FN)	15	
122	Folate receptor alpha	7	13
123	Fructose-bisphosphate aldolase (EC 4.1.2.13)	6, 15	
124	Galectin	15	
125	Galectin-1	12	
126	Gelsolin	6, 7, 15	
127	Glucose regulated protein 58 kDa	2	
128	Glucose-6-phosphate 1-dehydrogenase	10	
129	Glucose-6-phosphate isomerase (GPI) (EC 5.3.1.9)	15	
130	Glyceraldehyde-3-phosphate dehydrogenase (GAPDH)	2, 7, 9, 12, 15	
131	Glycerol-3-phosphate dehydrogenase 2	2	
132	Glycogen phosphorylase, liver form (EC 2.4.1.1)	15	
133	Glycoprotein 2		6, 13

Table 8.2 (continued)

Sr. number	Milk protein	Reference reporting increase in mastitis	Reference reporting decrease in mastitis
134	Glycosylation-dependent cell adhesion molecule-1 (Glycam-1) (Lactophorin)	4, 5, 8, 13	6, 15
135	GNAI2 protein	10	
136	Haptoglobin	5, 6, 7, 9, 10, 12, 15	
137	Heat shock protein	2, 6, 10, 15	
138	Heat-responsive protein 12	9	
139	Haemoglobin subunit alpha		12
140	Haemoglobin subunit beta	12	
141	Hemopexin	7, 9, 10, 15	
142	Hepatoma-derived growth factor (HDGF)	15	
143	Heterogeneous nuclear ribonucleoprotein K (hnRNP K)	15	
144	Hibernation-associated plasma protein HP-20-like	6	
145	High mobility group protein B1 (High mobility group protein 1) (HMG-1)	15	
146	High mobility group protein B2 (High mobility group protein 2) (HMG-2)	15	
147	Histidine triad nucleotide-binding protein 1	15	
148	Histidine-rich glycoprotein	15	
149	Histone	2, 5, 10, 15	
150	HRPE773-like (Uncharacterized protein)	15	
151	Hyaluronan and proteoglycan link protein 1	13	
152	Ig kappa chain or IGK protein	6	
153	Immunoglobulin heavy chain	5, 12, 13, 14	
154	Immunoglobulin lambda light chain variable region	6, 13, 14	12
155	Indolicidin (cathelicidin-4),	3, 5, 7, 10, 15	
156	Insulin-like growth factor-binding protein 6 (IBP-6)	15	
157	Integrin beta-2	10	
158	Interalpha (globin) inhibitor H4	6, 15	
159	Inter-alpha-trypsin inhibitor heavy chain (4 and H1)	5, 7, 10, 13, 15	
160	Intercellular adhesion molecule 1 (ICAM-1) (CD antigen CD54)	15	
161	Intercellular adhesion molecule 3	10	

(continued)

Table 8.2 (continued)

Sr. number	Milk protein	Reference reporting increase in mastitis	Reference reporting decrease in mastitis
162	Interleukin 1 receptor accessory protein (Uncharacterized protein)	15	
163	Interleukin-1 receptor antagonist protein (IL-1RN) (IL-1ra) (IRAP)	15	
164	Isoaspartyl peptidase/L-asparaginase (EC 3.4.19.5)	15	
165	Isocitrate dehydrogenase [NADP] cytoplasmic		12
166	KIAA1984 (Uncharacterized protein)		15
167	Kinesin-like protein	15	
168	Kininogen 1	5, 7, 10	
169	Kininogen 2	7	
170	Lactadherin (milk fat globule EGF factor 8 protein)	5, 7	3, 6, 13
171	Lactoperoxidase	7, 10	6, 14, 15
172	Lactotransferrin (lactoferrin)	4, 5, 6, 9, 10, 11, 13, 15	14
173	Leucine-rich alpha-2-glycoprotein 1 (Uncharacterized protein)	15	
174	Leukocyte elastase inhibitor (LEI) (Serpin B1)	9, 10, 15	
175	Lipocalin 2	12	
176	Lipopolysaccharide-binding protein	10, 15	
177	Lipoprotein lipase	7, 10	
178	L-Lactate dehydrogenase	6, 9, 15	
179	LOC788112 protein (Uncharacterized protein)	15	
180	L-Serine dehydratase/L-threonine deaminase (SDH) (EC 4.3.1.17)	15	
181	Lymphocyte cytosolic protein 1 (65K macrophage protein/L-plastin)	2, 10	
182	Lymphocyte-specific protein 1 (Uncharacterized protein)	15	
183	Macrophage migration inhibitory factor (MIF) (EC 5.3.2.1)	15	
184	Macrophage-capping protein	10, 15	
185	Manganous superoxide dismutase	12	
186	MARCKS-related protein (MARCKS-like protein 1)	15	
187	Matrix metalloproteinase-9 (MMP-9) (EC 3.4.24.35) (92 kDa gelatinase)	15	
188	Mitochondrial peptide methionine sulfoxide reductase (EC 1.8.4.11)	15	
189	Moesin	10, 15	
190	MPO protein	10	
191	Mucin 1	2	15

Table 8.2 (continued)

Sr. number	Milk protein	Reference reporting increase in mastitis	Reference reporting decrease in mastitis
192	Myeloid-associated differentiation marker	10	
193	Myoglobin		15
194	Myosin light polypeptide 6 (17 kDa myosin light chain) (LC17)	15	
195	Myosin regulatory light chain 12B	15	
196	Myozenin-1 (Calsarcin-2)		15
197	Na(+)/H(+) exchange regulatory cofactor NHE-RF1 (NHERF-1)	15	
198	Neutrophil cytosol factor 1	10	
199	Nuclease-sensitive element-binding protein 1	15	
200	Nucleobindin 2 (Uncharacterized protein)		15
201	Nucleobindin-1	7	6, 13, 15
202	Nucleoside diphosphate kinase B (NDK B) (NDP kinase B) (EC 2.7.4.6)	15	
203	Odorant-binding protein-like	13	
204	OLFM4 protein	10	
205	Osteoclast-stimulating factor 1	15	
206	Osteopontin	7, 10, 13	
207	Pantetheinase (EC 3.5.1.92) (Pantetheine hydrolase)	15	
208	Paraoxonase 1 (Uncharacterized protein)	15	
209	Pentaxin (Pentraxin)	15	
210	Pentraxin-related protein PTX3 (Pentaxin-related protein PTX3)	15	
211	Peptidoglycan recognition protein	2, 4, 5, 6, 7, 9, 10, 15	
212	Peptidyl-prolyl cis-trans isomerase A	10	12
213	Perilipin-2 (Adipophilin) (Adipose differentiation-related protein) (ADRP)	–	13, 15
214	Peripherin	12	
215	Peroxiredoxin-1 (EC 1.11.1.15)	15	
216	Peroxiredoxin-5, mitochondrial	9, 10, 15	
217	Peroxiredoxin-6	10, 15	
218	Peroxisome proliferator-activated receptor gamma coactivator 1 beta	2	
219	PHD finger protein 20-like protein 1		15
220	Phosphoglucomutase-1 (PGM 1) (EC 5.4.2.2) (Glucose phosphomutase 1)	15	
221	Phosphoglycerate kinase 1 (EC 2.7.2.3)	7, 15	

(continued)

Table 8.2 (continued)

Sr. number	Milk protein	Reference reporting increase in mastitis	Reference reporting decrease in mastitis
222	Phosphoglycerate mutase 1 (EC 3.1.3.13) (EC 5.4.2.11)	15	
223	Plasminogen	5, 6, 7, 15	
224	Platelet glycoprotein 4		10, 13, 15
225	Polymeric immunoglobulin receptor		3, 6, 14, 15
226	Primary amine oxidase, liver isozyme (EC 1.4.3.21)	15	
227	Procollagen-proline, 2-oxoglutarate 4-dioxygenase	2	
228	Profilin-1	6, 7, 15	
229	Prohibitin	2	
230	Prosaposin (Proactivator polypeptide)	15	
231	Prostaglandin D2 synthase		6
232	Prostaglandin reductase 1 (PRG-1) (EC 1.3.1.-)	15	
233	Prostaglandin-H2 D-isomerase (PTGDS)	10, 13	
234	Proteasome activator complex subunit 1	10, 15	
235	Proteasome activator complex subunit 2 (Proteasome activator 28)	15	
236	Protein disulfide-isomerase A3 (EC 5.3.4.1)	15	
237	Protein FAM49B	15	
238	Protein HP-20 homolog	15	
239	Protein HP-25 homolog 2	15	
240	Protein OS-9		15
241	Protein S100 (S100 calcium-binding protein)	15	
242	Protein S100-A12 (Calcium-binding protein in amniotic fluid 1) (CAAF1)	2, 3, 9, 12, 15	
243	protein S100-A2	12	
244	Protein S100-A8 (calgranulin A)	7, 6, 9, 10, 12, 13, 15	
245	Protein ZBED8 (Transposon-derived Buster3 transposase-like protein)	15	
246	Prothrombin	13, 15	
247	Pyruvate kinase (EC 2.7.1.40)	15	
248	Rab GDP dissociation inhibitor alpha (Rab GDI alpha)	15	
249	Ras suppressor protein 1 (Rsu-1)	15	
250	Ras-related protein Rab-1B	15	
251	Resistin	15	
252	Rho GDP dissociation I-beta	9	
253	Rho GDP-dissociation inhibitor 1 (Rho GDI 1) (Rho-GDI alpha)	15	

Table 8.2 (continued)

Sr. number	Milk protein	Reference reporting increase in mastitis	Reference reporting decrease in mastitis
254	Rho GDP-dissociation inhibitor 2 (Rho GDI 2)	15	
255	Ribonuclease pancreatic	13	
256	Ribonuclease UK114 (EC 3.1.-.-)	15	
257	Ribose-5-phosphate isomerase (EC 5.3.1.6)	15	
258	Ribosomal protein L7	2	
259	S100 calcium binding protein A11 (calgizzarin)	2, 12	
260	Selenoprotein 15 kDa		15
261	Serine–tRNA ligase, cytoplasmic (EC 6.1.1.11)	15	
262	Serotransferrin	1, 2, 3, 4, 5, 6, 7, 9, 10, 11, 13, 15	
263	Serotransferrin precursor		12
264	Serpin A3	13	
265	Serpin A3-1	10, 12, 15	
266	Serpin A3-2	12	
267	Serpin A3-3 (Endopin-1B) (Muscle endopin-1B) (mEndopin-1B)	13, 15	
268	Serpin A3-6	12, 13	
269	Serpin A3-8	13, 15	
270	Serpin peptidase inhibitor, clade A, member 1 (alpha-1 antiproteinase, antitrypsin)	6, 12	
271	SERPINB4 protein (Uncharacterized protein)	15	
272	SERPIND1 protein (Uncharacterized protein)	15	
273	Serum amyloid A 3	5, 6, 7, 10, 12, 15	
274	SH3 domain-binding glutamic acid-rich-like protein 3	15	
275	Similar to galactose-binding lectin	12	
276	Similar to lipocalin	9	
277	Sodium-dependent phosphate transporter	15	13
278	Spermadhesin-1 (Acidic seminal fluid protein) (ASFP)		15
279	Sulfhydryl oxidase (EC 1.8.3.2)	15	
280	Syndecan	15	
281	Syndecan-2 (SYND2) (CD antigen CD362)		15
282	Thioredoxin (Trx)	15	
283	THO complex subunit 4 (Tho4) (Ally of AML-1 and LEF-1)	15	

(continued)

Table 8.2 (continued)

Sr. number	Milk protein	Reference reporting increase in mastitis	Reference reporting decrease in mastitis
284	Thymosin beta-10 (Thymosin beta-9) [Cleaved into: Thymosin beta-8]	15	
285	Thymosin beta-4 (T beta-4)	15	
286	Thymosin, beta 4-like	12	
287	Thyroxine-binding globulin (Serpin A7) (T4-binding globulin)	15	
288	Transaldolase (EC 2.2.1.2)	15	
289	Transforming growth factor-beta-induced protein ig-h3	15	
290	Transgelin-2	15	
291	Transketolase (EC 2.2.1.1)	15	
292	Transthyretin	3, 6	
293	Triosephosphate isomerase (TIM) (EC 5.3.1.1) (Triose-phosphate isomerase)	15	
294	TTR	3	
295	Tubulin alpha-1C chain	15	
296	Tubulin beta-5 chain	9, 15	
297	Tubulin beta-6 chain	9, 15	
298	TWF2 protein (Uncharacterized protein)	15	
299	Ubiquitin carboxyl-terminal hydrolase 10 (EC 3.4.19.12)	–	15
300	Ubiquitin-S27a fusion protein	12	
301	Vasodilator-stimulated phosphoprotein (VASP)	15	
302	Vimentin	12, 15	
303	Vitamin D-binding protein	6, 7	13
304	WD repeat-containing protein 1	15	
305	Xanthine dehydrogenase/oxidase	7	6, 13, 15
306	Zinc phosphodiesterase ELAC protein 1 (EC 3.1.26.11) (ElaC homolog protein 1) (Ribonuclease Z 1)	15	

1 Hogarth et al. (2004)
2 Smolenski et al. (2007)
3 Boehmer et al. (2008)
4 Boehmer et al. (2010a)
5 Danielsen et al. (2010)
6 Ibeagha-Awemu et al. (2010)
7 Boehmer et al. (2010b)
8 Kim et al. (2011)
9 Alonso-Fauste et al. (2012)
10 Reinhardt et al. (2013)
11 Pongthaisong et al. (2016)

during episodes of inflammation also abound and may be considered as endogenous non-native proteases that could account for most of the proteolytic activity in high somatic cell count milk (Napoli et al. 2007). The proteolytic activities of enzymes in milk ultimately result in loss of milk caseins which compromises the quality and technological properties of milk such as in cheese formation (Larsen et al. 2010a). In this study, mass spectrometry-based peptidomics enabled the identification of a number of peptides in milk samples from which biomarkers relating to the process might be discovered.

In a recent study of milk from clinical cases of mastitis, Mansor et al. identified up to 31 polypeptides which, combined in a classification panel, could differentiate healthy from mastitic milk samples with 100% specificity and sensitivity (Mansor et al. 2013). A further set of 14 peptides was able to distinguish between cases of mastitis caused by different pathogens (*S. aureus* or with *E. coli*) responsible for infections with 100% sensitivity but a lower specificity of 75%. Rapid classification of the bacterial class of the pathogen causing mastitis would be of some value as it could direct more effective use of antimicrobial treatment so that a profile based on a peptidomic approach for this purpose would be valuable.

8.3.5 Metabolomic Investigation of Mastitis

Metabolomics is the study of the composition, relative abundances, interactions and dynamics of the metabolome in response to change of environment of metabolites within a biological system (Osorio et al. 2012). It entails the use of sophisticated analytical techniques in non-biased identification and quantification of all metabolites in a biological system (Dettmer et al. 2007). Metabolomics is aimed at characterising metabolic changes that result following the presence or absence of one or more factors in order to gain insights into systems biology and also to identify possible biomarkers of specific conditions (Courant et al. 2013).

Metabolomics has been valuable in several areas of study in the bovine species, in particular diagnostics of animal health and food safety as well as management practices geared to improvement of animal production. Numerous metabolomic studies have already been carried out in cattle and the bovine metabolome database (BMDB) is available at http://www.cowmetdb.ca/. This database comprises information on metabolites of dairy and beef cattle obtained by experiment on blood, meat, urine, milk and ruminal fluid (Hailemariam et al. 2014). Targeted evaluations of the metabolic profile (of known metabolites) in bovine samples such as urine, serum, plasma and milk have been carried out, however untargeted approaches that aid in detecting new metabolites are gaining importance especially with innovations in bioinformatics and mass spectrometric techniques.

A number of metabolomic studies have been conducted in the bovine species including studies in endocrinology by Rijk et al. who utilised an untargeted UPLC-TOF MS to identify biomarker candidates for the anabolic steroid prohormones (Rijk et al. 2009) dehydroepiandrosterone (DHEA) and pregnenolone in cattle urine. Similar studies were also carried out by Regal et al., this time using serum samples, for assessing two other anabolic steroids, estradiol-17β and progesterone,

(Regal et al. 2011). They utilised HPLC coupled to an Orbitrap spectrometer and found significant differences that discriminated between the use and non-use of these hormones. In the same area of research, markers of natural steroids and 4-androstenedione abuse in urine of cattle have been explored by Anizan et al. (2011, 2012). All these studies resulted in the detection of several compounds that were not previously recognised in the analytes, and which once properly validated, could serve as markers for screening of animals for steroid abuse.

Bender et al. observed significant differences in metabolites in follicular fluid of heifers compared to those of lactating cows using GC-MS, and also between dominant and subordinate follicles (Bender et al. 2010); these discrepancies were suggested to be able to give an insight into increasing incidences of low fertility and variances in fertility level between these two groups of cows. Metabolomics studies have also shown that differences in the concentration of up to 19 metabolites are potentially able to distinguish subclinical ketosis from normal serum samples, while up to 31 differentiated clinical ketosis from normal. Eight metabolites were also found to vary between subclinical ketotic and normal serum samples. These metabolites are thus potential biomarkers of ketosis in dairy cows (Zhang et al. 2013). The bovine ruminal fluid metabolome has been investigated by Saleem et al., using a combination of NMR spectroscopy and GC-MS (Saleem et al. 2012). A database containing the metabolites identified in this study has been made available at http://www.rumendb.ca.

Application of metabolomics in mastitis research has been relatively neglected. Eriksson and colleagues compared volatile metabolites present in the headspace of mastitic and normal milk samples using GC-MS technology and demonstrated that an electronic nose (gas-sensor array consisting semi-conductive metallic oxide sensors and metal oxide semi-conductive field effect transistors) can differentiate between normal milk and mastitis milk (Eriksson et al. 2005). Hettinga and colleagues successfully developed an Artificial Neural Network (ANN)-based multivariate classifier to differentiate the milk samples that were positive for five different pathogens (*S. aureus*, coagulase-negative *Staphylococci*, *Streptococcus dysgalactiae*, *S. uberis* or *E. coli*) or healthy control, based on the volatile metabolites present in the samples using GC/MS (Hettinga et al. 2008a). They also found the sources of the volatile metabolites in the mastitis samples, concluding that most of the volatile metabolites were products of distinct pathogens (Hettinga et al. 2009). Recently, Sundekilde and colleagues used an NMR spectroscopy method to compare metabolite profiles of milk with higher and lower SCC and identified significant relative increase in concentrations of lactate, acetate, isoleucine, butyrate and BHBA in samples with high SCC and corresponding significant relative decrease in lactose, hippurate and fumarate concentrations (Sundekilde et al. 2013b). They have also reviewed the application of NMR spectroscopy method in the metabolomics of milk (Sundekilde et al. 2013a). Oxylipids are a diverse group of lipid mediators of inflammation that are biosynthesised from the oxidation of polyunsaturated fatty acids (PUFAs), such as arachidonic acid, docosahexaenoic acid and eicosapentaenoic acid through enzymatic and free radical-mediated reactions (Stables and Gilroy 2011; Massey and Nicolaou 2013). Using a LC-MS/MS-based lipidomic approach, recent studies investigated the role of oxylipids in bovine coliform mastitis and *S. uberis* mastitis (Mavangira et al. 2015; Ryman et al. 2015). In coliform mastitis, lipoxygenase and cytochrome P450 derived oxylipids were the predominant fraction of total oxylipids present in both milk and plasma. Similarly, higher concentration of arachidonic acid and linoleic acid-derived oxylipids such as hydroxyoctadecadienoic acid and oxooctadecadienoic acid were reported from *S. uberis* mastitis. Thomas and colleagues recently reported an untargeted metabolomic analysis of milk obtained from an experimentally induced mastitis with a host-adapted strain of *S. uberis*, which is described in detail under section X.4 (Thomas et al. 2016a).

8.3.6 Microbiome Investigations in Mastitis

Bovine "microbiome" is the catalogue of microbes and their genes associated with bovines (Ursell et al. 2012). There has been an increased interest in the characterisation of the bovine microbiome during mastitis and its comparison between healthy and disease states (Addis et al. 2016). It is postulated that up to 99% of the microbes in the environment cannot be readily cultivated (Bhatt et al. 2012), and culture-based tests in mastitis fail to identify pathogenic organisms in about 30% of cases (Oikonomou et al. 2014). With the developments in the DNA analysis technologies, particularly with the arrival of the Next-Generation sequencing technologies, it has become possible to sequence and analyse the hypervariable regions within the 16S rRNA gene to study microbial diversity and to identify microbes in culture negative milk samples. Developments in milk microbiota from multiple host species including bovines and human have recently been reviewed (Quigley et al. 2013; Addis et al. 2016). Recently, studies using metagenomic sequencing of bacterial 16S rRNA genes from clinical and subclinical mastitis milk reported the presence of diverse microbial communities, including hitherto unknown anaerobic pathogens in mastitis milk (Bhatt et al. 2012; Oikonomou et al. 2012, 2014). Recent studies suggest Lactobacillus supplementation could reduce incidence of subclinical mastitis (Qiao et al. 2015; Ma et al. 2015; Bouchard et al. 2013).

8.4 An Integrated Omic Investigation of *Streptococcus uberis* Mastitis

Most investigations using proteomics or metabolomics have focussed on a single technology, whereas it is plain that changes occur in all biospheres simultaneously. To undertake a true systems approach it would be ideal to monitor each omic entity in the same samples to gain maximal insight into the response to a major pathogenic response as occurs in the mammary gland during mastitis. Indeed, with the ready availability of milk from an infected gland this could be described as the ideal experimental model to examine the total response of a mammalian system to bacteria.

In order to understand the pathology of mastitis at molecular level, a team at the University of Glasgow (led by Professor P. D. Eckersall and Professor R. N. Zadoks) have conducted an integrated polyomic study in milk in an experimental model of *Streptococcus uberis* mastitis (Thomas et al. 2016a, b; Mudaliar et al. 2016). The study was conducted on milk samples obtained from an intramammary challenge experiment with a host-adapted strain (FSL Z1-048) of *S. uberis* (Tassi et al. 2013). The integrated polyomic study of mastitis consisted of analysing global profiles of milk peptidome, proteome and metabolome over the course of the infection, inflammation and resolution. This comprehensive study analysed almost the complete spectrum of molecules in milk—the peptidomic analysis considered polypeptide fragments in the range of 380–6000 Da, the proteomic analysis included proteins with masses ranging up to 3 MDa and the metabolomics analysed small molecules with a mass less than 1500 Da. Based on the changes in clinical parameters, SCC and bacterial concentration in milk, six time points (0, 36, 42, 57, 81 and 312 h post-challenge) were selected for the high-throughput polyomic study, while all the milk samples collected over the time course (19 time points) were used to study the high-abundance proteins and three acute phase proteins—Hp, M-SAA3 and CRP. The high-throughput quantitative peptidomic, label-free quantitative proteomic and untargeted metabolomic data were generated using capillary electrophoresis-mass spectrometry (CE-MS), LC-MS/MS and LC-MS, respectively. A one-dimensional electrophoresis approach combined with LC-MS/MS-based protein identification was used to study the high-abundance proteins, and the acute phase proteins were measured using ELISA. The peptidomic study identified 460 peptides, of which 77 peptides could be used to classify pre- and post-infection time points. Most of the identified peptides belonged to caseins, while a few peptides belonged to serum amyloid A and Glycosylation-dependent cell adhesion molecule 1 (GDCAM). Correspondingly, there was reduction in the caseins, one of the high-abundance proteins, which was noted first at 36-h post-challenge and continued even after the elimination of infection, suggesting the role of host rather than bacterial proteases in protein degradation. The label-free quantitative proteomics study quantified 570 bovine proteins and reported their changes over the course of the infection. Particularly, the proteomics analysis showed more than thousand-fold increase in antimicrobial proteins (peptidoglycan recognition protein 1, cathelicidins) between 36 and 81 h post-challenge. The APPs—Hp, M-SAA3 and CRP, which were quantified in the proteomics analysis and also using immunoassay showed identical pattern of their expression levels in both techniques. The APPs started increasing by hundreds of folds at 30-hour post-challenge. Immunoassay showed maximum median concentration of Hp was 421 μg/mL at 72-h post-challenge, M-SAA3 was 9900 μg/mL at 96-h post-challenge, and CRP was 16,687 ng/mL at 72-h post-challenge. The untargeted metabolomics analysis putatively identified 690 metabolites. The temporal profile of the metabolites showed an increasing concentration of bile acids over the time course until 81-h post-challenge. The putatively identified bile acids that showed up-regulation in the pro-inflammatory phase include glycocholate ($C_{26}H_{43}NO_6$), taurocholic acid ($C_{26}H_{45}NO_7S$), taurochenodeoxycholic acid ($C_{26}H_{45}NO_6S$), glycodeoxycholate ($C_{26}H_{43}NO_5$) and cholate ($C_{24}H_{40}O_5$). The integrated analysis of the polyomics data showed the involvement of APR signalling, liver X receptor (LXR), retinoid X receptor (LXR) and Farnesoid X receptor (FXR) activation pathways. Particularly, the up-regulation of bile acids can be linked to mastitis via FXR activation pathway. This study also identified patterns of protein and metabolite changes in acute phase and resolving phase of the inflammatory process. The integrated analysis showed while the bacterial count reaches a peak within 36–42 h post challenge, most of the host responses do not reach a peak until a 57 or 81 h post challenge.

The data mountain generated by this combined omic approaches to examine the host response to mastitis required extensive bioinformatic expertise to fully assess the sequence of events leading to the multiple peptide, protein and metabolite disruption. However, this investigation gives a glimpse into the possibilities for demonstrating the complete picture of the mammary gland's response to mastitis.

8.5 Systems Biology in Mastitis Research: Current and Future Perspectives

As described in the preceding sections, advances in the omics technologies, the key technologies for systems biology investigations including Next-Generation sequencing, liquid chromatography, mass spectrometry and bioinformatics for proteomics and metabolomics have enabled in-depth investigations in bovine mastitis. These reductionist approaches have provided a breadth of understanding of molecular changes during mastitis with high precision and have illuminated mechanisms of host–pathogen interactions. Thus individual classes of pathogen-host responsiveness have been described with the identification of bacterial species causing the infection being determined with high specificity and sensitivity by genomic investigations. Similarly, individual omics have been able to categorise the responses to infection. Proteomics has demonstrated that proteins such as the acute phase proteins and cathelicidins respond within a few hours, peptidomics have characterised a rise in small peptides and metabolomics has documented the wholesale changes that occur in the metabolites of carbohydrate, lipid, nitrogen and nucleic acid metabolism during infection of the udder. Lately, synchronised changes in the transcriptome of mammary tissue and liver during IMI have been studied to the increase our understanding of the coordination between the mammary gland and the liver and the transcriptional network controlling inflammation in both these tissues (Moyes et al. 2016).

However, the majority of investigations have utilised a single technology which though highly informative may be limiting. Thus a system-wide approach integrating genomics, transcriptomics, proteomics, metabolomics from both host and pathogen systems could potentially lead to a better understanding of mastitis and provide improvements in diagnosing, managing and preventing mastitis (Ferreira et al. 2013). Towards this end, the mastitomics study (Mudaliar et al. 2016; Thomas et al. 2016a, b) is a small beginning but may presage a realisation that the host responses to a disease such as mastitis does not occur in isolated silos determined by the currently available technology platforms. While there are separate databases available in each omics area, there is a need to develop integrated systems biology resources for bovine mastitis. Integrated omics as a systems biology tool to investigate mastitis and other important areas of farm animal research should be an aim for the future and it is highly likely that with an integrated application of these modern disruptive approaches, our understanding of the biological processes could be revolutionised with a plethora of novel diagnostic and therapeutic approaches developed for dealing with the economically important disease of mastitis.

References

Abdel-Shafy H, Bortfeldt RH, Tetens J, Brockmann GA (2014) Single nucleotide polymorphism and haplotype effects associated with somatic cell score in German Holstein cattle. Genet Sel Evol 46:35. doi:10.1186/1297-9686-46-35

Addis MF, Tanca A, Uzzau S, Oikonomou G, Bicalho RC, Moroni P (2016) The bovine milk microbiota: insights and perspectives from -omics studies. Mol Biosyst 12(8):2359–2372. doi:10.1039/c6mb00217j

Akers RM, Nickerson SC (2011) Mastitis and its impact on structure and function in the ruminant mammary gland. J Mammary Gland Biol Neoplasia 16(4):275–289. doi:10.1007/s10911-011-9231-3

Åkerstedt M, Waller KP, Larsen LB, Forsbäck L, Sternesjö Å (2008) Relationship between haptoglobin and serum amyloid A in milk and milk quality. Int Dairy J 18(6):669–674. doi:10.1016/j.idairyj.2008.01.002

Akira S, Uematsu S, Takeuchi O (2006) Pathogen recognition and innate immunity. Cell 124(4):783–801. doi:10.1016/j.cell.2006.02.015

Albalat A, Mischak H, Mullen W (2011) Clinical application of urinary proteomics/peptidomics. Expert Rev Proteomics 8(5):615–629. doi:10.1586/epr.11.46

Almeida AM, Bassols A, Bendixen E, Bhide M, Ceciliani F, Cristobal S et al (2015) Animal board invited review: advances in proteomics for animal and food sciences. Animal 9(1):1–17. doi:10.1017/S1751731114002602

Alonso-Fauste I, Andres M, Iturralde M, Lampreave F, Gallart J, Alava MA (2012) Proteomic characterization by 2-DE in bovine serum and whey from healthy and mastitis affected farm animals. J Proteomics 75(10):3015–3030. doi:10.1016/j.jprot.2011.11.035

Anizan S, Bichon E, Di Nardo D, Monteau F, Cesbron N, Antignac JP et al (2011) Screening of 4-androstenedione misuse in cattle by LC-MS/MS profiling of glucuronide and sulfate steroids in urine. Talanta 86:186–194. doi:10.1016/j.talanta.2011.08.058

Anizan S, Bichon E, Duval T, Monteau F, Cesbron N, Antignac JP et al (2012) Gas chromatography coupled to mass spectrometry-based metabolomic to screen for anabolic practices in cattle: identification of 5alpha-androst-2-en-17-one as new biomarker of 4-androstenedione misuse. J Mass Spectrom 47(1):131–140. doi:10.1002/jms.2035

Baeker R, Haebel S, Schlatterer K, Schlatterer B (2002) Lipocalin-type prostaglandin D synthase in milk: a new biomarker for bovine mastitis. Prostaglandins Other Lipid Mediat 67(1):75–88

Barreiro JR, Braga PA, Ferreira CR, Kostrzewa M, Maier T, Wegemann B et al (2012) Nonculture-based identification of bacteria in milk by protein fingerprinting. Proteomics 12(17):2739–2745. doi:10.1002/pmic.201200053

Bassols A, Turk R, Roncada P (2014) A proteomics perspective: from animal welfare to food safety. Curr Protein Pept Sci 15(2):156–168

Bender K, Walsh S, Evans AC, Fair T, Brennan L (2010) Metabolite concentrations in follicular fluid may explain differences in fertility between heifers and lactating cows. Reproduction 139(6):1047–1055. doi:10.1530/REP-10-0068

Bendixen E, Danielsen M, Hollung K, Gianazza E, Miller I (2011) Farm animal proteomics—a review. J Proteomics 74(3):282–293. doi:10.1016/j.jprot.2010.11.005

Berg LC, Thomsen PD, Andersen PH, Jensen HE, Jacobsen S (2011) Serum amyloid A is expressed in histologically normal tissues from horses and cattle. Vet Immunol Immunopathol 144(1-2):155–159. doi:10.1016/j.vetimm.2011.06.037

Berning LM, Shook GE (1992) Prediction of mastitis using milk somatic cell count, N-acetyl-beta-D-glucosaminidase, and lactose. J Dairy Sci 75(7):1840–1848. doi:10.3168/jds.S0022-0302(92)77943-0

Bhatt VD, Ahir VB, Koringa PG, Jakhesara SJ, Rank DN, Nauriyal DS et al (2012) Milk microbiome signatures of subclinical mastitis-affected cattle analysed by shotgun sequencing. J Appl Microbiol 112(4):639–650. doi:10.1111/j.1365-2672.2012.05244.x

Bian Y, Lv Y, Li Q (2014) Identification of diagnostic protein markers of subclinical mastitis in bovine whey using comparative proteomics. Bull Vet Inst Pulawy 58(3). doi:10.2478/bvip-2014-0060

Bislev SL, Deutsch EW, Sun Z, Farrah T, Aebersold R, Moritz RL et al (2012) A Bovine PeptideAtlas of milk and mammary gland proteomes. Proteomics 12(18):2895–2899. doi:10.1002/pmic.201200057

Blum SE, Heller ED, Sela S, Elad D, Edery N, Leitner G (2015) Genomic and phenomic study of mammary pathogenic Escherichia coli. PLoS One 10(9):e0136387. doi:10.1371/journal.pone.0136387

Boehmer JL (2011) Proteomic analyses of host and pathogen responses during bovine mastitis. J Mammary Gland Biol Neoplasia 16(4):323–338. doi:10.1007/s10911-011-9229-x

Boehmer JL, Bannerman DD, Shefcheck K, Ward JL (2008) Proteomic analysis of differentially expressed proteins in bovine milk during experimentally induced Escherichia coli mastitis. J Dairy Sci 91(11):4206–4218. doi:10.3168/jds.2008-1297

Boehmer JL, DeGrasse JA, McFarland MA, Tall EA, Shefcheck KJ, Ward JL et al (2010a) The proteomic advantage: label-free quantification of proteins expressed in bovine milk during experimentally induced coliform mastitis. Vet Immunol Immunopathol 138(4):252–266. doi:10.1016/j.vetimm.2010.10.004

Boehmer JL, Ward JL, Peters RR, Shefcheck KJ, McFarland MA, Bannerman DD (2010b) Proteomic analysis of the temporal expression of bovine milk proteins during coliform mastitis and label-free relative quantification. J Dairy Sci 93(2):593–603. doi:10.3168/jds.2009-2526

Boss R, Cosandey A, Luini M, Artursson K, Bardiau M, Breitenwieser F et al (2016) Bovine Staphylococcus aureus: subtyping, evolution, and zoonotic transfer. J Dairy Sci 99(1):515–528. doi:10.3168/jds.2015-9589

Bouchard DS, Rault L, Berkova N, Le Loir Y, Even S (2013) Inhibition of Staphylococcus aureus invasion into bovine mammary epithelial cells by contact with live Lactobacillus casei. Appl Environ Microbiol 79(3):877–885. doi:10.1128/AEM.03323-12

Bradley A (2002) Bovine mastitis: an evolving disease. Vet J 164(2):116–128

Bradley AJ, Leach KA, Breen JE, Green LE, Green MJ (2007) Survey of the incidence and aetiology of mastitis on dairy farms in England and Wales. Vet Rec 160(8):253–257

Canavez FC, Luche DD, Stothard P, Leite KR, Sousa-Canavez JM, Plastow G et al (2012) Genome sequence and assembly of Bos indicus. J Hered 103(3):342–348. doi:10.1093/jhered/esr153

Ceciliani F, Ceron JJ, Eckersall PD, Sauerwein H (2012) Acute phase proteins in ruminants. J Proteomics 75(14):4207–4231. doi:10.1016/j.jprot.2012.04.004

Ceron JJ, Eckersall PD, Martynez-Subiela S (2005) Acute phase proteins in dogs and cats: current knowledge and future perspectives. Vet Clin Pathol 34(2):85–99

Cooray R, Waller KP, Venge P (2007) Haptoglobin comprises about 10% of granule protein extracted from bovine granulocytes isolated from healthy cattle. Vet Immunol Immunopathol 119(3-4):310–315. doi:10.1016/j.vetimm.2007.06.026

Costa EO, Ribeiro AR, Melville PA, Prada MS, Carciofi AC, Watanabe ET (1996) Bovine mastitis due to algae of the genus Prototheca. Mycopathologia 133(2):85–88

Courant F, Antignac JP, Monteau F, Le Bizec B (2013) Metabolomics as a potential new approach for investigating human reproductive disorders. J Proteome Res 12(6):2914–2920. doi:10.1021/pr400204q

Crawford RG, Leslie KE, Bagg R, Dick CP, Duffield TF (2005) The impact of controlled release capsules of monensin on postcalving haptoglobin concentrations in dairy cattle. Can J Vet Res 69(3):208–214

D'Amato A, Bachi A, Fasoli E, Boschetti E, Peltre G, Senechal H et al (2009) In-depth exploration of cow's whey proteome via combinatorial peptide ligand libraries. J Proteome Res 8(8):3925–3936. doi:10.1021/pr900221x

Dallas DC, Guerrero A, Khaldi N, Castillo PA, Martin WF, Smilowitz JT et al (2013) Extensive in vivo human milk peptidomics reveals specific proteolysis yielding protective antimicrobial peptides. J Proteome Res 12(5):2295–2304. doi:10.1021/pr400212z

Dallas DC, Guerrero A, Parker EA, Robinson RC, Gan J, German JB et al (2015) Current peptidomics: applications, purification, identification, quantification, and functional analysis. Proteomics 15(5-6):1026–1038. doi:10.1002/pmic.201400310

Danielsen M, Codrea MC, Ingvartsen KL, Friggens NC, Bendixen E, Rontved CM (2010) Quantitative milk proteomics—host responses to lipopolysaccharide-mediated inflammation of bovine mammary gland. Proteomics 10(12):2240–2249. doi:10.1002/pmic.200900771

Davies PL, Leigh JA, Bradley AJ, Archer SC, Emes RD, Green MJ (2016) Molecular epidemiology of Streptococcus uberis clinical mastitis in dairy herds: strain heterogeneity and transmission. J Clin Microbiol 54(1):68–74. doi:10.1128/JCM.01583-15

Dettmer K, Aronov PA, Hammock BD (2007) Mass spectrometry-based metabolomics. Mass Spectrom Rev 26(1):51–78. doi:10.1002/mas.20108

Dohoo IR, Smith J, Andersen S, Kelton DF, Godden S, Mastitis Research Workers' Conference (2011) Diagnosing intramammary infections: evaluation of definitions based on a single milk sample. J Dairy Sci 94(1):250–261. doi:10.3168/jds.2010-3559

Eckersall PD, de Almeida AM, Miller I (2012) Proteomics, a new tool for farm animal science. J Proteomics 75(14):4187–4189. doi:10.1016/j.jprot.2012.05.014

Eckersall PD, Young FJ, McComb C, Hogarth CJ, Safi S, Fitzpatrick JL et al (2001) Acute phase proteins in serum and milk from dairy cows with clinical mastitis. Vet Rec 148(2):35–41. doi:10.1136/vr.148.2.35

Eckersall PD, Young FJ, Nolan AM, Knight CH, McComb C, Waterston MM et al (2006) Acute phase proteins in bovine milk in an experimental model of Staphylococcus aureus subclinical mastitis. J Dairy Sci 89(5):1488–1501. doi:10.3168/jds.S0022-0302(06)72216-0

Eriksson A, Waller KP, Svennersten-Sjaunja K, Haugen JE, Lundby F, Lind O (2005) Detection of mastitic milk using a gas-sensor array system (electronic nose). Int Dairy J 15(12):1193–1201. doi:10.1016/J.Idairyj.2004.12.012

Ezzat Alnakip M, Quintela-Baluja M, Bohme K, Fernandez-No I, Caamano-Antelo S, Calo-Mata P et al (2014) The immunology of mammary gland of dairy ruminants between healthy and inflammatory conditions. J Vet Med 2014:659801. doi:10.1155/2014/659801

Ferreira AM, Bislev SL, Bendixen E, Almeida AM (2013) The mammary gland in domestic ruminants: a systems biology perspective. J Proteomics 94:110–123. doi:10.1016/j.jprot.2013.09.012

Galvao KN, Pighetti GM, Cheong SH, Nydam DV, Gilbert RO (2011) Association between interleukin-8 receptor-alpha (CXCR1) polymorphism and disease incidence, production, reproduction, and survival in Holstein cows. J Dairy Sci 94(4):2083–2091. doi:10.3168/jds.2010-3636

Gerardi G, Bernardini D, Azzurra Elia C, Ferrari V, Iob L, Segato S (2009) Use of serum amyloid A and milk amyloid A in the diagnosis of subclinical mastitis in dairy cows. J Dairy Res 76(4):411–417. doi:10.1017/S0022029909990057

Gilchrist TL, Smith DG, Fitzpatrick JL, Zadoks RN, Fontaine MC (2013) Comparative molecular analysis of ovine and bovine Streptococcus uberis isolates. J Dairy Sci 96(2):962–970. doi:10.3168/jds.2012-5705

Goertz I, Baes C, Weimann C, Reinsch N, Erhardt G (2009) Association between single nucleotide polymorphisms in the CXCR1 gene and somatic cell score in Holstein dairy cattle. J Dairy Sci 92(8):4018–4022. doi:10.3168/jds.2008-1536

Goldammer T, Zerbe H, Molenaar A, Schuberth HJ, Brunner RM, Kata SR et al (2004) Mastitis increases mammary mRNA abundance of beta-defensin 5, toll-like-receptor 2 (TLR2), and TLR4 but not TLR9 in cattle. Clin Diagn Lab Immunol 11(1):174–185

Goldstone RJ, Harris S, Smith DG (2016) Genomic content typifying a prevalent clade of bovine mastitis-associated Escherichia coli. Sci Rep 6:30115. doi:10.1038/srep30115

Green M, Bradley A (2013) The changing face of mastitis control. Vet Rec 173(21):517–521. doi:10.1136/vr.f7029

Gronlund U, Hallen Sandgren C, Persson WK (2005) Haptoglobin and serum amyloid A in milk from dairy cows with chronic sub-clinical mastitis. Vet Res 36(2):191–198. doi:10.1051/vetres:2004063

Guerrero A, Dallas DC, Contreras S, Bhandari A, Canovas A, Islas-Trejo A et al (2015) Peptidomic analysis of healthy and subclinically mastitic bovine milk. Int Dairy J 46:46–52. doi:10.1016/j.idairyj.2014.09.006

Gultiken N, Serhat A, Gul F, Elvan A, Findik M (2012) Evaluation of plasma and milk haptoglobin concentrations in the diagnosis and treatment follow-up of subclinical mastitis in dairy cows. Acta Vet Brno 62(2-3):271–279. doi:10.2298/avb1203271g

Gurjar A, Gioia G, Schukken Y, Welcome F, Zadoks R, Moroni P (2012) Molecular diagnostics applied to mastitis problems on dairy farms. Vet Clin North Am Food Anim Pract 28(3):565–576. doi:10.1016/j.cvfa.2012.07.011

Hailemariam D, Mandal R, Saleem F, Dunn SM, Wishart DS, Ametaj BN (2014) Identification of predictive biomarkers of disease state in transition dairy cows. J Dairy Sci 97(5):2680–2693. doi:10.3168/jds.2013-6803

Halasa T, Huijps K, Osteras O, Hogeveen H (2007) Economic effects of bovine mastitis and mastitis management: a review. Vet Q 29(1):18–31. doi:10.1080/01652176.2007.9695224

Heikkila AM, Nousiainen JI, Pyorala S (2012) Costs of clinical mastitis with special reference to premature culling. J Dairy Sci 95(1):139–150. doi:10.3168/jds.2011-4321

Hettinga KA, van Valenberg HJ, Lam TJ, van Hooijdonk AC (2008a) Detection of mastitis pathogens by analysis of volatile bacterial metabolites. J Dairy Sci 91(10):3834–3839. doi:10.3168/jds.2007-0941

Hettinga KA, van Valenberg HJ, Lam TJ, van Hooijdonk AC (2009) The origin of the volatile metabolites found in mastitis milk. Vet Microbiol 137(3-4):384–387. doi:10.1016/j.vetmic.2009.01.016

Hettinga KA, van Valenberg HJF, Lam TJGM, van Hooijdonk ACM (2008b) Detection of mastitis pathogens by analysis of volatile bacterial metabolites. J Dairy Sci 91(10):3834–3839. doi:10.3168/jds.2007-0941

Hillerton JE, Berry EA (2005) Treating mastitis in the cow—a tradition or an archaism. J Appl Microbiol 98(6):1250–1255. doi:10.1111/j.1365-2672.2005.02649.x

Hinz K, Larsen LB, Wellnitz O, Bruckmaier RM, Kelly AL (2012) Proteolytic and proteomic changes in milk at quarter level following infusion with Escherichia coli lipopolysaccharide. J Dairy Sci 95(4):1655–1666. doi:10.3168/jds.2011-4813

Hiss S, Mielenz M, Bruckmaier RM, Sauerwein H (2004) Haptoglobin concentrations in blood and milk after endotoxin challenge and quantification of mammary Hp mRNA expression. J Dairy Sci 87(11):3778–3784. doi:10.3168/jds.S0022-0302(04)73516-X

Hiss S, Weinkauf C, Hachenberg S, Sauerwein H (2009) Short communication: Relationship between metabolic status and the milk concentrations of haptoglobin and lactoferrin in dairy cows during early lactation. J Dairy Sci 92(9):4439–4443. doi:10.3168/jds.2008-1632

Hogarth CJ, Fitzpatrick JL, Nolan AM, Young FJ, Pitt A, Eckersall PD (2004) Differential protein composition of bovine whey: a comparison of whey from healthy animals and from those with clinical mastitis. Proteomics 4(7):2094–2100. doi:10.1002/pmic.200300723

Holland JW, Deeth HC, Alewood PF (2006) Resolution and characterisation of multiple isoforms of bovine kappa-casein by 2-DE following a reversible cysteine-tagging enrichment strategy. Proteomics 6(10):3087–3095. doi:10.1002/pmic.200500780

Hovinen M, Siivonen J, Taponen S, Hanninen L, Pastell M, Aisla AM et al (2008) Detection of clinical mastitis with the help of a thermal camera. J Dairy Sci 91(12):4592–4598. doi:10.3168/jds.2008-1218

http://may2009.archive.ensembl.org/Bos_taurus/Info/Index. Ensembl genome browser 54: B.taurus—Description—Search Archive EnsEMBL Cow (2009)

Hu ZL, Park CA, Reecy JM (2016) Developmental progress and current status of the Animal QTLdb. Nucleic Acids Res 44(D1):D827–D833. doi:10.1093/nar/gkv1233

Huang J, Luo G, Zhang Z, Wang X, Ju Z, Qi C et al (2014) iTRAQ-proteomics and bioinformatics analyses of mammary tissue from cows with clinical mastitis due to natural infection with Staphylococci aureus. BMC Genomics 15:839. doi:10.1186/1471-2164-15-839

Ibeagha-Awemu EM, Ibeagha AE, Messier S, Zhao X (2010) Proteomics, genomics, and pathway analyses of Escherichia coli and Staphylococcus aureus infected milk whey reveal molecular pathways and networks involved in mastitis. J Proteome Res 9(9):4604–4619. doi:10.1021/pr100336e

Ibeagha-Awemu EM, Peters SO, Akwanji KA, Imumorin IG, Zhao X (2016) High density genome wide genotyping-by-sequencing and association identifies common and low frequency SNPs, and novel candidate genes influencing cow milk traits. Sci Rep 6:31109. doi:10.1038/srep31109

Jacobsen S, Niewold TA, Kornalijnslijper E, Toussaint MJ, Gruys E (2005) Kinetics of local and systemic isoforms of serum amyloid A in bovine mastitic milk. Vet Immunol Immunopathol 104(1-2):21–31. doi:10.1016/j.vetimm.2004.09.031

Janosi S, Ratz F, Szigeti G, Kulcsar M, Kerenyi J, Lauko T et al (2001) Review of the microbiological, pathological, and clinical aspects of bovine mastitis caused by the alga Prototheca zopfii. Vet Q 23(2):58–61. doi:10.1080/01652176.2001.9695082

Jensen LE, Whitehead AS (1998) Regulation of serum amyloid A protein expression during the acute-phase response. Biochem J 334(Pt 3):489–503

Kalmus P, Simojoki H, Pyorala S, Taponen S, Holopainen J, Orro T (2013) Milk haptoglobin, milk amyloid A, and N-acetyl-beta-D-glucosaminidase activity in bovines with naturally occurring clinical mastitis diagnosed with a quantitative PCR test. J Dairy Sci 96(6):3662–3670. doi:10.3168/jds.2012-6177

Kamal RM, Bayoumi MA, Abd El Aal SFA (2014) Correlation between some direct and indirect tests for screen detection of subclinical mastitis. Int Food Res J 21(3):1249–1254

Kaşikçi G, Çetin O, Bingöl EB, Gündüz MC (2012) Relations between electrical conductivity, somatic cell count, California mastitis test and some quality parameters in the diagnosis of subclinical mastitis in dairy cows. Turk J Vet Anim Sci 36(1):49–55. doi:10.3906/vet-1103-4

Kawai K, Hayashi T, Kiku Y, Chiba T, Nagahata H, Higuchi H et al (2013) Reliability in somatic cell count measurement of clinical mastitis milk using DeLaval cell counter. Anim Sci J 84(12):805–807. doi:10.1111/asj.12136

Kempf F, Slugocki C, Blum SE, Leitner G, Germon P (2016) Genomic comparative study of bovine mastitis Escherichia coli. PLoS One 11(1):e0147954. doi:10.1371/journal.pone.0147954

Kim Y, Atalla H, Mallard B, Robert C, Karrow N (2011) Changes in Holstein cow milk and serum proteins during intramammary infection with three different strains of Staphylococcus aureus. BMC Vet Res 7:51. doi:10.1186/1746-6148-7-51

Kim YG, Lone AM, Saghatelian A (2013) Analysis of the proteolysis of bioactive peptides using a peptidomics approach. Nat Protoc 8(9):1730–1742. doi:10.1038/nprot.2013.104

Kitchen BJ (1976) Enzymic methods for estimation of the somatic cell count in bovine milk. 1. Development of assay techniques and a study of their usefulness in evaluating the somatic cell content of milk. J Dairy Res 43(2):251–258

Kosciuczuk EM, Lisowski P, Jarczak J, Krzyzewski J, Zwierzchowski L, Bagnicka E (2014) Expression patterns of beta-defensin and cathelicidin genes in parenchyma of bovine mammary gland infected with coagulase-positive or coagulase-negative Staphylococci. BMC Vet Res 10:246. doi:10.1186/s12917-014-0246-z

Kovacevic-Filipovic M, Ilic V, Vujcic Z, Dojnov B, Stevanov-Pavlovic M, Mijacevic Z et al (2012) Serum amyloid A isoforms in serum and milk from cows with Staphylococcus aureus subclinical mastitis. Vet Immunol Immunopathol 145(1-2):120–128. doi:10.1016/j.vetimm.2011.10.015

Lai IH, Tsao JH, Lu YP, Lee JW, Zhao X, Chien FL et al (2009) Neutrophils as one of the major haptoglobin sources in mastitis affected milk. Vet Res 40(3):17. doi:10.1051/vetres:2008055

Larsen LB, Hinz K, Jorgensen AL, Moller HS, Wellnitz O, Bruckmaier RM et al (2010a) Proteomic and peptidomic study of proteolysis in quarter milk after infusion with lipoteichoic acid from Staphylococcus aureus. J Dairy Sci 93(12):5613–5626. doi:10.3168/jds.2010-3409

Larsen T, Aulrich K (2012) Optimizing the fluorometric beta-glucuronidase assay in ruminant milk for a more precise determination of mastitis. J Dairy Res 79(1):7–15. doi:10.1017/S0022029911000720

Larsen T, Rontved CM, Ingvartsen KL, Vels L, Bjerring M (2010b) Enzyme activity and acute phase proteins in milk utilized as indicators of acute clinical E. coli LPS-induced mastitis. Animal 4(10):1672–1679. doi:10.1017/S1751731110000947

Larson MA, Weber A, McDonald TL (2006) Bovine serum amyloid A3 gene structure and promoter analysis: induced transcriptional expression by bacterial components and the hormone prolactin. Gene 380(2):104–110. doi:10.1016/j.gene.2006.05.019

Lawless N, Foroushani AB, McCabe MS, O'Farrelly C, Lynn DJ (2013) Next generation sequencing reveals the expression of a unique miRNA profile in response to a gram-positive bacterial infection. PLoS One 8(3):e57543. doi:10.1371/journal.pone.0057543

Le Marechal C, Seyffert N, Jardin J, Hernandez D, Jan G, Rault L et al (2011) Molecular basis of virulence in Staphylococcus aureus mastitis. PLoS One 6(11):e27354. doi:10.1371/journal.pone.0027354

Lee JW, Bannerman DD, Paape MJ, Huang MK, Zhao X (2006) Characterization of cytokine expression in milk somatic cells during intramammary infections with Escherichia coli or Staphylococcus aureus by real-time PCR. Vet Res 37(2):219–229. doi:10.1051/vetres:2005051

Leigh JA, Ward PN, Field TR (2004) The exploitation of the genome in the search for determinants of virulence in Streptococcus uberis. Vet Immunol Immunopathol 100(3-4):145–149. doi:10.1016/j.vetimm.2004.04.004

Leitner G, Shoshani E, Krifucks O, Chaffer M, Saran A (2000) Milk leucocyte population patterns in bovine udder infection of different aetiology. J Vet Med B Infect Dis Vet Public Health 47(8):581–589

Leyva-Baca I, Schenkel F, Martin J, Karrow NA (2008) Polymorphisms in the 5′ upstream region of the CXCR1 chemokine receptor gene, and their association with somatic cell score in Holstein cattle in Canada. J Dairy Sci 91(1):407–417. doi:10.3168/jds.2007-0142

Lindsay JA (2014) Staphylococcus aureus genomics and the impact of horizontal gene transfer. Int J Med Microbiol 304(2):103–109. doi:10.1016/j.ijmm.2013.11.010

Lippolis JD, Brunelle BW, Reinhardt TA, Sacco RE, Nonnecke BJ, Dogan B et al (2014) Proteomic analysis reveals protein expression differences in Escherichia coli strains associated with persistent versus transient mastitis. J Proteomics 108:373–381. doi:10.1016/j.jprot.2014.06.008

Lippolis JD, Reinhardt TA (2010) Utility, limitations, and promise of proteomics in animal science. Vet Immunol Immunopathol 138(4):241–251. doi:10.1016/j.vetimm.2010.10.003

Liu Y, Qin X, Song XZ, Jiang H, Shen Y, Durbin KJ et al (2009) Bos taurus genome assembly. BMC Genomics 10:180. doi:10.1186/1471-2164-10-180

Lomborg SR, Nielsen LR, Heegaard PM, Jacobsen S (2008) Acute phase proteins in cattle after exposure to complex stress. Vet Res Commun 32(7):575–582. doi:10.1007/s11259-008-9057-7

Luther DA, Almeida RA, Oliver SP (2008) Elucidation of the DNA sequence of Streptococcus uberis adhesion molecule gene (sua) and detection of sua in strains of Streptococcus uberis isolated from geographically diverse locations. Vet Microbiol 128(3–4):304–312. doi:10.1016/j.vetmic.2007.10.015

Ma C, Zhao J, Xi X, Ding J, Wang H, Zhang H et al (2016) Bovine mastitis may be associated with the deprivation of gut Lactobacillus. Benef Microbes 7:95–102. doi:10.3920/BM2015.0048

Malek dos Reis CB, Barreiro JR, Mestieri L, Porcionato MA, Dos Santos MV (2013) Effect of somatic cell count and mastitis pathogens on milk composition in Gyr cows. BMC Vet Res 9:67. doi:10.1186/1746-6148-9-67

Mansor R, Mullen W, Albalat A, Zerefos P, Mischak H, Barrett DC et al (2013) A peptidomic approach to biomarker discovery for bovine mastitis. J Proteomics 85:89–98. doi:10.1016/j.jprot.2013.04.027

Massey KA, Nicolaou A (2013) Lipidomics of oxidized polyunsaturated fatty acids. Free Radic Biol Med 59:45–55. doi:10.1016/j.freeradbiomed.2012.08.565

Mavangira V, Gandy JC, Zhang C, Ryman VE, Daniel Jones A, Sordillo LM (2015) Polyunsaturated fatty acids influence differential biosynthesis of oxylipids and other lipid mediators during bovine coliform mastitis. J Dairy Sci 98(9):6202–6215. doi:10.3168/jds.2015-9570

McDonald TL, Larson MA, Mack DR, Weber A (2001) Elevated extrahepatic expression and secretion of mammary-associated serum amyloid A 3 (M-SAA3) into colostrum. Vet Immunol Immunopathol 83(3-4):203–211

Menschaert G, Vandekerckhove TT, Baggerman G, Schoofs L, Luyten W, Van Criekinge W (2010) Peptidomics coming of age: a review of contributions from a bioinformatics angle. J Proteome Res 9(5):2051–2061. doi:10.1021/pr900929m

Meredith BK, Berry DP, Kearney F, Finlay EK, Fahey AG, Bradley DG et al (2013) A genome-wide association study for somatic cell score using the Illumina high-density bovine beadchip identifies several novel QTL potentially related to mastitis susceptibility. Front Genet 4:229. doi:10.3389/fgene.2013.00229

Meredith BK, Kearney FJ, Finlay EK, Bradley DG, Fahey AG, Berry DP et al (2012) Genome-wide associations for milk production and somatic cell score in Holstein-Friesian cattle in Ireland. BMC Genet 13:21. doi:10.1186/1471-2156-13-21

Metzner M, Sauter-Louis C, Seemueller A, Petzl W, Klee W (2014) Infrared thermography of the udder surface of dairy cattle: characteristics, methods, and correlation with rectal temperature. Vet J 199(1):57–62. doi:10.1016/j.tvjl.2013.10.030

Miglio A, Moscati L, Fruganti G, Pela M, Scoccia E, Valiani A et al (2013) Use of milk amyloid A in the diagnosis of subclinical mastitis in dairy ewes. J Dairy Res 80(4):496–502. doi:10.1017/S0022029913000484

Milner P, Page KL, Walton AW, Hillerton JE (1996) Detection of clinical mastitis by changes in electrical conductivity of foremilk before visible changes in milk. J Dairy Sci 79(1):83–86. doi:10.3168/jds.S0022-0302(96)76337-3

Mogensen TH (2009) Pathogen recognition and inflammatory signaling in innate immune defenses. Clin Microbiol Rev 22(2):240–273. doi:10.1128/CMR.00046-08

Molenaar AJ, Harris DP, Rajan GH, Pearson ML, Callaghan MR, Sommer L et al (2009) The acute-phase protein serum amyloid A3 is expressed in the bovine mammary gland and plays a role in host defence. Biomarkers 14(1):26–37. doi:10.1080/13547500902730714

Moyes KM, Drackley JK, Morin DE, Bionaz M, Rodriguez-Zas SL, Everts RE et al (2009) Gene network and pathway analysis of bovine mammary tissue challenged with Streptococcus uberis reveals induction of cell proliferation and inhibition of PPARgamma signaling as potential mechanism for the negative relationships between immune response and lipid metabolism. BMC Genomics 10:542. doi:10.1186/1471-2164-10-542

Moyes KM, Sorensen P, Bionaz M (2016) The impact of intramammary Escherichia coli challenge on liver and mammary transcriptome and cross-talk in dairy cows during early lactation using RNAseq. PLoS One 11(6):e0157480. doi:10.1371/journal.pone.0157480

Mudaliar M, Tassi R, Thomas FC, McNeilly TN, Weidt SK, McLaughlin M et al (2016) Mastitomics, the integrated omics of bovine milk in an experimental model of Streptococcus uberis mastitis: 2. Label-free relative quantitative proteomics. Mol Biosyst 12(9):2748–2761. doi:10.1039/c6mb00290k

Nagahata H, Saito S, Noda H (1987) Changes in N-acetyl-B-D-glucosaminidase and B-glucuronidase activities in milk during bovine mastitis. Can J Vet Res 51(1):126–134

Napoli A, Aiello D, Di Donna L, Prendushi H, Sindona G (2007) Exploitation of endogenous protease activity in raw mastitic milk by MALDI-TOF/TOF. Anal Chem 79(15):5941–5948. doi:10.1021/ac0704863

Nicholas RA (2011) Bovine mycoplasmosis: silent and deadly. Vet Rec 168(17):459–462. doi:10.1136/vr.d2468

Nielsen BH, Jacobsen S, Andersen PH, Niewold TA, Heegaard PM (2004) Acute phase protein concentrations in serum and milk from healthy cows, cows with clinical mastitis and cows with extramammary inflammatory conditions. Vet Rec 154(12):361–365

Nissen A, Bendixen E, Ingvartsen KL, Rontved CM (2013) Expanding the bovine milk proteome through extensive fractionation. J Dairy Sci 96(12):7854–7866. doi:10.3168/jds.2013-7106

Notcovich S, de Nicolo G, Williamson NB, Grinberg A, Lopez-Villalobos N, Petrovski KR (2016) The ability of four strains of Streptococcus uberis to induce clinical mastitis after intramammary inoculation in lactating cows. N Z Vet J 64(4):218–223. doi:10.1080/00480169.2016.1157049

O'Mahony MC, Healy AM, Harte D, Walshe KG, Torgerson PR, Doherty ML (2006) Milk amyloid A: correlation with cellular indices of mammary inflammation in cows with normal and raised serum amyloid A. Res Vet Sci 80(2):155–161. doi:10.1016/j.rvsc.2005.05.005

Oikonomou G, Bicalho ML, Meira E, Rossi RE, Foditsch C, Machado VS et al (2014) Microbiota of cow's milk; distinguishing healthy, sub-clinically and clinically diseased quarters. PLoS One 9(1):e85904. doi:10.1371/journal.pone.0085904

Oikonomou G, Machado VS, Santisteban C, Schukken YH, Bicalho RC (2012) Microbial diversity of bovine mastitic milk as described by pyrosequencing of metagenomic 16s rDNA. PLoS One 7(10):e47671. doi:10.1371/journal.pone.0047671

Olsson T, Sandstedt K, Holmberg O, Thore A (1986) Extraction and determination of adenosine 5′-triphosphate in bovine milk by the firefly luciferase assay. Biotechnol Appl Biochem 8(5):361–369

Opsal MA, Lien S, Brenna-Hansen S, Olsen HG, Vage DI (2008) Association analysis of the constructed linkage maps covering TLR2 and TLR4 with clinical mastitis in Norwegian Red cattle. J Anim Breed Genet 125(2):110–118. doi:10.1111/j.1439-0388.2007.00704.x

Osorio MT, Moloney AP, Brennan L, Monahan FJ (2012) Authentication of beef production systems using a metabolomic-based approach. Animal 6(1):167–172. doi:10.1017/S1751731111001418

Pachauri S, Varshney P, Dash S, Gupta M (2013) Involvement of fungal species in bovine mastitis in and around Mathura, India. Vet World 6(7):393. doi:10.5455/vetworld.2013.393-395

Pawlik A, Sender G, Kapera M, Korwin-Kossakowska A (2015) Association between interleukin 8 receptor alpha gene (CXCR1) and mastitis in dairy cattle. Cent Eur J Immunol 40(2):153–158. doi:10.5114/ceji.2015.52828

Pedersen LH, Aalbaek B, Rontved CM, Ingvartsen KL, Sorensen NS, Heegaard PM et al (2003) Early pathogenesis and inflammatory response in experimental bovine mastitis due to Streptococcus uberis. J Comp Pathol 128(2-3):156–164

Pepe G, Tenore GC, Mastrocinque R, Stusio P, Campiglia P (2013) Potential anticarcinogenic peptides from bovine milk. J Amino Acids 2013:939804. doi:10.1155/2013/939804

Petzl W, Gunther J, Muhlbauer K, Seyfert HM, Schuberth HJ, Hussen J et al (2016) Early transcriptional events in the udder and teat after intra-mammary Escherichia coli and Staphylococcus aureus challenge. Innate Immun 22(4):294–304. doi:10.1177/1753425916640057

Pezeshki A, Stordeur P, Wallemacq H, Schynts F, Stevens M, Boutet P et al (2011) Variation of inflammatory dynamics and mediators in primiparous cows after intramammary challenge with Escherichia coli. Vet Res 42:15. doi:10.1186/1297-9716-42-15

Pighetti GM, Kojima CJ, Wojakiewicz L, Rambeaud M (2012) The bovine CXCR1 gene is highly polymorphic. Vet Immunol Immunopathol 145(1-2):464–470. doi:10.1016/j.vetimm.2011.09.012

Polat B, Colak A, Cengiz M, Yanmaz LE, Oral H, Bastan A et al (2010) Sensitivity and specificity of infrared thermography in detection of subclinical mastitis in dairy cows. J Dairy Sci 93(8):3525–3532. doi:10.3168/jds.2009-2807

Pongthaisong P, Katawatin S, Thamrongyoswittayakul C, Roytrakul S (2016) Milk protein profiles in response to Streptococcus agalactiae subclinical mastitis in dairy cows. Anim Sci J 87(1):92–98. doi:10.1111/asj.12391

Pyorala S (2003) Indicators of inflammation in the diagnosis of mastitis. Vet Res 34(5):565–578. doi:10.1051/vetres:2003026

Pyorala S, Hovinen M, Simojoki H, Fitzpatrick J, Eckersall PD, Orro T (2011) Acute phase proteins in milk in naturally acquired bovine mastitis caused by different pathogens. Vet Rec 168(20):535. doi:10.1136/vr.d1120

Qiao J, Kwok L, Zhang J, Gao P, Zheng Y, Guo Z et al (2015) Reduction of Lactobacillus in the milks of cows with subclinical mastitis. Benef Microbes 6(4):485–490. doi:10.3920/BM2014.0077

Quigley L, O'Sullivan O, Stanton C, Beresford TP, Ross RP, Fitzgerald GF et al (2013) The complex microbiota of raw milk. FEMS Microbiol Rev 37(5):664–698. doi:10.1111/1574-6976.12030

Rainard P, Riollet C (2006) Innate immunity of the bovine mammary gland. Vet Res 37(3):369–400. doi:10.1051/vetres:2006007

Reese JT, Childers CP, Sundaram JP, Dickens CM, Childs KL, Vile DC et al (2010) Bovine Genome Database: supporting community annotation and analysis of the Bos taurus genome. BMC Genomics 11:645. doi:10.1186/1471-2164-11-645

Regal P, Anizan S, Antignac JP, Le Bizec B, Cepeda A, Fente C (2011) Metabolomic approach based on liquid chromatography coupled to high resolution mass spectrometry to screen for the illegal use of estradiol and progesterone in cattle. Anal Chim Acta 700(1-2):16-25. doi:10.1016/j.aca.2011.01.005

Reinhardt TA, Lippolis JD (2006) Bovine milk fat globule membrane proteome. J Dairy Res 73(4):406–416. doi:10.1017/S0022029906001889

Reinhardt TA, Sacco RE, Nonnecke BJ, Lippolis JD (2013) Bovine milk proteome: quantitative changes in normal milk exosomes, milk fat globule membranes and whey proteomes resulting from Staphylococcus aureus mastitis. J Proteomics 82:141–154. doi:10.1016/j.jprot.2013.02.013

Reyher KK, Dohoo IR, Scholl DT, Keefe GP (2012a) Evaluation of minor pathogen intramammary infection, susceptibility parameters, and somatic cell counts on the development of new intramammary infections with major mastitis pathogens. J Dairy Sci 95(7):3766–3780. doi:10.3168/jds.2011-5148

Reyher KK, Haine D, Dohoo IR, Revie CW (2012b) Examining the effect of intramammary infections with minor mastitis pathogens on the acquisition of new intramammary infections with major mastitis pathogens—a systematic review and meta-analysis. J Dairy Sci 95(11):6483–6502. doi:10.3168/jds.2012-5594

Rijk JC, Lommen A, Essers ML, Groot MJ, Van Hende JM, Doeswijk TG et al (2009) Metabolomics approach to anabolic steroid urine profiling of bovines treated with prohormones. Anal Chem 81(16):6879–6888. doi:10.1021/ac900874m

Roncada P, Piras C, Soggiu A, Turk R, Urbani A, Bonizzi L (2012) Farm animal milk proteomics. J Proteomics 75(14):4259–4274. doi:10.1016/j.jprot.2012.05.028

Rupp R, Boichard D (2003) Genetics of resistance to mastitis in dairy cattle. Vet Res 34(5):671–688. doi:10.1051/vetres:2003020

Russell CD, Widdison S, Leigh JA, Coffey TJ (2012) Identification of single nucleotide polymorphisms in the bovine Toll-like receptor 1 gene and association with health traits in cattle. Vet Res 43(1):17. doi:10.1186/1297-9716-43-17

Ryman VE, Pighetti GM, Lippolis JD, Gandy JC, Applegate CM, Sordillo LM (2015) Quantification of bovine oxylipids during intramammary Streptococcus uberis infection. Prostaglandins Other Lipid Mediat 121(Pt B):207–217. doi:10.1016/j.prostaglandins.2015.09.006

Sahana G, Guldbrandtsen B, Thomsen B, Holm LE, Panitz F, Brondum RF et al (2014) Genome-wide association study using high-density single nucleotide polymorphism arrays and whole-genome sequences for clinical mastitis traits in dairy cattle. J Dairy Sci 97(11):7258–7275. doi:10.3168/jds.2014-8141

Saleem F, Bouatra S, Guo AC, Psychogios N, Mandal R, Dunn SM et al (2012) The bovine ruminal fluid metabolome. Metabolomics 9(2):360–378. doi:10.1007/s11306-012-0458-9

Sargeant JM, Leslie KE, Shirley JE, Pulkrabek BJ, Lim GH (2001) Sensitivity and specificity of somatic cell count and California Mastitis Test for identifying intramammary infection in early lactation. J Dairy Sci 84(9):2018–2024. doi:10.3168/jds.S0022-0302(01)74645-0

Savijoki K, Iivanainen A, Siljamaki P, Laine PK, Paulin L, Karonen T et al (2014) Genomics and proteomics provide new insight into the commensal and pathogenic lifestyles of bovine- and human-associated Staphylococcus epidermidis strains. J Proteome Res. doi:10.1021/pr500322d

Schukken Y, Chuff M, Moroni P, Gurjar A, Santisteban C, Welcome F et al (2012) The "other" gram-negative bacteria in mastitis: Klebsiella, serratia, and more. Vet Clin North Am Food Anim Pract 28(2):239–256. doi:10.1016/j.cvfa.2012.04.001

Seker I, Risvanli A, Yuksel M, Saat N, Ozmen O (2009) Relationship between California Mastitis Test score and ultrasonographic teat measurements in dairy cows. Aust Vet J 87(12):480–483. doi:10.1111/j.1751-0813.2009.00520.x

Sharma A, Lee JS, Dang CG, Sudrajad P, Kim HC, Yeon SH et al (2015) Stories and challenges of genome wide association studies in Livestock—a review. Asian-Australas J Anim Sci 28(10):1371–1379. doi:10.5713/ajas.14.0715

Sharma RS, Misra DS (1966) Milk lactose and its importance in diagnosis of mastitis. Indian Vet J 43(2):154–159

Smolenski G, Haines S, Kwan FY, Bond J, Farr V, Davis SR et al (2007) Characterisation of host defence proteins in milk using a proteomic approach. J Proteome Res 6(1):207–215. doi:10.1021/pr0603405

Spittel S, Hoedemaker M (2012) Mastitis diagnosis in dairy cows using PathoProof real-time polymerase chain reaction assay in comparison with conventional bacterial culture in a Northern German field study. Berl Munch Tierarztl Wochenschr 125(11-12):494–502

Stables MJ, Gilroy DW (2011) Old and new generation lipid mediators in acute inflammation and resolution. Prog Lipid Res 50(1):35–51. doi:10.1016/j.plipres.2010.07.005

Sun J, Aswath K, Schroeder SG, Lippolis JD, Reinhardt TA, Sonstegard TS (2015) MicroRNA expression profiles of bovine milk exosomes in response to Staphylococcus aureus infection. BMC Genomics 16:806. doi:10.1186/s12864-015-2044-9

Sundekilde UK, Larsen LB, Bertram HC (2013a) NMR-based milk metabolomics. Metabolites 3(2):204–222. doi:10.3390/metabo3020204

Sundekilde UK, Poulsen NA, Larsen LB, Bertram HC (2013b) Nuclear magnetic resonance metabonomics reveals strong association between milk metabolites and somatic cell count in bovine milk. J Dairy Sci 96(1):290–299. doi:10.3168/Jds.2012-5819

Suojala L, Orro T, Jarvinen H, Saatsi J, Pyorala S (2008) Acute phase response in two consecutive experimentally induced E coli intramammary infections in dairy cows. Acta Vet Scand 50:18. doi:10.1186/1751-0147-50-18

Swanson KM, Stelwagen K, Dobson J, Henderson HV, Davis SR, Farr VC et al (2009) Transcriptome profiling of Streptococcus uberis-induced mastitis reveals fundamental differences between immune gene expression in the mammary gland and in a primary cell culture model. J Dairy Sci 92(1):117–129. doi:10.3168/jds.2008-1382

Szczubial M, Dabrowski R, Kankofer M, Bochniarz M, Komar M (2012) Concentration of serum amyloid A and ceruloplasmin activity in milk from cows with subclinical mastitis caused by different pathogens. Pol J Vet Sci 15(2):291–296

Takahashi E, Kuwayama H, Kawamoto K, Matsui T, Inokuma H (2009) Detection of serum amyloid A isoforms in cattle. J Vet Diagn Invest 21(6):874–877

Takeda K, Akira S (2005) Toll-like receptors in innate immunity. Int Immunol 17(1):1–14. doi:10.1093/intimm/dxh186

Tassi R, McNeilly TN, Fitzpatrick JL, Fontaine MC, Reddick D, Ramage C et al (2013) Strain-specific pathogenicity of putative host-adapted and nonadapted strains of Streptococcus uberis in dairy cattle. J Dairy Sci 96(8):5129–5145. doi:10.3168/jds.2013-6741

Tassi R, McNeilly TN, Sipka A, Zadoks RN (2015) Correlation of hypothetical virulence traits of two Streptococcus uberis strains with the clinical manifestation of bovine mastitis. Vet Res 46:123. doi:10.1186/s13567-015-0268-y

Tellam RL, Lemay DG, Van Tassell CP, Lewin HA, Worley KC, Elsik CG (2009) Unlocking the bovine genome. BMC Genomics 10:193. doi:10.1186/1471-2164-10-193

Thielen MA, Mielenz M, Hiss S, Zerbe H, Petzl W, Schuberth HJ et al (2007) Short communication: cellular localization of haptoglobin mRNA in the experimentally infected bovine mammary gland. J Dairy Sci 90(3):1215–1219. doi:10.3168/jds.S0022-0302(07)71609-0

Thomas FC, Mudaliar M, Tassi R, McNeilly TN, Burchmore R, Burgess K et al (2016a) Mastitomics, the integrated omics of bovine milk in an experimental model of Streptococcus uberis mastitis: 3. Untargeted metabolomics. Mol Biosyst 12(9):2762–2769. doi:10.1039/c6mb00289g

Thomas FC, Mullen W, Tassi R, Ramirez-Torres A, Mudaliar M, McNeilly TN et al (2016b) Mastitomics, the integrated omics of bovine milk in an experimental model of Streptococcus uberis mastitis: 1. High abundance proteins, acute phase proteins and peptidomics. Mol Biosyst 12(9):2735–2747. doi:10.1039/c6mb00239k

Thomas FC, Santana AM, Waterston M, Haining H, Eckersall PD (2016c) Effect of pre-analytical treatments on bovine milk acute phase proteins. BMC Vet Res 12(1):151. doi:10.1186/s12917-016-0769-6

Thomas FC, Waterston M, Hastie P, Haining H, Eckersall PD (2016d) Early post parturient changes in milk acute phase proteins. J Dairy Res 83(3):352–359. doi:10.1017/S0022029916000297

Thompson-Crispi KA, Sargolzaei M, Ventura R, Abo-Ismail M, Miglior F, Schenkel F et al (2014) A genome-wide association study of immune response traits in Canadian Holstein cattle. BMC Genomics 15:559. doi:10.1186/1471-2164-15-559

Tiezzi F, Parker-Gaddis KL, Cole JB, Clay JS, Maltecca C (2015) A genome-wide association study for clinical mastitis in first parity US Holstein cows using single-step approach and genomic matrix re-weighting procedure. PLoS One 10(2):e0114919. doi:10.1371/journal.pone.0114919

Tomasinsig L, De Conti G, Skerlavaj B, Piccinini R, Mazzilli M, D'Este F et al (2010) Broad-spectrum activity against bacterial mastitis pathogens and activation of mammary epithelial cells support a protective role of neutrophil cathelicidins in bovine mastitis. Infect Immun 78(4):1781–1788. doi:10.1128/IAI.01090-09

Tomasinsig L, Skerlavaj B, Scarsini M, Guida F, Piccinini R, Tossi A et al (2012) Comparative activity and mechanism of action of three types of bovine antimicrobial peptides against pathogenic Prototheca spp. J Pept Sci 18(2):105–113. doi:10.1002/psc.1422

Turk R, Piras C, Kovacic M, Samardzija M, Ahmed H, De Canio M et al (2012) Proteomics of inflammatory and oxidative stress response in cows with subclinical and clinical mastitis. J Proteomics 75(14):4412–4428. doi:10.1016/j.jprot.2012.05.021

Uhlar CM, Whitehead AS (1999) Serum amyloid A, the major vertebrate acute-phase reactant. Eur J Biochem 265(2):501–523

Ursell LK, Metcalf JL, Parfrey LW, Knight R (2012) Defining the human microbiome. Nutr Rev 70(Suppl 1):S38–S44. doi:10.1111/j.1753-4887.2012.00493.x

Viguier C, Arora S, Gilmartin N, Welbeck K, O'Kennedy R (2009) Mastitis detection: current trends and future perspectives. Trends Biotechnol 27(8):486–493. doi:10.1016/j.tibtech.2009.05.004

Waldmann P, Meszaros G, Gredler B, Fuerst C, Solkner J (2013) Evaluation of the lasso and the elastic net in genome-wide association studies. Front Genet 4:270. doi:10.3389/fgene.2013.00270

Wang X, Ma P, Liu J, Zhang Q, Zhang Y, Ding X et al (2015) Genome-wide association study in Chinese Holstein cows reveal two candidate genes for somatic cell score as an indicator for mastitis susceptibility. BMC Genet 16:111. doi:10.1186/s12863-015-0263-3

Wang XG, Ju ZH, Hou MH, Jiang Q, Yang CH, Zhang Y et al (2016) Deciphering transcriptome and complex alternative splicing transcripts in mammary gland tissues from cows naturally infected with Staphylococcus aureus mastitis. PLoS One 11(7):e0159719. doi:10.1371/journal.pone.0159719

Ward PN, Field TR, Ditcham WG, Maguin E, Leigh JA (2001) Identification and disruption of two discrete loci encoding hyaluronic acid capsule biosynthesis genes hasA, hasB, and hasC in Streptococcus uberis. Infect Immun 69(1):392–399. doi:10.1128/IAI.69.1.392-399.2001

Ward PN, Holden MT, Leigh JA, Lennard N, Bignell A, Barron A et al (2009) Evidence for niche adaptation in the genome of the bovine pathogen Streptococcus uberis. BMC Genomics 10:54. doi:10.1186/1471-2164-10-54

Watson CJ, Kreuzaler PA (2011) Remodeling mechanisms of the mammary gland during involution. Int J Dev Biol 55(7–9):757–762. doi:10.1387/ijdb.113414cw

Wellenberg GJ, van der Poel WH, Van Oirschot JT (2002) Viral infections and bovine mastitis: a review. Vet Microbiol 88(1):27–45

Wenz JR, Fox LK, Muller FJ, Rinaldi M, Zeng R, Bannerman DD (2010) Factors associated with concentrations of select cytokine and acute phase proteins in dairy cows with naturally occurring clinical mastitis. J Dairy Sci 93(6):2458–2470. doi:10.3168/jds.2009-2819

Whelehan CJ, Meade KG, Eckersall PD, Young FJ, O'Farrelly C (2011) Experimental Staphylococcus aureus infection of the mammary gland induces region-specific changes in innate immune gene expression. Vet Immunol Immunopathol 140(3-4):181–189. doi:10.1016/j.vetimm.2010.11.013

Wijga S, Bastiaansen JW, Wall E, Strandberg E, de Haas Y, Giblin L et al (2012) Genomic associations with somatic cell score in first-lactation Holstein cows. J Dairy Sci 95(2):899–908. doi:10.3168/jds.2011-4717

Winter P, Miny M, Fuchs K, Baumgartner W (2006) The potential of measuring serum amyloid A in individual ewe milk and in farm bulk milk for monitoring udder health on sheep dairy farms. Res Vet Sci 81(3):321–326. doi:10.1016/j.rvsc.2006.03.003

http://www.animalgenome.org/. Cattle QTL Database. 2016. http://www.animalgenome.org/cgi-bin/QTLdb/BT/index.

http://www.genome.gov/12512874. Bovine Genome Assembled. 2013 National Institutes of Health (NIH). http://www.genome.gov/12512874. Accessed 10/10/2013

Yamada T (1999) Serum amyloid A (SAA): a concise review of biology, assay methods and clinical usefulness. Clin Chem Lab Med 37(4):381–388. doi:10.1515/CCLM.1999.063

Yang YX, Zhao XX, Zhang Y (2009) Proteomic analysis of mammary tissues from healthy cows and clinical mastitic cows for identification of disease-related proteins. Vet Res Commun 33(4):295–303. doi:10.1007/s11259-008-9177-0

Youngerman SM, Saxton AM, Oliver SP, Pighetti GM (2004) Association of CXCR2 polymorphisms with subclinical and clinical mastitis in dairy cattle. J Dairy Sci 87(8):2442–2448. doi:10.3168/jds.S0022-0302(04)73367-6

Younis S, Javed Q, Blumenberg M (2016) Meta-analysis of transcriptional responses to mastitis-causing Escherichia coli. PLoS One 11(3):e0148562. doi:10.1371/journal.pone.0148562

Zadoks R, Fitzpatrick J (2009) Changing trends in mastitis. Irish Vet J 62(Suppl 4):S59–S70. doi:10.1186/2046-0481-62-S4-S59

Zadoks RN, Middleton JR, McDougall S, Katholm J, Schukken YH (2011) Molecular epidemiology of mastitis pathogens of dairy cattle and comparative relevance to humans. J Mammary Gland Biol Neoplasia 16(4):357–372. doi:10.1007/s10911-011-9236-y

Zhang H, Wu L, Xu C, Xia C, Sun L, Shu S (2013) Plasma metabolomic profiling of dairy cows affected with ketosis using gas chromatography/mass spectrometry. BMC Vet Res 9:186. doi:10.1186/1746-6148-9-186

Zhao X, Lacasse P (2008) Mammary tissue damage during bovine mastitis: causes and control. J Anim Sci 86(13 Suppl):57–65. doi:10.2527/jas.2007-0302

Zhou L, Wang HM, Ju ZH, Zhang Y, Huang JM, Qi C et al (2013) Association of novel single nucleotide polymorphisms of the CXCR1 gene with the milk performance traits of Chinese native cattle. Genet Mol Res 12(3):2725–2739. doi:10.4238/2013.July.30.10

Zimin AV, Delcher AL, Florea L, Kelley DR, Schatz MC, Puiu D et al (2009) A whole-genome assembly of the domestic cow, Bos taurus. Genome Biol 10(4):R42. doi:10.1186/gb-2009-10-4-r42

Laminitis: A Multisystems Veterinary Perspective with *Omics* Technologies

9

Richard R.E. Uwiera, Ashley F. Egyedy, and Burim N. Ametaj

Abstract

Laminitis and lameness-associated diseases in dairy cattle have plagued the dairy industry for decades. It is a costly illness with significant impact on dairy production and the animal's health and welfare. Laminitis in dairy cattle is complex and the initiation, progression, and severity of disease is affected by multiple factors including the individual animal's health, heritable traits, and its environment. As such, understanding the pathophysiology of bovine laminitis in dairy cattle requires research that explores both local events (i.e., histopathology alterations and mechanical injury) directly affecting the hoof and indirect pathophysiology perturbations (i.e., hyperinsulinemia) distant from the hoof. This provides a more generalized understanding of multiple physiologic systems—a systems veterinary approach. Presently, there is a relative paucity of information investigating the induction and progression of this disease in dairy cattle. Incorporating new *omics* technologies into the study of laminitis in cattle should shed light on the disease processes. Indeed, *omics* sciences, namely genomics, transcriptomics, proteomics, and metabolomics demonstrate that laminitis is a collection of complex changes in gene expression, protein translation, and the metabolism of components involved in the inflammatory process. Most certainly bovine laminitis is associated with increased production of pro-inflammatory cytokines, matrix metalloproteinases, and the metabolism of amino acids, carbohydrates, lipids, and energy producing molecules. As research advances, a comprehensive understanding of the pathophysiology of laminitis in dairy cattle will ensue and this could translate into better animal husbandry practices for the dairy industry and treatment modalities for mitigating the disease.

R.R.E. Uwiera, D.V.M., Ph.D. (✉) • A.F. Egyedy, M.Sc. • B.N. Ametaj, D.V.M., Ph.D., Ph.D.
Department of Agricultural, Food and Nutritional Science, University of Alberta, Edmonton, Alberta T6R 2J6, Canada
e-mail: richard.uwiera@ualberta.ca; egyedy@ualberta.ca; bametaj@ualberta.ca

B.N. Ametaj (ed.), *Periparturient Diseases of Dairy Cows*,
DOI 10.1007/978-3-319-43033-1_9

9.1 Introduction

The hoof (claw) is a specialized structure in cattle enabling the animal to support its weight and provide unimpeded movement. In order for cattle of any species to function, they must be able to support weight and to transition from one location to another, an important aspect for the animals' natural behavior, and a necessary quality for good animal welfare and health. Certainly the ability of animals to move freely is one tenant of the five freedoms of animal welfare, and free movement is a desired trait in the dairy industry as it translates into optimal animal production (Webster 1997). Dairy cattle are required to obtain feed from the feed bunk or pasture, transition from stanchions to the milking parlor, and in most cases stand for calving; all of these situations require sound supporting structure such as good hooves. That said, any process that impairs the ability of dairy cattle to stand or move (i.e., lameness) adversely affects productivity and lameness is a major cause of reduced productivity in the dairy industry. Lameness is pervasive with individual dairy farms experiencing nearly 25% of cattle presenting with an incidence of lameness within any given year (Whitaker et al. 2000). Lameness also increases culling within a herd, with cull rates in some herds reaching 6%. Lameness in dairy cattle also significantly impacts profitability with a subsequent inflation-adjusted reduction in profit by over $238 USD per animal (i.e., dairy cattle with sole ulcers; Cha et al. 2010). Clinical manifestations of lameness in cattle can vary markedly (Nordlund et al. 2004). Mild forms of lameness include slow walking cattle with shorter gaits while severe forms of lameness in dairy cattle can be exhibited as gaunt animals that partially weight bear with pronounced mobility difficulties. Importantly, 90% of all lameness in dairy cattle is associated with abnormalities of the hoof and the majority of these hoof-associated changes are linked to laminitis.

9.2 A Brief History of Lameness and Laminitis Throughout the Ages

Lameness in cattle is not unique and nor an exclusive ailment of the contemporary dairy industry, as it has plagued hooven animal species throughout history and has permeated all enterprises that employ farmed animals. In 3500 BC, Sumerian writers describe cattle doctors treating disorders in resident cattle (Dunlop and Williams 1996), and in 1350 BC Hittites described lameness in "beasts" being linked to overexercise and alterations in feed and watering practices. Lameness and its treatment in cattle was also described in agricultural treatises by Roman agricultural writers such as (Columella 2007) (55 AD) in the first century:

> If it's descended down to the feet, blood causes lameness. When this happens, the hoof is immediately inspected. When touched, it [seemingly the blood] presents heat, and the ox doesn't allow the affected part to be handled too roughly. But if the blood is still in the leg above the hooves, it's dispelled through constant rubbing. If this does no good, it's removed through scraping. On the other hand, if it's now in the hooves, you'll make a light opening between the toes with a knife. Afterwards, bandages moistened with salty vinegar are

applied, and the foot is shod in a broom sandal. Particular care should be taken to keep the ox from putting its foot into water and to make sure that it stables in a dry manner. If this blood is not expelled, it will create a swelling, and if this festers, the treatment will come too late.

Understanding of lameness and laminitis remained a mystery throughout the early Middle Ages and the treatments associated with the disease were still ineffective and not necessarily beneficial to the afflicted animal. For example, if bleeding to expel the "evil humors" was unsuccessful for treating lameness in livestock then castration was recommended as it was recognized that people with gout with corresponding difficulties in walking occurred less in eunuchs. Other less than desirable treatments of lameness and laminitis, including denuding of the sole of the hoof, persisted for centuries, and this practice was thankfully, discontinued in the nineteenth century under the direction of the Society for the Prevention of Cruelty to Animals (Smith 1976). Generally, the Middle Ages demonstrated a lack in effective treatments of laminitis, but changes in animal husbandry practices were evident in Western Europe and this may have improved livestock heath. Indeed, as written by Guibert de Nogent, the poor were shoeing "their oxen as though they were horses" (Severin 1989). Treatment and prevention of lameness and laminitis began to improve into the modern era, evidenced by improved shoeing practices, proper diets in concert with breeding, and employing animals for work that were more resistant to disease. This practice was emphasized by the US army quartermaster general in the nineteenth century, as they preferred mules and perhaps oxen to heavy military horses as these draft animals did not "overeat or overdrink" and had a significant reduction in the incidence of laminitis, a particularly desirable trait for the US cavalry (Kauffman 1996). During this period farmed animals were encouraged to tread on softer ground, and were provided pain control, as described with the eating "husks of the poppy seeds" to improve recovery. It was not until the late twentieth century (i.e., post-1980s) that medical treatment of lameness and laminitis became accepted and based on scientific evidence. During this time there was increased targeted use of anti-histamines and other compounds to reduce inflammation (Takahashi and Young 1981), a treatment that alleviated clinical disease, improved recovery time, and reduced the need to cull affected cattle.

9.3 Hoof Anatomy and Weight Transfer in Cattle

As mentioned, the ability to move freely is an essential requirement not only for the well-being of the cow but also for good animal production. The need for large animals to transfer concussive forces of heavy weight onto the ground requires adaptations present within the bovine hoof. The bovine claw acts as the intermediary between the lower limb and the surrounding environment, the ground. Although the bovine claw may not be as specialized as the horse hoof for movement and athletic performance, and it lacks specialized anatomical features namely bars, frogs, and secondary lamellae, the bovine claw still provides effective protection of the most distal structures of the bovine limb (Shively 1984). Anatomically, the bovine claw could be divided into the internal structures of the claw and the hoof capsule (Fig. 9.1). The internal support structures include the bone, joints, ligaments, and tendons. The third phalanx (pedal bone), distal aspect of the second phalanx and the distal sesamoid bone lie within the claw and are held together by interdigitating collateral, cruciate, and distal sesamoid ligaments and corresponding joint capsules. The pedal bone is also attached to the deep digital flexor tendon, and deep digital extensor tendon and these tendons are required for flexion and extension of the claw, respectively. The internal structures of the claw are in general surrounded by a collagen matrix and in the distal aspect of the claw three pads of collagen and fat (digital cushion) lie underneath the pedal bone adjacent to the flexor tendon (Räber et al. 2004). Their internal structures are also well innervated by digital nerves and nourished by blood from axial and abaxial digital arteries (Sisson et al. 1975). The hoof capsules are composed of specialized cornified structures categorized into four anatomic regions (Mason 2012). The perioplic segment is the most cranial aspect of the dorsal surface of the hoof capsule and interfaces with the dermis of the lower leg. The coronary segment is the largest region of the hoof capsule composed of the hard, thick horn wall and is the primary structure to protect the crucial internal structures of the claw. This area is also important for transferring a portion of the weight bearing forces to the ground. The white line connects the capsule and the

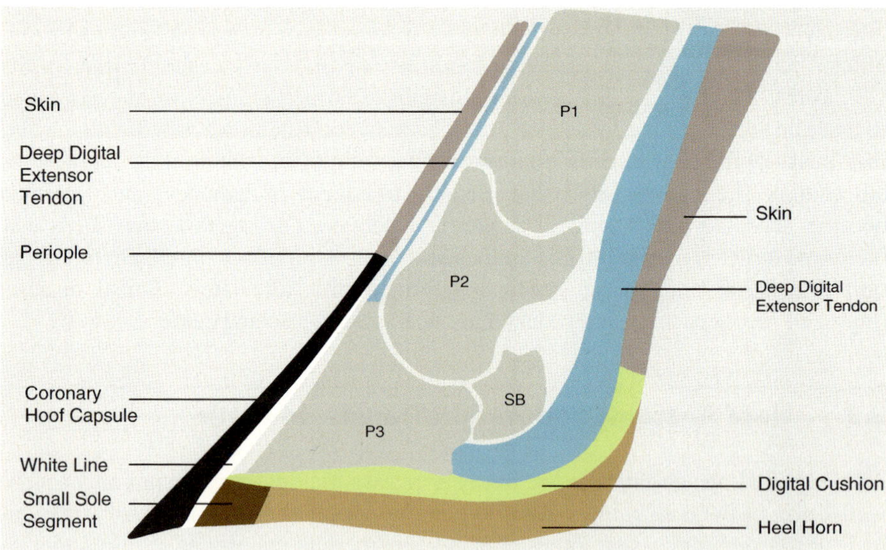

Fig. 9.1 The Bovine Hoof. The illustration identifies important anatomical structures of the bovine hoof. These include skin, deep digital flexor tendon, periople at the interface between the skin and capsule of the hoof, the coronary aspect of the hoof capsule, white line, firm small sole segment, softer heel horn, digital cushion, deep digital flexor tendon, and the internal bones of the hoof. P1, P2, P3, and SB represent the first phalanx, second phalanx, third phalanx, and the sesamoid bone, respectively

sole and interdigitates the laminae horn and laminae corium and is an important suspensory structure of the hoof capsule, pedal bone, and surrounding collagen. The sole comprises the small sole segment juxtaposed to the white line and larger heel horn which is further subdivided into the firm solar segment and softer bulbar heal.

In cattle, weight bearing and mechanical force transfer occurs in various components of the claw and the gross morphology, cellular architectural and vascular composition of the bovine hoof is critical in maintaining proper hoof function. The distal cushion is mainly involved in shock absorption while the coronary band with a network of complex dermo-microvascular structures is important in improving blood flow during movement and the stepping process (Räber et al. 2004).

The suspensory apparatus is critical for maintaining proper orientation of the pedal bone within the hoof capsule and transfers a portion of the mechanical forces to the hoof. The corium (dermis) that links the pedal bone to the hoof capsule is rich in collagen and elastic fibers and is composed of an intricate pegging architecture that interdigitates with the laminae of the epidermis, increasing the interconnecting surface area between the two structures and improving structural integrity. The corium is also filled with glycoproteolytic metalloproteinases, keratin proteins, angiogenic growth factors (Tomlinson et al. 2004; Hirschberg and Plendl 2005), and components that lead to the triggering and progression of laminitis.

9.4 Dairy Cow Laminitis

Laminitis in cattle was first coined in the early 1960s as a "diffuse aseptic inflammation of the dermal layer of the claw" and regardless of the type of initiating event, laminitis like all other forms of inflammation displays the classic hallmarks of inflammation, namely rubor (redness), tumor (swelling), dolor (pain), calor (heat), and functio laesa (loss of function) within the bovine hoof. Laminitis is also defined by the onset, duration, and severity of inflammation and these are classified into four groups: acute, subacute, chronic, and subclinical laminitis. Acute laminitis is a tissue response occurring within the first 24 h following an injurious event. Cattle presenting with acute laminitis have a stiff gait, prominent digital pulses, and discomfort. Markers of animal performance, such as milk production and feed intake, may not be affected during periods of acute laminitis. Subacute laminitis extends beyond 10 days post-injury and presents with milder forms of clinical disease compared to cattle with acute laminitis. Chronic laminitis persists longer than 6–8 weeks after initiation and is identified by the presentation of abnormal claw morphology. Finally, subclinical laminitis is a hidden form of laminitis. Dairy cows with subclinical laminitis have very few clinical manifestations and often have a normal gait. Indeed, the evidence of subclinical laminitis is usually discovered by observing subtle abnormalities in the hoof following routine hoof trimming.

The pathophysiology of laminitis is complex and involves changes in both vascular flow and disruption of normal cellular processes (Ossent and Lischer 1998). Although the etiology of bovine laminitis remains controversial, animal models such as the oligofructose overload model of lameness in cattle has provided insight into the pathophysiologic and morphologic changes involved in bovine hoof injury (Danscher et al. 2009, 2010; Thoefner et al. 2004). The induction and progression of bovine laminitis is believed to occur in three continuous phases (Ossent and Lischer 1998).

Phase 1 is associated with impairment of blood flow to the hoof caused by the release of vasoactive agents such as histamine, with subsequent stasis of blood flow and thrombosis within the microvasculature. There is effusion of a transudate into the surrounding tissue resulting in edema. The reduction in blood flow can also induce arteriovenous shunts that further redirect blood flow from the corium to deeper tissues. These early events are also associated with prominent morphologic changes within the hoof including stretching of the laminae with detachment of the laminae basement membrane following the degeneration and necrosis of the germinative cells (Thoefner et al. 2004). On occasion there is also modest perivascular infiltration of granulocytes with focal dermal hemorrhages and pooling of edema fluid.

During Phase 2 the compression of the underlying tissue of the pedal bone becomes evident. The dermis is also undergoing continued injury of the capillary and microvasculature which, similar to Phase 1, can be accompanied by perivascular inflammatory cell infiltrates, necrosis, and edema.

During Phase 3 of chronic bovine laminitis, many of the clinical manifestations to the hoof capsule appear between 6 and 8 weeks following injury. The changes to the hoof can include: increased softening and friability of the horn, evident corium hemorrhaging, increased necrosis and ulceration of the corium of the heel and sole with widening and yellow discoloration of the white line. In the most extreme presentations of severe bovine laminitis, there is distal rotation with penetration of the pedal bone through the sole of the hoof, an event that can lead to the humane destruction of the animal.

The cellular events associated with laminitis in cattle are complex and not as well delineated as the pathogenesis of laminitis in equine and as such, there is a relative paucity of information of laminitis in dairy cows. There are, however, some physiologic and biochemical processes that have been identified in the development and progression of bovine laminitis. Research shows various causes of lameness (i.e., sole ulcers as a manifestation of chronic bovine laminitis) can induce changes in the gene expression of inflammatory cell populations in plasma. Researchers have also shown sole ulcers can induce an elevated neutrophil lymphocyte ratio with enhanced cortisol and haptoglobin levels suggesting an inflammatory response in cattle with laminitis. Furthermore, sole injury not only increases leukocyte expression of the pro- and anti-inflammatory cytokines IL-1 and IL-10, the presence of sole ulcers also increases the expression of L-selectin, an important molecule resulting in the homing of granulocytes (i.e., neutrophils) to areas of inflammation. Finally, a considerable increase in the expression of matrix metalloproteinases, MMP-13, a large protein involved in the degradation of collagen and a potent activator of other matrix metalloproteinases, MMP-2 and MMP-9 also occurs; a process that induces further tissue injury within the hoof (O'Driscoll et al. 2015; Almeida et al. 2007). Other studies have shown alterations in cytokine profiles and growth factors in the hoof following injury to either the sole or hoof wall (Mills et al. 2009; Osorio et al. 2012). For instance, biopsies of the hoof dermis and epidermis in cattle presenting with lesions in the sole or hoof wall demonstrate increased pro-inflammatory cytokines, IL-1, IL-1 receptor, TNF, IL-18, and IL-12, an important inducer of TNF and INF-γ from T-cells. Furthermore, there is an induction of iNOS (inducible nitric oxide synthase) and keratinocyte growth factor in ulcerated tissue of the hoof sole and with a progressive increased expression of these compounds with prolonged periods of injury. Interestingly, a parallel increase in the expression of anti-inflammatory and restorative cytokines IL-10 and TGF-β in the bovine sole suggest that an intricate and balanced interaction between inflammation and repair occurs during episodes of chronic laminitis.

9.5 Omics Sciences Linking Multiple Physiologic Systems in Dairy Cattle

Investigating disease in dairy cattle may require a paradigm shift in the way diseases are examined. Classic reductionist investigations into etiology and progression of disease required examining tissues in isolation, often ignoring the animal as a whole. An integrated investigation studying multiple physiologic systems concurrently (i.e., systems veterinary approach) should be considered for laminitis as it will provide more comprehensive information replicating the process in a living organism and may unlock some of the challenging questions that have eluded researchers to date. Research into multisystem causes of bovine laminitis is limited, but some studies in equine species demonstrated changes in one body system, such as the endocrine system, can trigger induction of diseases in distant body structures, namely the equine hoof. This "metabolic theory" of induction of equine laminitis provides compelling evidence that multiple systems are associated with the manifestation of disease. Indeed, elevated insulin levels are potent activators of hoof inflammation in equine species. Prolonged hyperinsulinemia, for example, will trigger laminitis in horses by increasing the production of activated glycan end products, which subsequently induce increases in pro-inflammatory cytokine levels and reactive oxygen species within the hoof lamellae (de Laat et al. 2012). Moreover, hyperinsulinemia also causes histopathologic changes in the secondary lamellae of the hoof indicative of cellular degeneration and inflammation. This is exemplified by tissue injury characterized by dyskeratosis, disruption of epidermal basement membrane and neutrophil infiltrates within the surrounding stroma (Asplin et al. 2010). These few examples highlight that the induction and progression of laminitis is not a simple process associated only with injurious events within the hoof, but can be most certainly coupled to changes in components with biologic functions (i.e., hyperinsulinemia) distinct from the hoof. As such, broadening our thinking to include consideration of a systems veterinary approach will be required to gain a better understanding of the pathophysiology of laminitis in hooven animals. This will be particularly important for determining the etiology and progression of the disease in dairy cattle and will require the use of the latest analytical and diagnostic tools, including *omics* sciences.

Undoubtedly a rapid growth in biotechnology has occurred over the last several decades and increased use of sophisticated laboratory techniques to study many aspects of both basic and clinical research in animal science and veterinary medicine have resulted. The terms "*omics*" and "*omics* sciences" are relatively new fields of study and are an extension of advanced techniques used to study biological sciences. Most certainly, *omics* is linked to the field of biology, as the suffix "*ome*" denotes "objects or parts of having a specific nature" (Ome 2016) and when related to the study of various cellular constituents, the *ome* suffix then pertains to the four main cellular constituents as well as biochemical and metabolic functions of the cell: genome, proteome, transcriptome, and metabolome. On the other hand, *omics* was a term coined to describe techniques and analytical algorithms that could compile and synthesize large amounts of data generated from these advanced biology sciences. Indeed, *omics* is often characterized as a high-throughput technology that can simultaneously assess biological processes providing an integrative and comprehensive understanding of complex cellular processes. Similar to the suffix "*ome*" and its relation to biological cellular processes, the suffix "*omics*" is associated with the study of biological macromolecules, biochemical and metabolic processes and it is categorized into four major groups: genomics, proteomics, transcriptomics, and metabolomics.

In general, genomics is the oldest and arguably the most established of the *omics* technologies and examines the function, structure, and sequence of DNA within the genome. Transcriptomics analyzes RNA transcripts from DNA sequences and can be considered a subcategory of genomics. Proteomics studies all translational protein isoforms of mRNA, changes in protein structure, and examines protein interactions with other macromolecules within the cell. Metabolomics, the newest member of the *omics* technologies, measures the presence of small metabolites in biological fluids and tissues. The strength of *omics* technologies stands in addressing the whole organism in concert; then, the integration of all *omics* technologies provides the most comprehensive understanding of biological and cellular processes within the animal.

In the context of dairy cow lameness and laminitis, the uses of *omics* sciences to investigate the etiology and pathomechanism of the disease is a growing field of study. As such, there is currently a shortage of information generated with *omics* technologies to study this disease in cattle. The sections below highlight application of *omics* sciences and some observations associated with bovine lameness and laminitis.

9.5.1 Genomics and Transcriptomics

It is well known that leg conformation, claw disorders, and the development of laminitis are associated with varying patterns of genetic and phenotypic heritability. Indeed, there are strong genetic correlations related to incidence of dermatitis and heel erosions which are infectious claw disorders, while white line disease or sole ulcers are laminitis associated claw disorders (van der Spek et al. 2013; Ødegård et al. 2014). Understanding the development of traits in dairy cattle would require the use of *omics* techniques such as genomics and transcriptomics to investigate livestock diseases, such as lameness and laminitis. These *omics* technologies include comparative genomics, quantitative trait loci mapping, and epigenetic inheritance. These have improved the understanding of complicated and multifaceted periparturient diseases not associated with the limb or hoof. Unfortunately, these *omics* technologies have not been applied to understanding the pathophysiology of lameness or laminitis in dairy cattle. Presently, the genomic and transcriptomic technologies employed to investigate lameness and laminitis in cattle include genome-wide associated studies (GWAS), microsatellite markers, and single nucleotide polymorphism (SNP; Matukumalli et al. 2009; Cole et al. 2011). As an example, there is a large-scale comprehensive study using *omics* technologies to examine the heritability of laminitis induced claw disorders in cows and heifers from bulls presenting with claw abnormalities (van der Spek et al. 2015). The investigation showed there are 10 significant and 45 suggestive changes in the genetic codons on 20 different chromosomes within the Holstein-Friesian genome. These genes were spread over several chromosomes and each SNP had an impact on the clinical manifestation of hoof diseases. It was suggested that these SNPs may be responsible for nearly 5% of the phenotypic variance between the hoof of healthy cows and cattle presenting with abnormal claws. Interestingly, some of the SNPs have been identified on chromosomal regions involved in leg and hoof confirmation and the quality of bone structure.

The importance of differential expression of mRNA transcripts in the development and progression of laminitis in dairy cattle is underscored with enhanced expression of matrix metalloproteinases. Indeed, elevated expression of these proteins initiates degradation of collagens needed to suspend the internal structures of the hoof. As previously indicated several studies have demonstrated that mRNA for MMP-13 is increased in dairy cattle with foot lesions and it might be partially responsible for exacerbation of disease and subsequent distal rotation and penetration of the pedal bone through the sole (O'Driscoll et al. 2015; Almeida et al. 2007). Importantly, the study by Almeida et al. (2007) employed microarray analysis to determine variations in mRNA expression, an example of advanced transcriptomics techniques used to investigate biologic events occurring during laminitis in dairy cattle.

Future work in this area will surely enhance the understanding of genetic effects on the risk of developing the disease, severity of the clinical manifestations, and may even suggest target areas for treatment strategies.

9.5.2 Proteomics

Genomics and transcriptomics technology can provide valuable information on the expression in genes linked to the development of laminitis in dairy cattle. The expression of transcripts of mRNA from the gene sequences may not, however, consistently translate into high fidelity protein expression as mRNA constructs can be modified or degraded. Conversely, the examination of protein products provides a higher level of cellular functionality as the proteins are the active macromolecular products of gene expression (Debnath et al. 2010). Indeed, examining protein profiles within the cell or other biologic networks will generate data that more accurately resembles cellular events.

Proteomic technologies have advanced and are a collection of protein separation techniques, integrated with bioinformatics (protein content; Mankowski and Graham 2008), and qualitative information (protein isoforms). Techniques that separate proteins have improved from simple 1- and 2D polyacrylamide gel electrophoresis (PAGE), to more advanced reversed phase high performance liquid chromatography (HPLC) and immune-affinity assays to very sophisticated Fourier transform infrared spectroscopy, X-ray crystallography, nuclear magnetic resonance (NMR), and mass spectrometry (MS). These techniques combined with powerful data analysis programs can provide information on the repertoire of proteins within the cell or biologic system.

Good and predictive proteomic information requires reliable genomic data and as such proteomic data is often an extension of accurate information gathered from genomic and transcriptomics investigations. Therefore, it is often recommended that investigations assessing the mechanism of disease should examine changes to the genome or transcriptome prior to measuring differences in protein content. Similar to genomic and transcriptomics, there are few investigations that examine adjustments in the proteome in dairy cattle with laminitis. Several early investigations, however, demonstrated laminitis can induce changes in protein production within the hooves of dairy cattle. Galbraith et al. (2006) used 2D-SDS PAGE to show differential protein expression of 169 proteins within the laminae of inflamed hooves. Similarly, another study using liquid chromatography MS in concert with multiple database analyses showed increased production of proteins in the coronary segment, corium, and lamellae (Tølbøll et al. 2012). In this study, over 440 proteins were identified with approximately 20% of those proteins being detected in all three tissues. These proteins had a diverse range of function and were involved in important processes associated with the induction and progression of bovine laminitis, namely cell structural integrity, metabolism, apoptosis, angiogenesis, immunity, and inflammation. Lastly, a large study examined protein changes in the plasma in dairy cattle afflicted with laminitis. The research showed dairy cows with clinical symptoms of disease had increased levels of proteins associated with energy metabolism (isocitrate dehydrogenase 1) and lipid metabolism (apolipoprotein A-I and A-IV, 3-hydroxy-3-methylglutaryl-coenzyme A reductase), voltage ion regulation (glycerol-3-phospahte dehydrogenase 1-like protein) oxidative stress, immune function regulation and inflammation (haptoglobin and conglutinin). Interestingly, proteins such as complement component 9 precursor and complement component 4 binding protein that are involved in complement activation, an important constituent of both innate and adaptive immunity, were downregulated in cattle with laminitis. These observations highlight the complex and coordinated nature of pro- and anti-inflammatory response in the context of bovine laminitis, underscoring that laminitis in dairy cattle is indeed not a simple inflammatory event but a highly regulated inflammatory process (Dong et al. 2015).

9.5.3 Metabolomics

Metabolomics is the newest *omics* science and will surely provide even more insight into the understanding of biological functions associated with bovine laminitis. This new technology has a great potential for providing the best predictive biological markers of diseases including bovine laminitis. Examination of the metabolome in dairy cattle is a new field of study and although previous studies have shown changes in rumen content (Saleem et al. 2013) and serum metabolites in dairy cattle with ruminal acidosis following alterations in the diet (Ametaj et al. 2010), there is currently only a single small-scale study that specifically examined alterations and the serum metabolomics metabolites in dairy cattle stricken with bacterial induced hoof inflammation (i.e., foot rot; Zheng et al. 2016). Disruptions within the metabolome of cattle afflicted with "foot rot" demonstrated changes of 21 different metabolomic compounds. These compounds were indicators of perturbations in metabolic pathways involved in aliphatic and hydroxyl amino acids and glycolipid metabolism, ketone bodies formation, oxidative protection, pyruvate metabolism, glycolysis, functioning of the tricarboxylic acid cycle, and gluconeogenesis. From the study, nonetheless, it is difficult to ascertain which altered metabolic pathways have the greatest impact on the manifestation of laminitis in dairy cattle. Some information from this investigation, however, is in-line with observations from other *omics* investigations demonstrating not only the utility of metabolic technology to study active diseases but complexity of biological responses associated with inflammation in the bovine hoof.

Metabolism is a complex process and alterations in these processes can produce a composite of metabolites measurable in both serum and tissues. In as such these metabolites can be used as predictors of concurrent disease. This raises the question whether changes in metabolite and other biomarker profiles can be used to predict development of disease prior to the onset of clinical manifestations. A recent study showed significant changes in serum markers of inflammation, lipid, and carbohydrate metabolism occurs in the transition period of dairy cattle prior to the onset of lameness. The investigators also demonstrated marked increases in acute phase proteins (serum amyloid A, haptoglobin), pro-inflammatory cytokines (IL-6, TNF), the carbohydrate metabolite (lactate) and lipid metabolites (non-esterified fatty acids and β-hydroxybutyrate), several weeks prior to the development of clinical lameness in dairy cattle (Zhang et al. 2015).

9.5.4 A Hypothetical Multisystems Model *of Laminitis*

The previous sections identified several *omics* technologies used to investigate disease process in the hooves of dairy cattle and questions whether these technologies could be used to investigate the induction and progression of disease in a multisystem approach. Can we extend this question to develop a hypothesis that examines events in tissues independent from the bovine hoof induced laminitis in cattle? As an example, could a hypothetical model involving histamine and/or LPS be applied to investigate whether the release of these products into circulation induces laminitis in dairy cattle (Fig. 9.2)? How would *omics* technologies be applied to measure changes in the physiological processes in cattle? Would this provide a multisystem approach to study hoof inflammation? It is well known that ruminal acidosis (Ametaj et al. 2010), mastitis (Burvenich et al. 2003), and metritis (Magata et al. 2015) are associated with the release of LPS and on occasion histamine and other vasoactive peptides into circulation (Stalberger and Kersting 1998). Moreover, several biogenic amines released in the rumen during feeding of high grain diets like methylamine, putrescine, ethanolamine, nitrosodimethylamine, and phenylacetylglycine also might be involved in the pathobiology of the disease (Ametaj et al. 2010;

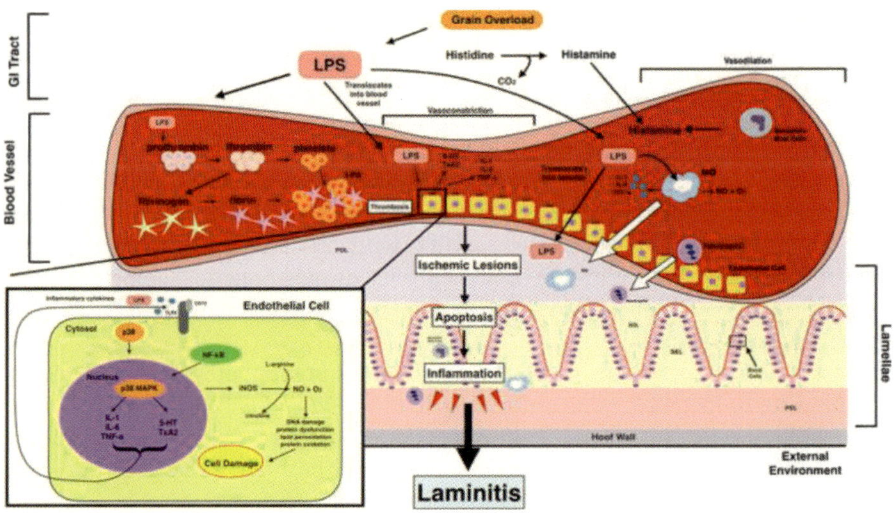

Fig. 9.2 Summary of the main factors involved in laminitis in dairy cattle. Grain overload in the GI tract results in the release of LPS (lipopolysaccharide) that translocates into the systemic circulation. LPS in the blood converts prothrombin into thrombin, then triggers aggregation of platelets. Platelets along with fibrin trigger thrombosis. LPS also acts on the endothelial cells binding to TLR-4 and CD14 complex resulting in activation of the p38 MAPK system which generates proinflammatory cytokines IL-1, IL-6, TNF-α, and also 5-HT (5-hydroxytryptamine) and TxA2 (thromboxane) which are potent vasoconstrictors. The NF-kB also is activated upon binding of LPS and triggers iNOS (inducible nitric oxide synthase) resulting in NO (nitric oxide) production and cell damage. Ischemic lesions result from damage of endothelial cells in the blood vessels resulting in apoptosis in the lamellae and inflammation. LPS also activates macrophages to produce pro-inflammatory cytokines and reactive oxygen species, which translocate into the lamellae. Another event during grain overload is the conversion of histidine into histamine that translocates into the systemic circulation. Basophils and mast cells within the blood vessels also produce histamine. The production of histamine results in vasodilation of the blood vessels further contributing to inflammation (swelling) within the hoof

Saleem et al. 2012). In summary, these molecules can induce marked alteration in blood flow at distant tissues, including the bovine hoof and thus initiating Phase 1 for the induction of acute inflammation of bovine laminitis. Lipopolysaccharides can then induce the production of reactive pro-inflammatory cytokines such as TNF and IL-1. These cytokines subsequently activate potent matrix metalloproteinases that degrade collagen and weaken the suspensory apparatus of the hoof causing rotation and possible penetration of the third phalanx into the underlying sole, a hallmark of the progression of Phase 2 bovine laminitis. Finally, as inflammation of the hoof progresses near 6–8 weeks post-injury, the clinical manifestations of chronic hoof disease becomes evident, an indicator of Phase 3 of the inflammatory process.

Although each *omics* technology investigates a specific *ome* of the host, it is evident that bovine laminitis is a complex disease and perhaps it is overly simplistic to only examine each *ome* in isolation. Understandably, this venture of examining the genome, transcriptome, proteome, and metabolome, would require access to the proper equipment and could be quite cost prohibitive. In an ideal setting, however, employing genomics, transcriptomics, proteomics, and metabolomic technologies in concert would indeed provide the most complete understanding of the induction and progression of disease.

As a simple hypothetical example, *omics* technologies could be employed to examine the early stages (Phase 1) of bovine laminitis in dairy cattle. For instance, genomics technologies could be used to identify whether the genetic background of the breed of cattle predisposes animals to increased incidence of laminitis or lameness induced hoof injury. Certain dairy cattle breeds such as Jersey cattle are more prone to developing laminitis than others (Edwards 1972). Transcriptomics would then determine the expression of mRNA of pro-inflammatory cytokines and matrix metalloproteinases and correlate the expression of these mRNA transcripts to the production of functional proteins with proteomic techniques. Lastly, metabolomics would determine the impact of these early inflammatory events on key pathways involved in amino acid, carbohydrate, and lipid metabolism, and products involved in innate and adaptive immune function and inflammation. Moreover, this strategy on the application of *omics* technologies could also be applied to determine mechanisms of disease as it progresses through Phases 2 and 3. From the example, it would therefore, seem reasonable that *omic* technologies could be applied, preferably in concert, to examine periparturient diseases such as laminitis in dairy cattle; employing a multisystems approach that examines the animal as whole.

Conclusions

Laminitis in cattle has a storied history that has permeated animal husbandry practices and the essence of civilization for millennia. Early writings describe the clinical manifestations of bovine laminitis and early forms of treatment that by today's standards were nothing short of barbaric. Over the centuries, however, the treatment of the disease has improved; progressing to modern-day evidence based veterinary practices. Current literature in veterinary anatomy and medicine describe in detail the anatomic structures of the bovine hoof. There is a growing body of evidence involving clinical and morphologic changes that manifest in the hooves of cattle with inflammation. Recent evidence in equine laminitis demonstrates laminitis can be induced by changes in biologic factors (i.e., hyperinsulinemia) distinct from local events (i.e., localized hoof infections, and mechanical injury) occurring at the hoof (de Laat et al. 2012). These observations suggest that in order to gain a more robust understanding of laminitis, disease investigations should not only study changes in the hoof in isolation, but should also examine the animal as a whole, a more holistic multisystems approach. As such, multisystem investigations should be considered when studying laminitis in dairy cattle. Although there is a relative paucity of information on cellular events during episodes of bovine laminitis in dairy cattle, new *omics* technologies are beginning to provide insight into the pathophysiology of the disease. During periods of inflammation there are increases in the production of pro-inflammatory cytokines, increased matrix metalloproteinase activity, and changes in metabolic pathways associated with amino acid, carbohydrate, lipid metabolism, and energy production. By using both *omics* technologies and considering laminitis as a complex disease influenced by both local events at the hoof and products produced from distant tissues or organs, a better understanding of the pathophysiology of this disease seems likely. A better understanding of bovine laminitis may translate into better husbandry practices for the dairy industry and possibly generate new therapies to mitigate disease.

Acknowledgments Thank you to Dr. Christopher S. MacKay, Professor of History and Classics at the University of Alberta, Canada, for help translating and interpreting the writings of Columella. Thank you as well to Dr. Trina C. Uwiera, Associate Professor of Surgery at the University of Alberta, Canada, for creating Fig. 9.1; to delineate the anatomy of the bovine hoof, a picture is truly worth a thousand words.

References

Almeida PE, Weber PSD, Burton JL et al (2007) Gene expression profiling of peripheral mononuclear cells in lame dairy cows with foot lesions. Vet Immunol Immunopathol 120:234–245

Ametaj BN, Zebeli Q, Saleem F et al (2010) Metabolomics reveals unhealthy alterations in rumen metabolism with increased proportion of cereal grain in the diet of dairy cows. Metabolomics 6:583–594

Asplin KE, Patterson-Kane JC, Sillence MN et al (2010) Histopathology of insulin-induced laminitis in ponies. Equine Vet J 42(8):700–706

Burvenich C, Van Merris V, Mehrzad J et al (2003) Severity of *E. coli* mastitis is mainly determined by cow factors. Vet Res 34:521–564

Cha E, Hertl JA, Bar D, et al (2010) The cost of different types of lameness in dairy cows calculated by dynamic programming. Prev Vet Med 97(1):1–8.

Cole JB, Wiggans GR, Ma L et al (2011) Genome-wide association analysis of thirty one production, health, reproduction and body conformation traits in contemporary US Holstein cows. BMC Genomics 12:408

Columella 2007, De re rustica, 6. 12. Translated by Christopher S. Mackay of the Department of History and Classics of the University of Alberta

Danscher AM, Enemark JMD, Telezhenko E et al (2009) Oligofructose overload induces lameness in cattle. J Dairy Sci 92:607–616

Danscher AM, Toelboell TH, Wattle O (2010) Biomechanics and histology of bovine claw suspensory tissue in early acute laminitis. J Dairy Sci 93:53–62

de Laat MA, Kyaw-Tanner MT, Sillence MN et al (2012) Advanced glycation endoproducts in horses with insulin-induced laminitis. Vet Immunol Immunopathol 145:395–401

Debnath M, Prasad GBKS, Bisen PS (2010) Molecular diagnostics: promises and possibilities. Springer, Dordrecht, p 527

Dong SW, Zhang SD, Wang DS et al (2015) Comparative proteomics analysis provide novel insight into laminitis in Chinese Holstein cows. BMC Vet Res 11:161

Dunlop RH, Williams DJ (1996) Veterinary medicine: an illustrated history. Mosby, St. Louis, p 53

Edwards GD (1972) Hereditary laminitis in Jersey cattle. In: Proceedings of the VIIth World Buiatrics Congress, London, pp 663–668

Galbraith H, Flannigan S, Swan L et al (2006) Proteomic evaluation of tissues at functionally important sites in the bovine claw. Cattle Pract 14:127–137

Hirschberg RM, Plendl J (2005) Pododermal angiogenesis and angioadaptation in the bovine claw. Mircosc Res Tech 66:145–155

Kauffman KD (1996) The U.S. Army as a rational economic agent: the choice of draft animals during the civil war. East Econ J 22(3):333–343

Magata F, Ishida Y, Miyamoto A et al (2015) Comparison of bacterial endotoxin lipopolysaccharide concentrations in the blood, ovarian follicular fluid and uterine fluid: a clinical case of bovine metritis. J Vet Med Sci 77:81–84

Mankowski JL, Graham DR (2008) Potential proteomic-based strategies for understanding laminitis: Predictions and pathogenesis. J Equine Vet Sci 28(8):484–487

Mason S (2012) The anatomy of the bovine hoof. http://dairyhoofhealth/40-hoof-anatomy/new-functional-foot-care-video. Accessed 10 Sept 2016

Matukumalli LK, Lawley CT et al (2009) Development and characterization of a high density SNP genotyping assay for cattle. PLoS One 4(4):1–13

Mills JA, Zarlenga DS et al (2009) Age, segment, and horn disease affect expression of cytokines, growth factors, and receptors in the epidermis and dermis of the bovine claw. J Dairy Sci 92:5977–5987

Nordlund KV, Cook NB, Oetzel GR (2004) Investigation strategies for laminitis problem herds. J Dairy Sci 87(Suppl E):E2–E35

Ødegård C, Svendsen M, Heringstad B (2014) Genetic correlations between claw health and feet and leg conformation in Norwegian Red Cows. J Dairy Sci 97:4522–4529

O'Driscoll K, McCabe M, Earley B (2015) Differences in leukocyte profile, gene expression, and metabolite status of dairy cows with or without sole ulcers. J Dairy Sci 98:1685–1695

Osorio JS, Fraser BC et al (2012) Corium tissue expression of genes associated with inflammation, oxidative stress, and keratin formation in relation to lameness in dairy cows. J Dairy Sci 95:6388–6396

Ossent P, Lischer C (1998) Bovine laminitis: the lesions and their pathogenesis, 4th edn. Vet Form, pp 415–427

Ome (2016) Oxford Dictionary. http://en.oxforddictionaries.com/definition/us/-ome. Accessed 10 Sept 2016

Räber M, Lischer CJ et al (2004) The bovine digital cushion—a descriptive anatomical study. Vet J 167:258–264

Saleem F, Ametaj BN, Bouatra S, Mandal R, Zebeli Q, Dunn SM, Wishart DS (2012) A metabolomics approach to uncover the effects of grain diets on rumen health in dairy cows. J Dairy Sci 95:6606–6623

Saleem F, Bouatra S, Guo AC et al (2013) The Bovine ruminal fluid metabolome. Metabolomics 9:360–378

Severin T (1989) Retracting the first crusade. Natl Geogr 176(3):326–365

Shively MJ (1984) Veterinary anatomy: basic, comparative and clinical. Texas A&M University Press, College Station, p 561

Sisson S, Grossman JD, Getty R (1975) The anatomy of the domestic animals, vol 1, 5th edn. WB Saunders Company, London, pp 788–789

Smith F (1976) The early history of veterinary literature, vol 2. JA Allen, London

Stalberger RJ, Kersting KW (1998) Peracute toxic coliform mastitis. Iowa State Univ Vet 50(1):48–53

Takahashi K, Young BA (1981) Effects of grain overfeeding and histamine injection on physiological responses related to acute bovine laminitis. Jpn Vet Sci 1(43):375–385

Thoefner MB, Pollitt CC et al (2004) Acute bovine laminitis: a new induction model using alimentary oligofructose overload. J Dairy Sci 87:2932–2940

Tølbøll TH, Danscher AM, Anderson PH et al (2012) Proteomics: a new tool in bovine claw disease research. Vet J 193:694–700

Tomlinson DJ, Mülling CH, Faklet TM (2004) Invited review: formation of keratins in the bovine claw: roles of hormones, minerals, and vitamins in functional claw integrity. J Dairy Sci 87:797–809

van der Spek D, van Arendonk JAM, AAA V, Bovenhuis H (2013) Genetic parameters of claw disorders and the effect of preselecting cows for trimming. J Dairy Sci 96:6070–6078

van der Spek D, van Arendonk JAM, Bovenhuis H (2015) Genome wide association study for claw disorders and trimming status in dairy cattle. J Dairy Sci 98:1286–1295

Webster AJE (1997) Review: farm animal welfare: the five freedoms and the free market. Vet J 161:229–237

Whitaker DA, Kelly JM, Smith S (2000) Disposal and disease rates in 340 British dairy herds. Vet Rec 146:363–367

Wishart DS (2016) Emerging applications of metabolomics in drug discovery and precision medicine. Nat Rev Drug Discov 15:473–484

Zhang G, Hailemariam D, Dervishi E, Deng Q, Goldansaz SA, Dunn SA, Ametaj BN (2015) 90 Alterations of innate immunity reactants in transition dairy cows before clinical signs of lameness. Animals 5:717–747

Zheng J, Lingwell SUN, Shi SHU et al (2016) Nuclear magnetic resonance-based serum metabolic profiling of dairy cows with footrot. J Vet Med Sci 78:1421–1428

Ketosis Under a Systems Veterinary Medicine Perspective

10

Guanshi Zhang and Burim N. Ametaj

Abstract

Ketosis (hyperketonemia) is a prevalent metabolic disease of transition dairy cows that affects ~30–40% of the cows during early lactation. Cows with ketosis have lower milk yield and reproductive performance, greater risk of other periparturient diseases, and higher culling rates. Ketosis is characterized by an excess level of circulating ketone bodies with blood concentration of beta-hydroxybutyrate (BHBA) been recognized as a golden standard for diagnosis of the disease. However, the cutoff value for serum BHBA for diagnosis of ketosis appears to be somewhat arbitrary. Negative energy balance at early lactation is the primary hypothesis for the explanation of the pathobiology of ketosis. Treatment strategies are focused on maintenance of glucose and ketone body homeostasis. The exact causes and the etiopathology of ketosis remain incompletely understood. In some recent metabolomics studies, data show that the number of metabolites altered in the plasma/serum or milk and metabolic pathways perturbed during ketosis are numerous. It is obvious that ketone bodies and glucose metabolism are not the only perturbed metabolites during ketosis. Mounting evidence indicates that multiple alterations also occur at proteome, transcriptome, and genome levels and other component networks. Integration of all the knowledge generated

G. Zhang, D.V.M., Ph.D. (✉)
Department of Agricultural, Food and Nutritional Science, University of Alberta, Edmonton, AB T6G 2P5, Canada

Center for Renal Translational Medicine, School of Medicine, Institute of Metabolomic Medicine, University of California, La Jolla, San Diego, CA 92093, USA
e-mail: guz016@ucsd.edu

B.N. Ametaj, D.V.M., Ph.D., Ph.D.
Department of Agricultural, Food and Nutritional Science, University of Alberta, Edmonton, AB T6G 2P5, Canada
e-mail: burim.ametaj@ualberta.ca

B.N. Ametaj (ed.), *Periparturient Diseases of Dairy Cows*,
DOI 10.1007/978-3-319-43033-1_10

by genomics, transcriptomics, proteomics, metabolomics, and lipidomics can help detect perturbations of biological information on distinct tiers and give insight on the causes and pathobiology of ketosis in dairy cows.

10.1 Introduction

10.1.1 Current Understanding of Ketosis

Ketosis (i.e., hyperketonemia) is a prevalent metabolic disease in transition dairy cows during early lactation. It is strongly associated with elevated levels of circulating ketone bodies [i.e., β-hydroxybutyrate (BHBA), acetone (Ac), and acetoacetate (AcAc)], hypoglycemia, and hyperinsulinemia or insulin resistance (Oetzel 2007). Cows with ketosis have lower milk yield and reproductive performance, greater risk of other periparturient diseases, and higher culling rates. The incidence of ketosis at herd-level is extremely high ranging from 26.4 to 55.7% for the subclinical form of ketosis (SCK), and varying from 2 to 15% for clinical ketosis (CK) (Gordon et al. 2013; Oetzel 2013). It is estimated that a single case of ketosis costs a dairy producer between CAD $50 to $100, which include treatment costs, decreased milk yield, impaired fertility, and occurrence of concurrent diseases (Duffield 2000). Based on the high morbidity of ketosis, a moderate cost per animal can lead to very large economic losses to the dairy industry.

Currently, there are two types of classification schemes used for ketosis by researchers. The first type of categorization scheme divides ketosis cases into SCK and CK based on blood concentration of BHBA and the presence of clinical symptoms. The second classification strategy classifies ketosis into three types: type I ketosis, type II ketosis, and ketosis related to butyric acid originating from silage. This latter classification is based on different causes of the pathology and timing of hyperketonemia (Herdt 2000; Holtenius and Holtenius 1996; Oetzel 2007). Type I ketosis, which is aligned with type I diabetes mellitus in humans, usually occurs between 3 and 6 weeks postpartum when milk energy outflow reaches its peak (Holtenius and Holtenius 1996). Type II ketosis is the metabolic counterpart of type II diabetes mellitus, and occurs immediately after parturition and is concurrent with fatty liver (Holtenius and Holtenius 1996). Butyric acid silage ketosis is related to intake of feedstuff high in ketogenic precursors such as ketogenic silages (Tveit et al. 1992).

The etiology of ketosis is related mostly to negative energy balance (NEB) or glucose deficiency during early lactation (Fig. 10.1). Treatment strategies also have been focused mostly on the maintenance of glucose and ketone body homeostasis.

10.1.2 New Challenges

Ketosis is of complexity and the current hypothesis of NEB is insufficient for interpretation of the disease. The occurrence of the disease involves dietary factors and

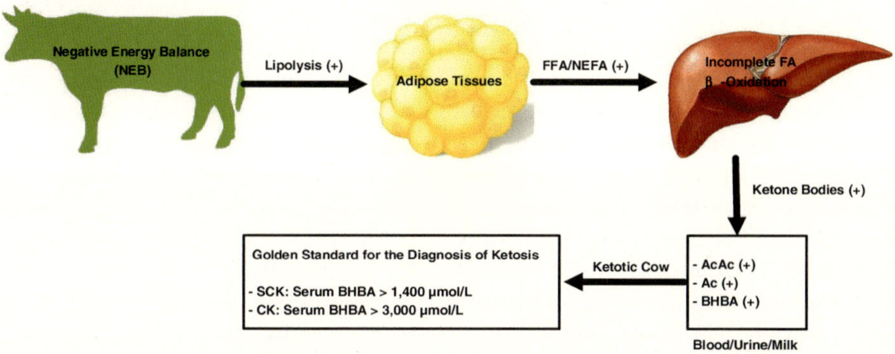

Fig. 10.1 Conventional understanding of ketosis by reductionist approach. *Abbreviations*: *Ac* acetone, *AcAc* acetoacetate, *BHBA* β-hydroxybutyric acid, *CK* clinical ketosis, *FA* fatty acid, *FFA* free fatty acids, *NEFA* non-esterified fatty acids, *SCK* subclinical ketosis

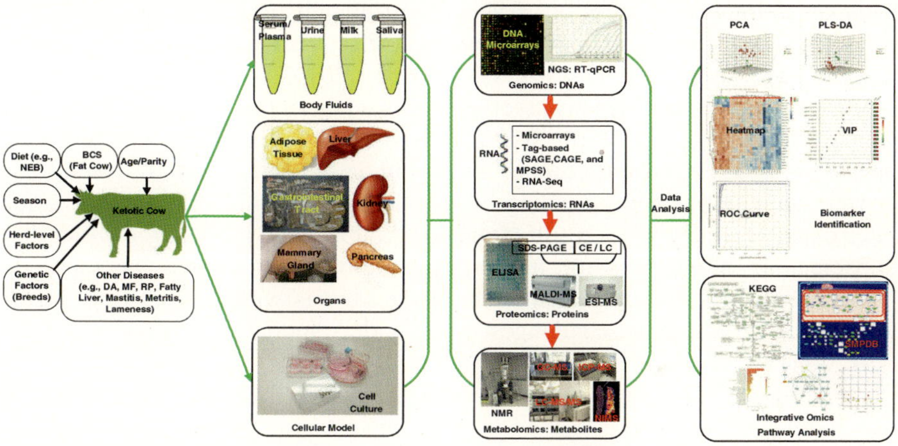

Fig. 10.2 Understanding ketosis using "omics" based systems biology approaches. *Abbreviations*: *BCS* body condition score, *CAGE* cap analysis of gene expression, *CE* capillary electrophoresis, *DA* displaced abomasum, *ELISA* enzyme-linked immunosorbent assay, *ESI-MS* electrospray ionization-mass spectrometry, *GC-MS* gas chromatography-mass spectrometry, *ICP-MS* inductively coupled plasma–mass spectrometry, *KEGG* Kyoto Encyclopedia of Genes and Genomes, *LC* liquid chromatography, *LC-MS/MS* liquid chromatography-tandem mass spectrometry, *MALDI-MS* matrix-assisted laser desorption/ionization-mass spectrometry, *MF* milk fever, *MPSS* massively parallel signature sequencing, *NEB* negative energy balance, *NGS* next-generation DNA sequencing, *NIMS* nanostructure imaging mass spectrometry, *PCA* principal component analysis, *PLS-DA* partial least squares-discriminant analysis, *RNA-Seq* RNA sequencing, *ROC* receiver operating characteristic, *RP* retained placenta, *RT-qPCR* quantitative reverse transcription polymerase chain reaction, *SAGE* serial analysis of gene expression, *SDS-PAGE* sodium dodecyl sulfate-polyacrylamide gel electrophoresis, *SMPDB* The Small Molecule Pathway Database, *VIP* variable importance in projection

inputs from various organ systems including the endocrine system, liver, adipose tissue, mammary gland, gastrointestinal tract, and immune system. Causal factors of ketosis and the workflow of "omics" based system biology approaches are shown in Fig. 10.2.

In order to better understand the disease, the first step is the acquisition of global sets of omics-oriented data from multiple hierarchical levels (i.e., DNA sequences, RNA expression, protein or lipid abundance, and metabolite variation). Data integration from several levels will result in formulation of detailed mathematical or graphical models that are important to monitor systems perturbation and evaluate the progress of the disease. Comprehensive quantification of alterations on different hierarchical levels data help build more accurate models to explain the systems-level properties of ketosis. Ultimately, ketosis investigators or veterinarians can utilize these models to accomplish at least four tasks not feasible before: (1) understand the onset and progress of ketosis based on given perturbations; (2) predict ketosis by identified screening biomarkers; (3) redesign or make interventions on the molecular networks to prevent the occurrence of ketosis; and (4) evaluate the cow body's response to appropriate intervention such as a new diet or medication.

In this chapter, we are not going to be comprehensive. Various topics of ketosis like its impact on other diseases, current diagnostic approaches, and cow-side tests have been extensively reviewed elsewhere (Duffield et al. 2009; Ospina et al. 2010; Oetzel 2007; Zhang et al. 2012). This chapter begins by reviewing omics based systems biology approaches and the associated bioinformatics tools for each platform. We then discuss a series of omics studies carried out in previous and recent years on ketosis in dairy cows or nonruminants. The current gaps in our understanding of ketosis will be summarized. Then, some preliminary metabolomics results from our group, which indicate the possibility of screening biomarkers for further evaluation of ketosis will be reported. Mechanistic insights and speculations about ketosis will be discussed. In the end, we will outline the challenges in the understanding of ketosis and future directions for ketosis studies.

10.1.3 Current Gaps in the Understanding of Ketosis

Ketosis is a typical example of the application of reductionist approach on the etiopathology of a disease. The reductionist methodology is focused on dividing complex problems into smaller, simpler, and more tractable units and understanding each basic unit separately. Ketosis in cows was reported first as hypoglycemia and hyperketonemia since 1929, and later NEB was proposed as the primary cause of the disease (Shaw 1956). Current strategies to diagnose ketosis rely heavily on concentrations of ketone bodies in the blood, urine, or milk and the "golden standard" for diagnosing ketosis is the measurement of blood concentration of BHBA.

Hyperketonemia, especially the SCK form, is a chronic condition that develops asymptomatically during the transition period, presenting itself clinically in a smaller number of transition cows during the early lactation and progression of disease. Currently, diagnosis of ketosis is based on a reductionist approach, focusing on a few metabolites, namely, concentration of ketone bodies (i.e., BHBA, Ac, and AcAc) in various body fluids.

Restoring ketone bodies and glucose homeostasis has been a central dogma for the treatment of ketosis. However, there are multiple challenges facing dairy producers and veterinary practitioners with regard to prevention rather than treatment of ketosis in dairy cows. Moreover, although herd-level diagnostic methods and prophylactic strategies are applied by dairy producers, the incidence rate of ketosis is still exceptionally high and the incidence of SCK can reach as high as 80%, in some dairy herds (Duffield 2000).

This state of ketosis has raised a few questions that need answers like: Is the hypothesis of NEB and hyperketonemia sufficient to explain the pathogenesis of ketosis? Are ketone bodies the only perturbed metabolites during ketosis? What is the contribution of perturbed DNA and RNA networks as well as altered proteins in the pathobiology of ketosis? Is inflammation and immunity involved during ketosis and if yes, why? Are there alterations before subclinical or clinical appearance of ketosis and can we prevent development of ketosis? If ketone bodies are not the only metabolites altered during ketosis, but multiple networks and pathways are involved, then, what is the real cause of ketosis?

If we accept the hypothesis that ketosis is caused by a single factor (i.e., NEB), which leads to excessive adipose tissue lipolysis for precursors of gluconeogenesis and ketone bodies to meet energy requirements, then, the simple phenotype of ketosis at the biochemical level should be the elevated levels of ketone bodies only. However, it has been reported that animals with severe hyperketonemia may not show clinical symptoms and animals with low levels of ketone bodies may be noticeably sick. Indeed, Herdt (2000) indicated that cows have individual differences in their ability to process and tolerate ketone bodies. Therefore, hyperketonemia in ketotic cows may not be the only reason that triggers lower milk production, decreased reproductive performance, and higher susceptibility to other diseases. It would be logical and impelling to ask whether the concept of ketosis can be simply delineated by hyperketonemia and whether ketosis can be merely diagnosed by concentrations of ketone bodies. The current diagnosis method for ketosis, which has overly relied on ketone bodies, is facing a crisis. The complexity of ketosis is deeper than our current understanding and multilevels analysis are crucial to the monitoring and pre-diagnosis of ketosis as well as in development of new prevention strategies. Several other factors to be considered are screening biomarkers, dietary interventions, age-related risk factors, body condition score, epigenetics of the individual and parents, genetics, transcriptomics, proteomics, metabolomics, lipidomics, and other systems biology sciences.

10.2 Omics-Based Systems Biology Approaches and Their Application in Biomarker Identification for Ketosis

10.2.1 Genomics and Transcriptomics Identify Genes and mRNA Biomarkers for Ketosis

Genome-wide studies have revealed that the heritability of ketosis in cattle varies from 0.01 to 0.16 (Heringstad et al. 2005; Kadarmideen et al. 2000; Uribe et al.

1995). It should be pointed out that a limited number of genomics and transcriptomics studies have been undertaken on ketosis in dairy cows. Recently, an excellent genome-wide association study (GWAS) was carried out by Tetens et al. (2015) based on the previously identified milk metabolite biomarkers [i.e., glycerophosphocholine (GPC), phosphocholine (PC), and the ratio of GPC to PC in milk] for ketosis. In this study, cows were genotyped using the Illumina BovineSNP50 BeadChip (Illumina, San Diego, CA), which comprises 54,001 single nucleotide polymorphisms (SNPs). Results showed that both the milk level of GPC and the GPC/PC ratio are of high heritability at $h^2 = 0.43$ and 0.34, respectively. GWAS data also revealed that four SNPs located on cattle chromosome 25 (BTA25) reached genome-wide significance, two of which were simultaneously associated with both milk GPC levels and the GPC/PC ratio.

In addition, the apolipoprotein B-receptor (*APOBR*) gene was located and analyzed as a candidate gene that is associated with the development of ketosis in dairy cows (Tetens et al. 2015). Particularly, polymorphisms within the *APOBR* gene are highly associated prognostic biomarkers (i.e., GPC and the GPC/PC ratio) for ketosis in the milk of dairy cows (Tetens et al. 2015). A study involving hepatic gene expression during ketosis revealed that fibroblast growth factor 21 (*FGF21*) and its co-receptor *KLB* mRNA expression was upregulated in the liver of ketotic cows compared with CON ones (Akbar et al. 2015). The 41-fold upregulation in *FGF21* mRNA expression in ketotic cows confirms that this gene is a biomarker of NEB in dairy cows (Schoenberg et al. 2011). And the data from the study revealed the potential roles of liver-derived FGF21 in the onset of ketosis or inflammation, stimulation of fatty acid oxidation, and helping cows adapt to disturbances in energy balance (Akbar et al. 2015).

An interesting genomics study on ketosis was conducted by Loor et al. (2007). A bovine cDNA microarray platform which contains 13,257 annotated oligonucleotides was applied to investigate alterations of hepatic gene networks underlying nutrition-induced ketosis in dairy cows. Ingenuity pathway analysis (IPA) indicated that genes associated with cholesterol metabolism, fatty acid desaturation, growth hormone signaling, oxidative phosphorylation, protein ubiquitination, proton transport, and ubiquinone biosynthesis were downregulated in the liver of periparturient dairy cows with ketosis. Genes that were upregulated during ketosis included those associated with cytokine signaling, fatty acid uptake/transport, and fatty acid oxidation (Loor et al. 2007).

Additionally, Loor's lab (2007) reported that mRNA expression of several nuclear receptors and transcription factors (e.g., *HNF4A*, *PPARA*, and *PPARGC1A*) in the liver are essential to maintain ketogenesis and fatty acid oxidation. Hepatic mRNA expression of angiopoietin-like 4 (*ANGPTL4*, a hepatokine), and other genes associated with gluconeogenesis and ketogenesis were also greater during a state of NEB and ketosis. More importantly, some potentially novel genes (e.g., *LPIN1*, *LPIN3*, and *ANGPTL4*) were identified to be useful markers for hepatic metabolic adaptations to NEB, and the altered physiological state, such as hepatic fatty acid oxidation and adipose tissue lipolysis, during the transition period (Loor et al. 2007).

10.2.2 Proteomics Identifies Protein Biomarkers for Ketosis

A few studies on proteomic analysis of liver, milk, urine, and serum of ketotic cows have been reported. The 2-DE gels coupled with MALDI-TOF-TOF MS/MS were used to compare proteomics profile in the liver between ketotic and healthy cows (Xu and Wang 2008; Xu et al. 2008). Numerous molecular pathways were activated in the liver of cows with ketosis. Results showed that 38 hepatic proteins of which 19 were upregulated (e.g., alpha-enolase, creatinine kinase M-type, fast-twitch myosin light chain 1, hypothetical protein MGC128326, myoglobin, similar to myosin light chain 2, similar to myosin light chain 1, and tropomyosin 3), and 19 were downregulated (e.g., acetyl-Coenzyme A acetyltransferase 2 (ACAT2), 3-hydroxyacyl-CoA dehydrogenase type-2 (HCDH), arginase-1, and elongation factor Tu (EF-Tu)] in ketotic cows and most of them play pivotal roles in various cellular functions and metabolic pathways including amino acid metabolism, antioxidation, carbohydrate degradation, cell structure, energy metabolism, fatty acid metabolism, glycolysis, nucleotide metabolism, and protein metabolism (Xu and Wang 2008). Findings from the study also indicated that fatty acid β-oxidation, ketogenesis, and protein synthesis were depressed, however, gluconeogenesis and energy production were elevated in ketotic cows (Xu et al. 2008).

The same research group (Xu et al. 2015a) performed surface-enhanced laser desorption/ionization time-of-flight (SELDI-TOF) MS-based proteomics to screen urine proteomic profiles of dairy cows diagnosed with CK. Urine concentrations of 11 proteins including amyloid precursor protein, apolipoprotein (Apo) C-III, cystatin C, fibrinogen, hepcidin, human neutrophil peptides, osteopontin, VGF (non-acronymic) protein, serum amyloid A (SAA), the C1 inhibitor (C1INH), and transthyretin were decreased in ketotic cows (Xu et al. 2015a). The relationship between the urine protein profile and the other three parameters (i.e., the inflammatory response, the clinical symptoms of depression, and lipid metabolism) were discussed by the authors in more detail. These 11 identified urine proteins might be used as novel protein biomarkers for ketosis and help to better understand the physiological perturbations and pathobiology of ketosis.

In the aspect of assembly of very low-density lipoprotein (VLDL), some apolipoproteins such as Apo A-I, Apo B-100, and Apo C-III were reported lower in ketotic cows (Oikawa et al. 1997; Yamamoto et al. 2001). Nakagawa-Ueta and Katoh (2000) also reported that the activity of lecithin:cholesterol acyltransferase was decreased before the occurrence of ketosis. In a recently published short communication, an enzyme-linked immunosorbent assay (ELISA) based proteomics approach revealed that serum FGF-21 can be used as a sensitive biomarker for diagnosis of ketosis in dairy cows; this finding is consistent with previous results from genomics studies in regard to hepatic *FGF21* mRNA expression (Akbar et al. 2015; Xu et al. 2016a).

Another proteomics study was also conducted in dairy cows diagnosed with fatty liver, which is a concurrent disorder of type II ketosis in dairy cows (Kuhla et al. 2009). In that study, liver specimens were analyzed by 2-DE and MALDI-TOF-TOF MS/MS for measurement of 5′-AMP-activated protein kinase (AMPK) and regulatory/regulated proteins. Results of this study showed that among 34 differentially expressed proteins downregulation of proteins associated with fatty acid oxidation [e.g., Ac-CoA acetyltransferase 2 (thiolase), acyl-CoA dehydrogenase, aldehyde dehydrogenase, aldo-keto reductase family 1 member C1, fatty acid binding protein 1 (FABP1), peroxiredoxin 6, and sterol carrier protein 2 (SCP2)], carbohydrate metabolism [(e.g., 6-phosphofructokinase), aldehyde dehydrogenase 2, enolase 1, fructose-bisphosphate aldolase B, sorbitol dehydrogenase, and triosephosphate isomerase], electron transfer (e.g., cytochrome B5 and electron transfer flavoprotein-β), protein degradation and antigen processing [e.g., a member of the heat-shock protein-70 (HSP70) family, dihydrodipicolinate synthase (DHDPS), heat-shock 70 kDa protein 5 (HSPA5), proteasome 26S subunit, protein disulfide-isomerase-related protein 5 (Erp57), and ubiquitin carboxyl-terminal esterase L3], as well as cytoskeletal rearrangement (e.g., an isoform of cytokeratin 8 and cofilin) were identified.

Proteins upregulated during fatty liver included two enzymes of the urea cycle (i.e., arginase-1 and argininosuccinate synthetase), fatty acid or cholesterol transport proteins (e.g., Apo-AI and sterol carrier protein 2), an inhibitor of glycolysis (i.e., parathymosin), and previously unknown changes in calcium signaling network (e.g., annexin IV, calcium binding protein SPEC 2D, and regucalcin) (Kuhla et al. 2009). These identified proteins might be used as important indicators for metabolic adaptation to feed restriction and development of fatty liver, as well as subsequently development of type II ketosis in dairy cows.

Changes in milk proteome, especially in the milk fat globule membrane (MFGM) proteins, and their association with NEB (usually involving ketosis-related parameters) in postparturient dairy cows were reported by Lu et al. (2013). Results indicated that cows with NEB during the early lactation, the MFGM had greater amounts of proteins related to acute inflammatory and immune response [e.g., compliment C3, lipopolysaccharide binding protein (LBP), fibronectin (FN1), prothrombin (F2), alpha-1-acid glycoprotein (ORM1), inter-alpha-trypsin inhibitor heavy chain H4 (ITIH4), alpha-2-HS-glycoprotein (AHSG), complement component C9, beta-2-microglobulin (B2M), MHC class I antigen (Man8), zinc-alpha-2-glycoprotein (AZGP1), peroxiredoxin-1 (PRDX1), syntaxin binding protein 2 (STXBP2), IGL protein, and IGK protein].

In a recent study from our group, we used ELISA-kit based immune-assays to do proteomics analysis of serum cytokines and acute phase proteins (APPs) in cows with ketosis and healthy cows (Zhang et al. 2016). Results showed that ketotic cows had greater levels of interleukin (IL)-6, tumor necrosis factor (TNF), and serum amyloid A (SAA) in the serum in comparison with the CON cows. Most importantly, serum concentrations of IL-6 and TNF were greater starting at −8 and −4 weeks prior to parturition in pre-ketotic cows.

10.2.3 Metabolomics Identifies Metabolite Biomarkers for Ketosis

Metabolites are considered as the end products of biochemical pathways that indicate specific perturbations in the development of disease. Nuclear magnetic resonance (NMR) and mass spectrometry (MS) based metabolomics are the most popular "omics" approaches that have been used for ketosis studies by dairy researchers. Blood (i.e., serum or plasma), urine, and milk are most commonly used body fluids in these metabolomics based ketosis studies in dairy cows.

For example, a GC-MS based metabolomics approach was used to compare plasma metabolic profiles in cows that were diagnosed with CK, SCK, or were clinically healthy (CON) (Zhang et al. 2013). In this study, 40 plasma biomarkers (i.e., amino acids, carbohydrates, fatty acids, sitosterol, vitamin E isomers, and others) that can distinguish CK, SCK, and CON groups were identified, among which 25 metabolites were similar in both CK and SCK cows. Univariate analysis data showed that 9 of the 25 metabolites, namely, 2,3,4-trihydroxybutyric acid, galactose, glucose, glucuronic acid, glycolic acid, lactate, L-alanine (L-Ala), pyroglutamic acid, and ribitol were decreased in both CK and SCK cows compared to CON cows. Whereas the other 16 metabolites including aminomalonic acid, α-aminobutyric acid, L-isoleucine (L-Ile), glycine (Gly), 3-hydroxybutyric acid, palmitic acid, heptadecanoic acid, stearic acid, trans-9-octadecenoic acid, myristic acid, cis-9-hexadecenoic acid, 2-piperidinecarboxylic acid, 3-hydroxyvaleric acid, 3-hydroxy-3-methylglutaric acid, α-tocopherol, and sitosterol were increased in both CK and SCK groups compared to the CON group. These results suggest that the development and progression of ketosis involves complex perturbations of metabolic profiles and multi-biochemical pathways such as amino acid metabolism, fatty acids metabolism, gluconeogenesis, glycolysis, and pentose phosphate pathway (Zhang et al. 2013).

Another recent study using LC-MS identified several other plasma metabolites in cows affected by CK (Li et al. 2014). This group of researchers reported 13 plasma metabolites that were different between the CK and CON cows. More specifically, plasma levels of five metabolites including Gly, glycocholic acid, palmitoleic acid, tetradecenoic acid, and valine (Val) were elevated in cows with CK when compared with CON cows. On the other hand, nine plasma metabolites like arginine (Arg), aminobutyric acid, creatinine, leucine (Leu)/isoleucine (Ile), lysine (Lys), norcotinine, tryptophan (Trp), and undecanoic acid were decreased in CK cows. These 13 plasma metabolites of CK are involved in various metabolic pathways including amino acid metabolism, fat metabolism, gluconeogenesis, nerve signal transduction, and vitamin metabolism.

As a complementary platform to MS technologies, ^{1}H-NMR-based metabolomics was also applied to measure the metabolic profiles in the plasma of the three groups of cows including cows diagnosed with CK, SCK, and CON cows (Sun et al., 2014). Twenty-five metabolites, including acetate, AcAc, Ac, BHBA, choline, citrate, creatine, formate, glucose, glutamic acid (Glu), glutamine (Gln), Gly, Ile, lactate, Leu, Lys, phenylalanine (Phe), proline (Pro), tyrosine (Tyr), and Val were different among the 3 groups. In particular, plasma concentrations of acetate, AcAc, Ac, BHBA, and n-acetyl glycoprotein were increased in both CK and SCK cows compared with CON cows. Plasma levels of choline, creatine, Gly, Leu, Ile, and Val were elevated only in cows with SCK. To the contrary, some metabolites such as Gln, Glu, glucose, histidine (His, e.g., 1-methyl histidine and 3-methyl histidine), lactate, Lys, and Phe were lower in plasma of cows with ketosis. And the plasma levels of Ala, citrate, formate, low-density lipoprotein (LDL), *myo*-inositol, Pro,

Tyr, and VLDL were decreased only in cows with CK. More importantly, the authors developed OPLS-DA models that can diagnose CK with a very high sensitivity and specificity of 100%. In the case of SCK, the OPLS-DA models have a sensitivity of 97.0% and specificity of 95.7%, respectively.

The combination of ^{1}H-NMR and multivariate analyses have been used to effectively distinguish the plasma metabolic profiles among cows with type I ketosis, type II ketosis, and healthy control cows (Xu et al. 2015b). The results of this study indicated remarkable differences in plasma metabolites among the three groups. Specifically, comparison between type I ketosis and CON group indicated that cows with type I ketosis had greater concentrations of Ac and BHBA, and lower levels of Ala, α-glucose, β-glucose, citrate, creatine, formate, Gln, Glu, Gly, His, Lys, *myo*-inositol, O-acetyl glycoprotein, PC, Phe, and Tyr. Type II ketotic cows had higher concentrations of plasma Ac, BHBA, and lactate, and decreased concentrations of plasma Ala, creatine, Lys, and Tyr than CON cows. Moreover, different metabolic profiles were also determined between type I ketosis and type II ketosis. When compared with type II ketosis, cows affected by type I ketosis had greater levels of Ac, acetate, BHBA, Ile, Leu, LDL, Val, and VLDL, and lower levels of α-glucose, β-glucose, citrate, creatine, formate, Gln, Glu, Gly, His, Lys, O-acetyl glycoprotein, PC, Phe, and Tyr (Xu et al. 2015b). Results from the study indicated different etiopathologies of type I and type II ketosis in dairy cows.

Milk is an easily accessible and noninvasively collected bio-fluid, which has been investigated by dairy science researchers using metabolomics technology for identification of diagnostic biomarkers for ketosis in different breeds (i.e., Brown Swiss, Holstein-Friesian, and Simmental Fleckvieh) and for animals in different lactations (Klein et al. 2012). After acquisition of 1D ^{1}H-NMR and 2D ^{1}H-^{13}C NMR spectra for each milk sample, statistical analyses revealed that the milk GPC/PC ratio can be used as a reliable prognostic biomarker for risk of ketosis. Particularly, higher concentrations of milk GPC can be used as an indicator of a low risk of ketosis incidence throughout lactation. The cutoff value of 2.5 (i.e., $\geq$2.5) for milk GPC/PC ratio indicates that the cow has a very low risk for developing ketosis (Klein et al. 2012).

Another milk metabolomics study was conducted on metabolites associated with the metabolic status (e.g., known biomarkers for energy metabolism such as Ac and BHBA for ketosis) of dairy cows by the same research group (Klein et al. 2010). In the study, NMR (i.e., 1D and 2D) and GC-MS were utilized for measurement of a total of 44 milk metabolites (23 by NMR and 25 by GC-MS, 4 overlaps from two approaches) in milk specimens collected from individual dairy cows of two different breeds during early and late lactation. Results indicated that there were significant differences present in numerous milk metabolites, especially known biomarkers of ketosis such as Ac and BHBA in different breeds (Holstein-Friesian, Brown Swiss, and Simmental Fleckvieh) of cows during the early lactation (Klein et al. 2010).

So far, there are no specific lipidomics based studies performed on cows affected by ketosis. Our group (Ametaj's) is conducting a comprehensive NMR and MS based metabolomics study to identify early biomarkers for monitoring, diagnosis, and prediction of several periparturient diseases in dairy cows by screening serum, urine, and milk specimens. We also have specifically screened for biomarkers of ketosis in three body fluids including serum, urine, and milk, respectively. In a DI/LC-MS/MS based metabolomics study, 128 metabolites including amino acids, glycerophospholipids, sphingolipids, acylcarnitines, biogenic amines, and hexose were identified and quantified in the serum of pre-ketotic, ketotic, and post-ketotic cows and CON ones. Results indicate that cows with ketosis had numerous perturbations of serum metabolites related to amino acid, carbohydrate, and lipid metabolism before, during, and after the occurrence of the disease (Zhang et al. 2017a). One of the most significant findings of the study was that we identified a seven-metabolite (i.e., Lys, , lysophosphatidylcholine acyl (lysoPC a) C17:0, lysoPC a C18:0, lysoPC a C16:0, Ile, kynurenine, and Leu) signature set for early diagnosis of ketosis at –8 wks prepartum (i.e., 9–11 wks prior to traditional diagnosis of ketosis through measurement of serum BHBA) (Zhang et al. 2017a). In a more recently published paper, inductively coupled plasma mass spectrometry (ICP-MS) metallotyping was performed in the serum and urine of ketotic and healthy cow. Overall results revealed major alterations of mineral elements (e.g., aluminum, iron, manganese, arsenic, boron, calcium, phosphorus, potassium, and magnesium) in both serum and urine of preketotic (i.e., at –8 wks or –4 wks prepartum) and postketotoc cows (i.e., at +4 wks or +8 wks postpartum) (Zhang et al. 2017b).

10.3 Network Models and Pathways Involved in Ketosis

In consideration of all the "omics" studies conducted on the ketosis in dairy cows, enormous alterations of genes, proteins (e.g., enzymes), and metabolites occur during ketosis. There is no exception that almost all investigators involved in bovine ketosis have used one "omics" platform only for studying the pathobiology of the disease. However, no single "omics" technology can completely unravel the complexities of ketosis. Integration of multiple tiers of "omics" data is indispensable to acquire a precise image of ketosis and build network models for better understanding of the disease pathobiology. Correlations between all levels of biological information require to be clarified. All levels of omics data from mRNA, proteins, and metabolites are affected by genetic variants, epigenetic, and environmental factors (Petersen et al. 2014; Shin et al. 2014; Zierer et al. 2015). It is feasible to put the results of ketosis "omics" studies in a systems biology context by integrating the variables of interest (i.e., ketosis-related genes, proteins, or metabolites) and metabolic pathways into known biological networks. Some of the metabolic pathways affected during ketosis are outlined below. The major metabolic pathways involved in ketosis based on omics data (i.e., selected DNAs, mRNAs, proteins, and metabolites) from the liver and blood are summarized in Fig. 10.3.

10.3.1 Lipid Metabolism

Perturbations in lipid metabolism, fatty acid transport, or metabolic process in cows with ketosis have been previously determined by associated gene expression. A microarray and qPCR based transcriptomics study provided evidence for

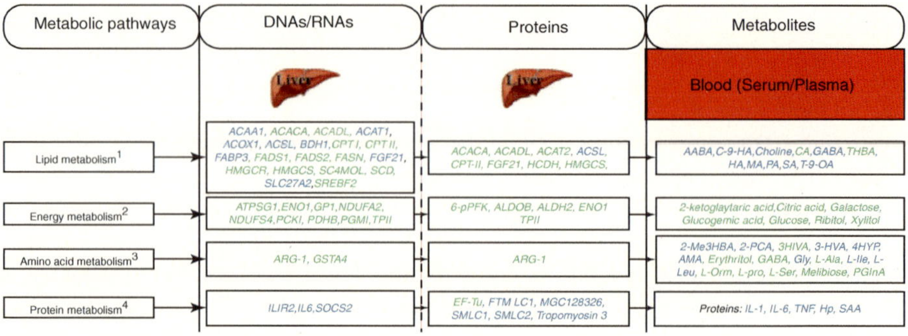

Fig. 10.3 Contribution of omics sciences to better understanding of ketosis

upregulation or downregulation of specific lipid catabolism-related genes in the liver of periparturient dairy cows with nutrition-induced ketosis (Loor et al. 2007). Data from this study revealed some lipid metabolism associated genes including acetyl-coenzyme A acyltransferase 1 (*ACAA1*), acetyl-coenzyme A acetyltransferase 1 (*ACAT1*), acyl-CoA oxidase 1-palmitoyl (*ACOX1*), 3-hydroxybutyrate dehydrogenase-type 1 (*BDH1*), fatty acid binding protein 3 (*FABP3*), and solute carrier family 27 (fatty acid transporter)-member 2 (*SLC27A2*) were upregulated in ketotic cows. *ACAT1* gene encodes the enzyme acetyl-coenzyme A acetyltransferase 1, which is responsible for the last step of ketone body breakdown. Similarly, *BDH1* is also an important gene regulating ketolysis (Mahendrean et al. 2013). Upregulated expression of *ACAT1* and *BDH1* might be a defense mechanism towards hyperketonemia. *ACAA1* and *ACOX1* are both important in the process of fatty acid β-oxidation (Baes and Van Veldhoven 2012). Upregulation of *FABP3* and *SLC27A2* might be associated with the large influx of NEFA during the state of NEB. Loor et al. (2007) speculated that upregulation of *FABP3* might channel fatty acids such as palmitic and oleic acids towards peroxisomal β-oxidation in cows with ketosis. In the gene expression study, it was also reported that genes associated with de novo fatty acid synthesis [e.g., fatty acid synthase (*FASN*), acetyl-coenzyme A carboxylase (*ACACA*)], fatty acid desaturation [e.g., stearoyl-CoA desaturase (*SCD*), fatty acid desaturase 1 (*FADS1*), fatty acid desaturase 2 (*FADS2*)], and cholesterol synthesis [e.g., 3-hydroxy-3-methylglutaryl-coenzyme A reductase (*HMGCR*), sterol-C4-methyl oxidase-like (*SC4MOL*)] were downregulated during ketosis (Loor et al. 2007). Most of these lipid metabolism genes are regulated by a central transcription factor [i.e., sterol regulatory element binding transcription factor 2 (SREBF2)], which also was downregulated. Protein levels of ACACA were also proved to be decreased in the liver of ketotic cows (Li et al. 2012). Both genes *FASN* and *ACACA* play a pivotal role in de novo lipogenesis, which generates fatty acids from surplus glucose (Okuda and Morita 2012; Postic and Girard 2008). Several fatty acids (especially C18:1) are endogenously synthesized and serve as major substrates for hepatic TAG synthesis (Loor et al. 2007). Downregulation of *FASN* and *ACACA* might be related with hypoglycemia during ketosis and hyperketonemia might suppress expression of two genes to prevent further development of steatosis. *FADS1* and *FADS2* genes encodes two rate-limiting enzymes, which desaturate linoleic acid [LA; 18:2(n-6) to arachidonic acid [ARA; 20:4(n-6)], and α-linolenic acid [ALA; 18:3(n-3)] to eicosapentaenoic acid [EPA; 20:5(n-3)] or docosahexaenoic acid [DHA; 22:6(n-3)], respectively (Xie and Innis 2008). Downregulation of hepatic *FADS1* and *FADS2* suggested polyunsaturated fatty acid (PUFA) synthesis (e.g., n-3 and n-6 fatty acids) is impaired in cows with ketosis (Loor et al. 2007). Similarly, downregulation of *SCD* indicated a substantial impairment of de novo synthesis of the major monounsaturated fatty acid (MUFA, i.e., oleic acid) in cellular membranes (Loor et al. 2007). MUFA synthesized endogenously are crucial for TAG, phospholipids, and cholesteryl esters synthesis (Scaglia and Igal 2005). Results from Loor et al.'s (2007) study suggested that downregulation of *SCD* during ketosis might contribute to liver lipidosis by decreased synthesis of oleic acid and subsequently impaired VLDL synthesis and TAG secretion.

Expression of fatty acid metabolic enzymes was also studied by comparative proteomic analysis of livers, plasma, and urine in ketotic cows. Two enzymes ACAT2 and HCDH, which play pivotal roles in fatty acid β-oxidation pathway, were reported lowered in the liver tissues of cows with ketosis, suggesting impaired liver function in utilization of fatty acids (Xu and Wang 2008; Xu et al. 2008). Thereafter, accumulation of fatty acids occurs in hepatic cells. Decreased levels of ACAT2 and HCDH in ketotic cows might be a negative feedback induced by high levels of ketone bodies. Another study conducted by Kuhla et al. (2009) demonstrated consistent results showing ACAT2 and acyl-CoA dehydrogenase (ACAD) were both downregulated in cows with fatty liver, suggesting a diminished fatty acid β-oxidation might be a factor contributing to hepatic lipidosis, which subsequently lead to ketosis. FGF21, an important regulator of lipid metabolism, has been studied by several researchers (Badman et al. 2007; Gälman et al. 2008). Xu and Wang (2008) reported that FGF21 is negatively correlated with BHBA and can be used independently as a diagnostic biomarker of ketosis. FGF21 potentially stimulates glucose uptake and inhibits lipolysis (Chen et al. 2011; Kharitonenkov et al. 2005). An explanation for decreased FGF21 in ketotic cows might be lower glucose concentration in blood, which also leads to excessive lipid mobilization from adipose tissues.

In another study, six important liver enzymes involved in fatty acid metabolism were measured on levels of mRNA abundance and protein content in postpartum cows with ketosis (Li et al. 2012). Real-time PCR and ELISA analyses revealed that mRNA and protein levels of acyl-CoA synthetase long-chain (ACSL) were greater in ketotic cows compared with the CON group. By contrast, mRNA levels of carnitine palmitoyltransferase I (CPT I), and both mRNA and protein levels of ACACA, acyl-CoA dehydrogenase long chain (ACADL), 3-hydroxy-3-methylglutaryl-CoA synthase (HMGCS), and carnitine palmitoyltransferase II (CPT II) were decreased in the liver of ketotic cows (Li et al. 2012). Upregulation of *ACSL* gene and increased ACSL enzyme suggested cows with ketosis require elevated oxidation of long-chain fatty acids (LCFA), which are mainly derived from lipolysis of adipose tissue, to make up the energy requirement. Increased expression of *ACSL* also can produce greater levels of fatty acyl-CoA, which might be esterified to TAG and lead to severe liver lipodosis during ketosis (Li et al. 2012). Enzymes CPT I and CPT II together with carnitine-acylcarnitine translocase mediate the mitochondrial translocation of acyl-groups (Dann and Drakley 2005). ACADL is an important enzyme catalyzing the first step of the β-oxidation of LCFA in the mitochondria. Decreased expression of *ACADL* gene and protein indicated the function of fatty acid β-oxidation in hepatocyte is impaired during ketosis. HMGCS is the rate-limiting enzyme of the ketogenic pathway, controlling ketogenesis under several physiological conditions such as a high-fat diet, fasting, and prolonged exercise (Li et al. 2012). One speculation for downregulation of *HMGCS* and diminished levels of HMGCS enzyme might be related to feedback inhibition of hyperketonemia in cows with ketosis (Ness and Chambers 2000).

Disturbances of metabolites involved in fatty acid metabolism have been reported by numerous metabolomics or lipidomics studies. Plasma metabolic profiling by GC-MS revealed that cows affected by ketosis experienced altered metabolites involved in fatty acid metabolism (Zhang et al. 2013). Concentrations of glycolic acid (GA) and 2,3,4-trihydroxybutyric acid (THBA) in plasma were decreased, whereas α-aminobutyric acid (AABA), palmitic acid (PA), heptadecanoic acid (HA), stearic acid (SA), trans-9-octadecenoic acid (T-9-OA), myristic acid (MA), cis-9-hexadecenoic acid (C-9-HA), 4-aminobutyric acid (GABA), and xylitol were elevated in ketotic cows. The most dominant species of the nine increased metabolites were NEFA, which confirmed that excessive lipolysis during hypoglycemia may lead to ketosis. In another NMR-based plasma profiling study, Sun et al. (2014) reported that cows with SCK had greater levels of choline in the plasma. Choline, as a component of lecithin, plays an important role in transportation of fatty acids in the liver and promoting fatty acid β-oxidation to avoid excessive lipid deposition in hepatocytes (Gao et al. 2008). Increased concentrations of choline in plasma of SCK cows suggested that choline might be involved in a protection mechanism to prevent further development of fatty liver during ketosis.

NMR-based metabolomics analyses of milk showed alterations of phospholipid pathway in ketotic cows (Klein et al. 2012). Greater levels of milk GPC indicated lower incidence of ketosis and milk GPC/PC ratio greater than 2.5 showed lower risk of ketosis. Raised GPC and GPC/GC ratios in milk of ketotic cows might be caused by a higher rate of blood phosphatidylcholine breakdown due to higher enzyme concentration or activities of phospholipase A_2 and lysophospholipase, which are used as alternative energy source during hypoglycemia in cows with ketosis. Alterations of phospholipid pathway during ketosis were further investigated on levels of gene expression based on the previously identified screening biomarker (i.e., the ratio of milk GPC and PC) for the risk of ketosis (Tetens et al. 2015). Results showed a major quantitative trait locus on BTA25 is highly heritable with the milk level of GPC and milk GPC/PC ratio. Specifically, the *APOBR* gene coding for the apolipoprotein B receptor was identified as a risk biomarker for ketosis.

10.3.2 Energy Metabolism: Carbohydrate Metabolism, Glycolysis, Gluconeogenesis, and the Pentose Phosphate Pathway

It has been reported that ketotic cows have downregulations of genes associated with glycolysis, gluconeogenesis, and tricarboxylic acid (TCA) cycle such as α-enolase 1 (*ENO1*), pyruvate dehydrogenase (lipoamide) beta (*PDHB*), phosphoglycerate kinase 1 (*PCK1*), triosephosphate isomerase 1 (*TPI1*), glucose phosphate isomerase (*GPI*), and phosphoglucomutase 1 (*PGM1*) in the liver (Loor et al. 2007). Additionally, more than 80% of genes associated with functions in oxidative phosphorylation (e.g., *ATP5G1*), mitochondrial electron transport, and ubiquinone biosynthesis (e.g., *NDUFA2*, *NDUFS4*) were downregulated in cows with ketosis. PCK1 has long been considered as a rate-limiting enzyme catalyzing gluconeogenesis (Rognstad 1979). It was reported that the lower activities of crucial gluconeogenic enzymes and failure of hepatic gluconeogenesis to generate adequate glucose for body needs and milk production may lead to the occurrence of ketosis (Herdt 2000; Murondoti et al. 2004). In the nutrition-induced ketosis study, gene expression results indicate that the most affected pathways by ketosis are oxidative phosphorylation, protein ubiquitination, and ubiquinone synthesis with 97, 125, and 41 genes altered, respectively (Loor et al. 2007).

Proteome analysis of nutritional-derived fatty liver, which is a common complication of ketosis, revealed that glycolytic and gluconeogenic enzymes, such as 6-phosphofructokinase (6-pPFK), ENO1, TPI1, fructose-bisphosphate aldolase B (ALDOB), sorbitol dehydrogenase (SDH), and aldehyde dehydrogenase 2 (ALDH2), were less expressed on the protein level under a restricted feeding model (Kuhla et al. 2009). Furthermore, parathymosin, an inhibitor of glycolytic enzymes, is decreased in cows suffering from hypoglycemia.

The levels of another important hepatic enzyme, EF-Tu, which is important for global protein biosynthesis (i.e., transforming amino acids into proteins), were reported lower in the livers of ketotic cows. It was speculated that most glucogenic amino acids are involved in gluconeogenesis and the TCA cycle, for ATP production, during the state of ketosis (Xu et al. 2008). Since propionate, glycerol, lactate, and glucogenic amino acids are the substrates for gluconeogenesis, comprehensive analysis of these substrates in ketotic cows is required.

Numerous metabolomics studies have revealed significant alterations in the metabolites involved in energy metabolism. Citric acid, a TCA intermediate, was reported lower in plasma of cows with type I ketosis, suggesting that the efficiency of the TCA cycle was affected in ketotic cows (Xu et al. 2015b). Additionally, Zhang et al. (2013) reported that both CK and SCK cows had decreased plasma glucogenic amino acids, which are important for gluconeogenesis. Moreover, other metabolites (e.g., glucuronic acid, ribitol, galactose, and glucose), which can enter glycolysis via the pentose phosphate pathway, were also decreased in the plasma of ketotic cows. A precursor of xylulose 5-phospahte (i.e., xylitol), which is an intermediate of the pentose phosphate and glycolytic pathways, was increased in cows with CK. Moreover, 2-ketogluytaric acid, which is involved in the TCA cycle, was found to be downregulated in cows affected by SCK. Therefore, altered concentrations of plasma carbohydrates, glucogenic precursors (e.g., lactic acid and L-Ala), and vitamin C (a precursor of glucuronic acid) may play a pivotal role in the development of ketosis (Zhang et al. 2013). When comparing type II ketosis with type I ketosis, it was reported that cows with type II ketosis had less glucogenic amino acids than type I ketotic cows, which indicated that different forms of ketosis (i.e., SCK vs CK, type I vs type II ketosis) had different metabolic profiles, which gave insights that ketosis need to be carefully investigated (Xu et al. 2015b; Zhang et al. 2013). Given the evidence that cows with ketosis experience perturbations of intermediates involved in TCA cycle, glycolysis, and gluconeogenesis, more research is required for better understanding the full picture of energy metabolism during ketosis.

10.3.3 Amino Acid Metabolism

Besides serving as building blocks for protein synthesis, amino acids play important roles as a source of energy and in immune functions (Li et al. 2007; Moriwaki et al. 2004). Three types of amino acids (i.e., glucogenic, ketogenic, or both) can be utilized as precursors for gluconeogenesis or ketogenesis, and 10 to 15% of energy production in humans are from catabolism of amino acids (D'Mello 2003; Pasquale 2007). Moreover, amino acids also serve as building blocks for synthesis of enzymes (catalyzing different pathways) and innate immunity reactants (e.g., pro-inflammatory cytokines, APPs, and antibodies) during bacterial infection or inflammation. Based on the concept of NEB and our speculation for a potential role of bacterial endotoxins in the pathobiology of ketosis, new data on amino acids provide considerable information for better understanding of the disease.

Alterations of various amino acid associated metabolic pathways in cows with ketosis have been previously reported. Xu and Wang (2008) demonstrated that a key enzyme (arginase-1), which catalyzes the first step of arginine degradation through the urea cycle, was decreased in the livers of ketotic cows. Downregulation of arginase-1 suggested that arginine degradation was suppressed and arginine might be utilized by other pathways such as gluconeogenesis.

Plasma concentrations of numerous metabolites that are involved in different amino acid metabolic pathways such as Ala-aspartate (Asp) metabolism (e.g., L-Ala), glutathione metabolism (e.g., pyroglutamic acid), Gly-serine (Ser)-threonine (Thr) metabolism (e.g., Gly, L-Ser), Val-Leu-Ile degradation (e.g., L-Ile), Val-Leu-Ile biosynthesis [e.g., 3-hydroxyvaleric acid (3HVA), 2-methyl-3-hydroxybutyric acid (2Me3HBA)], Lys metabolism [e.g., 2-piperidinecarboxylic acid (2PCA)], and Arg-Pro metabolism (e.g., melibiose, L-Pro) are altered in cows with SCK and CK. Several other amino acids and their catabolic products such as L-Ile (both ketogenic and glucogenic), Gly (originated from Ser), aminomalonic acid (AMA, a constituent of proteins before hydrolysis), and 2PCA (Lys metabolism) are upregulated in ketotic cows (D'Mello 2003; Hinko et al. 1996; Marsh et al. 1993; Marvin and Francis 1949), implying that proteolysis is enhanced to meet body energy requirements in ketotic cows. In particular, upregulation of 2PCA, a catabolic product of Lys (a ketogenic amino acid), in ketotic cows suggest that 2PCA might indirectly contribute to production of ketone bodies (Hinko et al. 1996).

Despite similarities between metabolic profiles of cows with CK or SCK, various different alterations of metabolites regarding amino acid metabolism were identified between the two groups (Zhang et al. 2013). 2Me3HBA, a metabolite that participates in Ile catabolism, ketogenesis, and fatty acid β-oxidation, was increased in cows with CK compared with CON ones (Salway 2004). In contrast, some metabolites including GABA from L-glutamic acid (L-Glu) metabolism (Watanabe et al. 2002), 3-hydroxyisovaleric acid (3HIVA) catabolized from Leu (a ketogenic amino acid) (Ferreira et al. 2007), and erythritol, a precursor of L-Ser (a glucogenic amino acid) and fructose 6- phosphate (Munro et al. 1998; Hamana et al. 2010) were decreased in CK cows. Perturbed amino acids and their precursors or catabolites suggest that they might be important indicators of CK. In the comparisons between SCK and CON, different alterations of amino acid catabolites were observed. Upregulated metabolites in SCK cows included L-Leu and 4-hydroxyproline (4HYP, catabolized from Pro hydroxylation) (Siddiqi et al. 2002; Zhang et al. 2013). Whereas L-ornithine, which is involved in the urea cycle, was decreased in cows with SCK (Motyl and Barej 1986). Therefore, SCK and CK might have their own specific biomarkers and it is obvious that the mere ketone body concentration such as blood BHBA is insufficient to distinguish the different forms of ketosis in dairy cows.

Two other metabolomics studies have shown that cows with ketosis (both SCK and CK, and both type I and type II ketosis) had lower levels of glucogenic amino acids in the plasma (Sun et al., 2014; Zhang et al. 2013). It was speculated that a shortage of gluconeogenic substrates might be an important factor in the pathogenesis of ketosis. Contrasting results, however, were found in our recent study, which showed that ketotic cows had greater concentrations of gluconeogenic amino acids (unpublished data). We hypothesized that bacterial toxic products and activation of systemic inflammation might contribute to the pathogenesis of ketosis. Amino acids also play important roles in immune functions and increased mobilization of muscle protein might generate substrates (i.e., amino acids) for production of immunity reactants during activation of immune response.

10.3.4 Protein Metabolism

The homeostasis of body protein stores and tissue mass in normal dairy cows is maintaned through a balance between the rates of synthesis and degradation of cell proteins (Lecker et al. 2006). Genes associated with protein synthesis were among the most affected ones (i.e., downregulated) in cows with ketosis. Moreover, canonical signaling pathway analysis indicated that protein ubiquitination contained the largest number (42) of upregulated genes and many protein ubiquitination associated genes downregulated (Loor et al. 2007). Results indicated that protein degradation through the ubiquitin-proteasome pathway is more excessive in ketotic cows. Excessive proteolysis or muscle wasting can be triggered by a variety of pathologic conditions such as low fasting insulin levels, or cytokines (e.g., TNF) during sepsis (Lecker et al. 2006). Besides lipolysis from adipose tissue, proteolysis is also an alternative way to produce precursors (i.e., amino acids) for hepatic gluconeogenesis or other synthesis of proteins (Mitch and Goldberg 1996).

Xu and Wang (2008) demonstrated that alterations of cell structural proteins in the liver might be involved in the process of ketosis in dairy cows. In the study, various cytoskeleton proteins including fast-twitch myosin light chain 1, hypothetical protein MGC128326, myoglobin, similar to myosin light chain 1 similar to myosin light chain 2, and tropomyosin 3 were elevated in the liver of ketotic cows. Inversely, the expression of hepatic EF-Tu, which is an important enzyme for protein biosynthesis, was decreased in cows with ketosis (Xu et al. 2008). In comparison with the previously mentioned elevated proteolysis, decreased protein biosynthesis further disturbs homeostasis of body protein stores, which might be a factor that contributes in ketosis. In a proteomics study, numerous enzymes (e.g., a member of the heat-shock protein-70 family, heat-shock 70 kDa protein 5, proteasome 26S subunit, protein disulfide-isomerase-related protein 5, and ubiquitin carboxyl-terminal esterase L3) that participate in protein degradation were downregulated in cows with fatty liver (a concurrent disease with ketosis) (Kuhla et al. 2009). Thus, it is speculated that during the later stages of ketosis, a protection mechanism might be activated to prevent the cow's muscle's mass from further protein degradation.

10.3.5 Other Metabolic Pathways

Apart from the aforementioned alterations of metabolic pathways in cows with ketosis, more perturbations have been previously reported. For example, expression of proteins that are involved in antioxidation, cell structure, nervous signal transduction, nucleotide metabolism, and vitamin metabolism were different between ketotic and CON cows (Xu and Wang 2008; Xu et al. 2014). Please refer to the cited research articles for more information about other metabolic pathways that are altered in ketotic cows (Loor et al. 2007; Sun et al. 2014; Xu and Wang 2008; Xu et al. 2008, 2015b, 2016b; Zhang et al. 2013).

10.4 Conclusions and Future Directions

The limitations of the reductionism approach to biology are becoming increasingly obvious. Hyperketonemia, which is the only phenotypic characteristic of ketosis, is not sufficient for explanation of the pathobiology of the disease as well as for its prevention or treatment. The cutoff value of serum BHBA for defining ketosis, especially SCK, appears to be somewhat arbitrary. Concentrations of BHBA in the blood are changing over time, and different cutoff values during different periods (e.g., dry-off and early lactation) for diagnosis of ketosis need to be determined. The importance of management during the whole dry-off period also needs to be addressed. Based on published literature, ketosis should be characterized as perturbation of the overall metabolic status of the cow around calving resulting from an interplay of internal and external factors. The aforementioned advances in omics-based systems veterinary approach are driving the evolution of our knowledge on ketosis in dairy cows. As per the overview described above, significant alterations have been observed by different omics sciences (i.e., genomics, transcriptomics, proteomics, and metabolomics), which suggest that ketosis is a complex disease. Besides considering contribution of individual factors to the pathobiology of disease, the interconnection and interdependence among genes, proteins, and metabolite networks have to be taken into account. Integration of data obtained from multiple omes studies help to construct a complete picture of how ketosis is initiated and develops. There are still several areas that have remained ignored with regard to understanding of ketosis, such as the presence of subclinical cases of other diseases and their interactions with ketosis.

Even though high-throughput omics technologies (i.e., genomics, transcriptomics, proteomics, metabolomics, and lipidomics) are advancing and increasing our knowledge the amount of data related to bovine ketosis is increasing at a fast rate, and requires integration of different tiers of omics data. There is no exception that all the current published omics studies on ketosis in dairy cows have applied single platforms, focusing only on genes, proteins, or metabolites and missing vast amount of information. The same samples collected from a group of cows needs to be analyzed by different omics platforms simultaneously for better interpretation of ketosis under a systems veterinary approach. Furthermore, high-throughput and reliable bioinformatics tools need to be developed to help integrate multiple omics datasets based on the ontology, metabolic pathways, biological networks, or empirical correlations. It was also noticed that conflicting data have been obtained from different studies, which might be attributed to the fact that samples were collected at different stages of ketosis. Therefore, in future studies of ketosis, it is important to conduct longitudinal studies with repeated samplings from the same cows at different stages (i.e., before, during, and after the diagnosis of ketosis) for further understanding the onset and progression of the disease. Second, isotopic labeling techniques could be used in proteomics and metabolomics studies of bovine ketosis to track molecular pathways and biological processes of the targeted proteins or metabolites, helping to determine the source and end use of the molecules of interest. Third, high sensitivity and specificity risk biomarkers for ketosis need to be identified and general biomarker models to be validated with regard to ketosis. Fourth, new preventive and therapeutic strategies combining multi-target interventions need to be developed.

References

Akbar H, Batistel F, Drackley JK et al (2015) Alterations in hepatic FGF21, co-regulated genes, and upstream metabolic genes in response to nutrition, ketosis and inflammation in peripartal Holstein cows. PLoS One 10:e0139963

Badman MK, Pissios P, Kennedy AR et al (2007) Hepatic fibroblast growth factor 21 is regulated by PPARα and is a key mediator of hepatic lipid metabolism in ketotic States. Cell Metab 5:426–437

Baes M, Van Veldhoven PP (2012) Mouse models for peroxisome biogenesis defects and β-oxidation enzyme deficiencies. Biochim Biophys Acta 1822:1489–1500

Chen W, Hoo RL, Konishi M et al (2011) Growth hormone induces hepatic production of fibroblast growth factor 21 through a mechanism dependent on lipolysis in adipocytes. J Biol Chem 286:34559–34566

Dann HM, Drackley JK (2005) Carnitine palmitoyltransferase I in liver of periparturient dairy cows: Effects of prepartum intake, postpartum induction of ketosis, and periparturient disorders. J Dairy Sci 88:3851–3859

D'Mello JPF (2003) Amino acids as multifunctional molecules. In: D'Mello JPF (ed) Amino acids in animal nutrition, 2nd edn. CABI Publishing, Oxfordshire, pp 1–14

Duffield TF (2000) Subclinical ketosis in lactating dairy cattle. Vet Clin North Am Food Anim Pract 16:231–253

Duffield TF, Lissemore KD, McBride BW et al (2009) Impact of hyperketonemia in early lactation dairy cows on health and production. J Dairy Sci 92:571–580

Ferreira G, Weiss WP, Willett LB (2007) Changes in measures of biotin status do not reflect milk yield responses when dairy cows are fed supplemental biotin. J Dairy Sci 90:1452–1459

Gao H, Dong B, Liu X (2008) Metabonomic profiling of renal cell carcinoma: High-resolution proton nuclear magnetic resonance spectroscopy of human serum with multivariate data analysis. Anal Chim Acta 624:269–277

Gälman C, Lundåsen T, Kharitonenkov A et al (2008) The circulating metabolic regulator FGF21 is induced by prolonged fasting and PPARα activation in man. Cell Metab 8:169–174

Gordon JL, LeBlanc SJ, Duffield TF (2013) Ketosis treatment in lactating dairy cattle. Vet Clin North Am Food Anim Pract 29:433–445

Hamana M, Ohtsuka H, Oikawa M et al (2010) Blood free amino acids in the postpartum dairy cattle with left displaced abomasums. J Vet Med Sci 72:1355–1358

Herdt TH (2000) Ruminant adaptation to negative energy balance - Influences on the etiology of ketosis and fatty liver. Vet Clin North Am Food Anim Pract 16:215–230

Heringstad B, Chang YM, Gianola D et al (2005) Genetic analysis of clinical mastitis, milk fever, ketosis, and retained placenta in three lactations of Norwegian Red Cows. J Dairy Sci 88:3273–3281

Hinko CN, Crider AM, Kliem MA et al (1996) Anticonvulsant activity of novel derivatives of 2- and 3-piperidinecarboxylic acid in mice and rats. Neuropharmacology 35:1721–1735

Holtenius P, Holtenius K (1996) New aspects of ketone bodies in energy metabolism of dairy cows: a review. Zentralbl Veterinarmed A 43:579–587

Kadarmideen HN, Thompson R, Simm G (2000) Linear and threshold model genetic parameters for disease, fertility and milk production in dairy cattle. Anim Sci 71:411–419

Kharitonenkov A, Shiyanova TL, Koester A et al (2005) FGF-21 as a novel metabolic regulator. J Clin Invest 115:1627–1635

Klein MS, Almstetter MF, Schlamberger G et al (2010) Nuclear magnetic resonance and mass spectrometry-based milk metabolomics in dairy cows during early and late lactation. J Dairy Sci 93:1539–1550

Klein MS, Buttchereit N, Sebastian P et al (2012) NMR metabolomic analysis of dairy cows reveals milk glycerophosphocholine to phosphocholine ratio as prognostic biomarker for risk of ketosis. J Proteome Res 11:1373–1381

Kuhla B, Albrecht D, Kuhla S et al (2009) Proteome analysis of fatty liver in feed-deprived dairy cows reveals interaction of fuel sensing, calcium, fatty acid, and glycogen metabolism. Physiol Genomics 37:88–98

Lecker SH, Goldgerger AL, Mitch WE (2006) Protein degradation by the ubiquitin–proteasome pathway in normal and disease states. J Am Soc Nephrol 17:1807–1819

Li P, Lin Y, Li D et al (2007) Amino acids and immune function. Br J Nutr 98:237–352

Li P, Li XB, Fu SX et al (2012) Alterations of fatty acid β-oxidation capability in the liver of ketotic cows. J Dairy Sci 95:1759–1766

Li Y, Xu C, Xia C et al (2014) Plasma metabolic profiling of dairy cows affected with clinical ketosis using LC/MS technology. Vet Q 34:152–158

Loor JJ, Everts RE, Bionaz M et al (2007) Nutrition-induced ketosis alters metabolic and signaling gene networks in liver of periparturient dairy cows. Physiol Genomics 32:105–116

Lu J, Antunes Fernandes E, Páez Cano AE et al (2013) Changes in milk proteome and metabolome associated with dry period length, energy balance, and lactation stage in postparturient dairy cows. J Proteome Res 12:3288–3296

Mahendran Y, Vangipurapu J, Cedergerg H et al (2013) Association of ketone body levels with hyperglycemia and type 2 diabetes in 9,398 Finnish men. Diabetes 62:3618–3626

Marsh DC, Vreugdenhil PK, Mack VE et al (1993) Glycine protects hepatocytes from injury caused by anoxia, cold ischemia and mitochondrial inhibitors, but not injury caused by calcium ionophores or oxidative stress. Hepatology 17:91–98

Marvin DA, Francis B (1949) A metabolic study of α-aminobutyric acid. J Biol Chem 180:1059–1063

Mitch WE, Goldberg AL (1996) Mechanisms of muscle wasting: the role of the ubiquitin-proteasome system. N Engl J Med 335:1897–1905

Moriwaki H, Miwa Y, Tajika M et al (2004) Branched-chain amino acids as a protein- and energy-source in liver cirrhosis. Biochem Biophys Res Commun 313:405–409

Motyl T, Barej W (1986) Plasma amino acid indices and urinary 3-methyl histidine excretion in dairy cows in early lactation. Ann Rech Vet 17:153–157

Munro IC, Berndt WO, Borzelleca JF et al (1998) Erythritol: an interpretive summary of biochemical, metabolic, toxicological and clinical data. Food Chem Toxicol 36:1139–1174

Murondoti A, Jorritsma R, Beynen AC et al (2004) Unrestricted feed intake during the dry period impairs the postpartum oxidation and synthesis of fatty acids in the liver of dairy cows. J Dairy Sci 87:672–679

Nakagawa-Ueta H, Katoh N (2000) Reduction in serum lecithin: cholesterol acyltransferase activity prior to the occurrence of ketosis and milk fever in cows. J Vet Med Sci 62:1263–1267

Ness GC, Chambers CM (2000) Feedback and hormonal regulation of hepatic 3-hydroxy-3-methylglutaryl coenzyme A reductase: the concept of cholesterol buffering capacity. Proc Soc Exp Biol Med 224:8–19

Oetzel GR (2007) Herd-level ketosis—diagnosis and risk factors. Paper presented at the 40th Annual Conference—American Association of Bovine Practitioners, Vancouver, BC, Canada, 19 September 2007

Oetzel GR (2013) Understanding the impact of subclinical ketosis. Paper presented at the 24th Florida Ruminant Nutrition Symposium, University of Florida, Gainesville, 5–6 February 2013

Oikawa S, Katoh N, Kawawa F et al (1997) Decreased serum apolipoprotein B-100 and A-I concentrations in cows with ketosis and left displacement of the abomasums. Am J Vet Res 58:121–125

Okuda T, Morita N (2012) A very low carbohydrate ketogenic diet prevents the progression of hepatic steatosis caused by hyperglycemia in a juvenile obese mouse model. Nutr Diabetes 2:e50

Ospina PA, Nydam DV, Stokol T et al (2010) Evaluation of nonesterified fatty acids and beta-hydroxybutyrate in transition dairy cattle in the northeastern United States: Critical thresholds for prediction of clinical diseases. J Dairy Sci 93:546–554

Pasquale MGD (2007) Energy metabolism. In: Amino acids and proteins for the athlete: the anabolic edge, 2nd edn. CRC Press, Boca Raton, pp 107–133

Petersen AK, Zeilinger S, Kastenmüller G et al (2014) Epigenetics meets metabolomics: an epigenome-wide association study with blood serum metabolic traits. Hum Mol Genet 23:534–545

Postic C, Girard J (2008) The role of the lipogenic pathway in the development of hepatic steatosis. Diabetes Metab 34:643–648

Rognstad R (1979) Rate-limiting steps in metabolic pathways. J Biol Chem 254:1875–1878

Salway JG (2004) Metabolism at a glance, 3rd edn. Wiley-Blackwell, Hoboken

Scaglia N, Igal RA (2005) Stearoyl-CoA desaturase is involved in the control of proliferation, anchorage-independent growth, and survival in human transformed cells. J Biol Chem 280:25339–25349

Schoenberg KM, Giesy SL, Harvatine KJ et al (2011) Plasma FGF21 is elevated by the intense lipid mobilization of lactation. Endocrinology 152:4652–4661

Shaw JC (1956) Ketosis in dairy cattle. A review. J Dairy Sci 39:402–434

Shin SY, Fauman EB, Petersen AK et al (2014) An atlas of genetic influences on human blood metabolites. Nat Genet 46:543–550

Siddiqi NJ, Alhomida AS, Pandey VC (2002) Hydroxyproline distribution in the plasma of various mammals. J Biochem Mol Biol Biophys 6:159–163

Sun LW, Zhang HY, Wu L et al (2014) ^{1}H-Nuclear magnetic resonance-based plasma metabolic profiling of dairy cows with clinical and subclinical ketosis. J Dairy Sci 97:1552–1562

Tetens J, Heuer C, Heyer I et al (2015) Polymorphisms within the *APOBR* gene are highly associated with milk levels of prognostic ketosis biomarkers in dairy cows. Physiol Genomics 47:129–137

Tveit B, Lingass F, Svendsen M et al (1992) Etiology of acetonemia in Norwegian cattle. 1. Effect of ketogenic silage, season, energy level, and genetic factors. J Dairy Sci 75:2421–2432

Uribe HA, Kennedy BW, Martin SW et al (1995) Genetic parameters for common health disorders of Holstein cows. J Dairy Sci 78:421–430

Watanabe M, Maemura K, Kanbara K et al (2002) GABA and GABA receptors in the central nervous system and other organs. Int Rev Cytol 213:1–47

Xie L, Innis SM (2008) Genetic variants of the FADS1 FADS2 gene cluster are associated with altered (n-6) and (n-3) essential fatty acids in plasma and erythrocyte phospholipids in women during pregnancy and in breast milk during lactation. J Nutr 138:2222–2228

Xu C, Wang Z (2008) Comparative proteomic analysis of livers from ketotic cows. Vet Res Commun 32:263–273

Xu C, Wang Z, Liu G et al (2008) Metabolic characteristic of the liver of dairy cows during ketosis based on comparative proteomics. Asian-Australas J Anim Sci 21:1003–1010

Xu C, Shu S, Xia C et al (2014) Investigation on the relationship of insulin resistance and ketosis in dairy cows. J Vet Sci Technol 5:162

Xu C, Shu S, Xia C et al (2015b) Mass spectral analysis of urine proteomic profiles of dairy cows suffering from clinical ketosis. Vet Q 35:133–141

Xu C, Li Y, Xia C et al (2015a) ^{1}H NMR-based plasma metabolic profiling of dairy cows with Type I and Type II ketosis. Pharm Anal Acta 6:1000328

Xu C, Xu Q, Chen Y et al (2016b) FGF-21: promising biomarker for detecting ketosis in dairy cows. Vet Res Commun 40:49–54

Xu C, Sun LW, Xia C et al (2016a) ^{1}H-nuclear magnetic resonance-based plasma metabolic profiling of dairy cows with fatty liver. Asian-Australas J Anim Sci 29:219–229

Yamamoto M, Nakagawa-Ueta H, Katoh N et al (2001) Decreased concentration of serum apolipoprotein C-III in cows with fatty liver, ketosis, left displacement of the abomasum, milk fever and retained placenta. J Vet Med Sci 63:227–231

Zierer J, Menni C, Kastenmüller G et al (2015) Integration of 'omics' data in aging research: from biomarkers to systems biology. Aging Cell 14:933–944

Zhang G, Hailemariam D, Dervishi E et al (2016) Dairy cows affected by ketosis show alterations in innate immunity and lipid and carbohydrate metabolism during the dry off period and postpartum. Res Vet Sci 107:249–256

Zhang G, Dervishi E, Dunn SM et al (2017a) Metabotyping reveals distinct metabolic alterations in ketotic cows and identifies early predictive serum biomarkers for the risk of disease. Metabolomics 13:43

Zhang G, Dervishi E, Mandal R et al (2017b) Metallotyping of ketotic dairy cows reveals major alterations preceding, associating, and following the disease occurrence. Metabolomics 13:97

Zhang H, Wu L, Xu C et al (2013) Plasma metabolomic profiling of dairy cow affected with ketosis using gas chromatography mass spectrometry. BMC Vet Res 9:186

Zhang Z, Liu G, Wang H et al (2012) Detection of subclinical ketosis in dairy cows. Pak Vet J 32:156–151

The Omics Side of Fatty Liver: A Holistic Approach for a Commonly Occurring Peripartal Disease

11

Mario Vailati-Riboni, Valentino Palombo, and Juan J. Loor

Abstract

Compared with other stages of the lactation cycle, the transition period (from 3 weeks before to 3 weeks after calving) is critically important to health, production, and profitability of dairy cows. Due to the multifaceted physiological adaptations to lactation, dairy cows are at serious risk of developing metabolic disorders in the immediate weeks postpartum. Fatty liver (i.e., hepatic lipidosis) is a major metabolic disorder afflicting dairy cows in early lactation and is associated with decreased health status and reproductive performance. Fatty liver develops when the hepatic uptake of fatty acids exceeds the capacity of the organ to oxidize and/or export them as very-low density lipoproteins them. These events are often preceded by high concentrations of fatty acids in plasma mobilized from adipose tissue. Excess triacylglycerol accumulation in the liver can diminish physiologic functions of the liver. Studies of similar pathologies in humans and model organisms (e.g. mouse and rat) have revealed the complexity of the disease that goes beyond a simple disruption of lipid metabolic pathways. Advancements in computational biology, genome sequencing, and high-throughput technologies in the last decade have provided the tools to understand this metabolic disease using a systems rather than a reductionist approach. The application of "omics" technologies has enhanced understanding of the pathophysiology of fatty liver disease in peripartal dairy cows. This chapter focuses on the use of high-throughput technologies and bioinformatics to enhance our understanding of hepatic lipidosis, focusing on major breakthroughs generated using functional information from "omics" datasets. The goal is to provide specific examples of how these combined approaches have been used to advance our understanding of the pathophysiology of this commonly occurring metabolic disease.

M. Vailati-Riboni, Ph.D. • V. Palombo, Ph.D. • J.J. Loor, Ph.D. (✉)
Division of Nutritional Sciences, Department of Animal Sciences, University of Illinois at Urbana-Champaign, Urbana, IL 61801, USA
e-mail: vailati2@illinois.edu; valentino.palombo@gmail.com; jloor@illinois.edu

B.N. Ametaj (ed.), *Periparturient Diseases of Dairy Cows*,
DOI 10.1007/978-3-319-43033-1_11

11.1 Introduction

The transition period, defined as the last 3 weeks prepartum through the first 3 weeks postpartum (Grummer 1995), is one of the most challenging for dairy cattle management. A successful transition into lactation is a primary determinant of optimum production, reproduction, and health. Metabolic disorders and health problems are common during this time, especially in the first week postpartum (Ingvartsen 2006), and can easily erase a farm's profit potential. According to the USDA National Animal Health Monitoring System the major contributors to morbidity in dairy cows are clinical mastitis, lameness, infertility, retained placenta, milk fever, reproductive problems, and displaced abomasum, with at least 53% of all culled cows removed from the heard for one or more of the above diseases (National Animal Health Monitoring System 2008).

Besides milk fever and displaced abomasum, other metabolic diseases such as ketosis and fatty liver (i.e., hepatic lipidosis) tend to resolve themselves or pass unnoticed in their subclinical form, with only the severe clinical cases being recorded, thus contributing to nationwide statistics. Furthermore, fatty liver can only be diagnosed by liver biopsy, rendering the estimate of its incidence very difficult. However, in the case of fatty liver it has been estimated that up to 50% of all cows experience some degree of accumulation of triacylglycerol (TAG) in liver in the first 4 weeks after calving (Jorritsma et al. 2000, 2001), the period of primary occurrence (Grummer 1993). Because the disorder is associated with decreased health status, well-being, productivity, and reproductive performance of cows, its presence will increase veterinary costs, lengthen calving intervals, and decrease the average lifetime of a cow.

With an estimate national annual cost of over $60 million, understanding the intricate details of fatty liver physiologic mechanisms to design efficient treatments and preventive strategies is a research primary goal to enhance industry productivity, improving animal welfare and well-being. Due to the complexity of the metabolic disorder, spanning from impaired metabolism to inflammatory and apoptotic events, omics technologies are well suited to the study of fatty liver (Table 11.1.). Thanks to their broad and holistic nature, they can generate complete data to uncover different physiologic networks, and help in the discovery of more precise biomarkers to detect otherwise undetectable subclinical cases, or screen at-risk cows for preventive treatment.

11.2 Fatty Liver Pathophysiology

A key area of the biology of periparturient cows relates to lipid metabolism. The primary challenge faced by cows around calving is a sudden and marked increase of nutrient requirements for milk production, at a time when dry matter intake (DMI), and, thus, nutrient supply, lags far behind (Bell 1995). This discrepancy induces a period of negative energy balance (NEB) that lasts until the cow DMI gradually increases in the days after parturition. During times of NEB, adipose tissue mobilization contributes fuels that would fill the tissues metabolizable energy gap between intake and requirements of nutrients. These fuels come in the form of non-esterified fatty acids (NEFA) derived from the hydrolysis of TAG stored in the fat depots. If not taken up from the blood stream by the mammary gland to contribute to milk fat content, NEFA are taken up by the liver at a rate directly correlated with their plasma concentration (Bell 1995; Drackley 1999). They are then delivered to the hepatocyte's mitochondria by the activity of the shuttle carnitine palmitoyltransferase I (CPT-1), where they are completely or partially oxidized to CO_2 or ketone bodies, respectively. An alternate pathway for hepatic oxidation of NEFA is present in peroxisomes, subcellular organelles in most organs of the body, where they are only partially oxidized to acetyl-CoA (Drackley 1999).

Table 11.1 Summary of "omics" approaches aimed to understand fatty liver physiology in dairy cows

Technology	Model	Main findings	Reference
Transcriptomics	Ketosis	Transcriptome analysis revealed downregulation of genes associated with oxidative phosphorylation, cholesterol metabolism, growth hormone signaling, proton transport, fatty acid desaturation, protein ubiquitination, and ubiquinone biosynthesis with ketosis. Other molecular adaptations included upregulation of genes and nuclear receptors associated with cytokine signaling, fatty acid uptake/transport, and fatty acid oxidation	Loor et al. (2007)
Transcriptomics	Feed restriction	Feed restriction induced changes in expression of genes associated with lipid metabolism, connective tissue development and function, cell signaling, and cell cycle. It also reduced expression of transcription activators and signal transducers that regulate the expression of genes and gene networks associated with cell signaling and tissue repair. These alterations were linked with increased expression of abnormal cell cycle and cellular proliferation-associated pathways	McCarthy et al. (2010)
Transcriptomics	Feed restriction	The bioinformatics functional analysis of genes affected by feed restriction in mid-lactation uncovered biosynthesis of cholesterol and energy generation by mitochondrial respiration as the most relevant and inhibited functions. The data also indicated an increase of flux toward gluconeogenesis	Akbar et al. (2013)

(continued)

Table 11.1 (continued)

Technology	Model	Main findings	Reference
Proteomics	Feed restriction	Differentially expressed proteins due to feed restriction are related to energy and nucleotide metabolism and cellular stress. These proteins may be new candidate molecules that are potentially involved in signaling for maintaining energy homeostasis	Kuhla et al. (2007)
Proteomics	Ketosis	Proteins whose concentration was affected by ketosis play a role in energy metabolism, carbohydrate degradation, fatty acid metabolism, amino acid metabolism, antioxidation, cell structure, nucleotide metabolism, and protein metabolism	Xu and Wang (2008)
Proteomics	Feed restriction	Feed restriction downregulated proteins associated with fatty acid oxidation, glycolysis, electron transfer, protein degradation, and antigen processing, as well as cytoskeletal rearrangement. Upregulated proteins instead included enzymes of the urea cycle, fatty acid or cholesterol transport proteins, an inhibitor of glycolysis, and previously unknown changes in calcium signaling network	Kuhla et al. (2009)
Metabolomics	Ketosis	The development of clinical and subclinical ketosis involved disturbances in many metabolic pathways, mainly including fatty acid metabolism, amino acid metabolism, glycolysis, gluconeogenesis, and the pentose phosphate pathway. Results also identified potential biomarkers, including carbohydrates, fatty acids, amino acids, sitosterol, and vitamin E isomers	Zhang et al. (2013)
Metabolomics	Fatty liver	Metabolomics analysis yielded 29 metabolites (amino acids, phosphatidylcholine, and sphingomyelin) that, in conjunction, were able to distinguish between dairy cows with no hepatic lipidosis and those displaying different stages of the disorder	Imhasly et al. (2014)

Table 11.1 (continued)

Technology	Model	Main findings	Reference
Metabolomics	Ketosis	Compared with control cows, the levels of valine, glycine, glycocholic acid, tetradecenoic acid, and palmitoleic acid increased significantly in clinical ketosis. On the other hand, the levels of arginine, aminobutyric acid, leucine/isoleucine, tryptophan, creatinine, lysine, norcotinine, and undecanoic acid decreased markedly	Li et al. (2014b)
Metabolomics	Ketosis	The results indicated that plasma NMR-based metabolomics, coupled with pattern recognition analytical methods, not only has the sensitivity and specificity to distinguish cows with clinical and subclinical ketosis from healthy controls, but also has the potential to be developed into a clinically useful diagnostic tool that could contribute to a further understanding of the disease mechanisms	Sun et al. (2014)
Metabolomics	Sickness	Results from the univariate analysis indicated increased concentrations of multiple plasma amino acids in the sick cows as compared to healthy controls at—4 week before parturition. Altered profiles of plasma amino acids and sphingolipids in the sick cows during the transition period suggested an association among disease state and the dynamics of amino acid and sphingolipid metabolism providing insights on the etiopathology of the disease state	Hailemariam et al. (2014b)
Metabolomics	Sickness	Carnitine, propionyl carnitine, and lysophosphatidylcholine acyl C14:0 were elevated in diseased cows as early as 4 weeks before parturition, and used to develop a biomarker profile that predicted development of peripartal diseases with a sensitivity of 87% and a specificity of 85%. Furthermore, phosphatidylcholine acyl-alkyl C42:4 and phosphatidylcholine diacyl C42:6 could be used to discriminate diseased cows 1 week before parturition	Hailemariam et al. (2014a)

(continued)

Table 11.1 (continued)

Technology	Model	Main findings	Reference
Metabolomics	Fatty liver	Compared with the control group, the primary differences in the fatty liver group included increases in beta-hydroxybutyric acid, acetone, glycine, valine, trimethylamine-N-oxide, citrulline, and isobutyrate, and decreases in alanine, asparagine, glucose, gamma-aminobutyric acid glycerol, and creatinine	Xu et al. (2016)
Metabolomics	Fatty liver	Phosphatidylcholine plasma concentrations were lower in cows with fatty liver disease. The concentration of different bile acids tended to be increased instead. In addition, two metabolites related to inflammation, resolvin E1 and palmitoyl-ethanolamine (PEA), were detected and authors underscored the need for further investigation into their specific roles	Gerspach et al. (2017)

Cows often mobilize NEFA during periods of NEB in surplus of the negative gap of energy, exceeding the oxidation capacity of hepatocytes. The liver then repacks the NEFA into TAG and excretes them to the bloodstream as very-low density lipoproteins (VLDL) to be stored again, upon release of fatty acids via lipoprotein lipase, in the adipocytes. The greatest factor contributing to hepatic lipid accumulation and the onset of fatty liver disease may be the ruminant's inability to export TAG as VLDL due to the evolutionary adaptation to a diet poor in lipids (i.e., grass). When comparing in vitro TAG excretion capacity of monogastric hepatocytes (e.g., rat) to that of ruminant's (e.g., goats), up to a 25-fold difference in favor of the first was reported. This lower excretion capacity has been confirmed in vivo in early-lactating dairy cows, probably due to the lower expression and synthesis of apolipoprotein B100 (Bernabucci et al. 2004; Gruffat et al. 1997), the major protein of VLDL, indispensable for their assembly.

In summary, extreme rates of lipid mobilization early postpartum lead to increased uptake of NEFA by liver and increased TAG accumulation. If this lipid infiltration becomes severe, the syndrome of hepatic lipidosis or fatty liver may result, which can lead to prolonged recovery from other disorders, increased incidence of health problems, and development of "downer cows" that may die (Drackley 1999). Fatty liver can only be evaluated either by chemical or histological analyses of liver samples for liver TAG or total lipid concentration (Woitow et al. 1991), with normal, mild, moderate, and severe lipid infiltration determined at <1% (TAG % on a wet weight basis), 1–5%, 5–10%, and >10%, respectively.

Bobe et al. (2004) concluded based on an extensive review of the literature that around 5–10% of all cows in a herd experience severe fatty liver in the first month of lactation, and that 30–40% experience moderate fatty liver. This indicates that up to 50% of periparturient dairy cows are at a higher risk for diseases and reproductive problems. In fact, fatty liver can exacerbate the outcome of other metabolic diseases, in particular displaced abomasum and ketosis, as it decreases glucose availability for peripheral tissues (Veenhuizen et al. 1991). Furthermore, increased liver TAG concentration has been associated with increased incidence and severity of laminitis, mastitis, milk fever, retained placenta, and metritis (Herdt 1991). The early postpartum period is strongly and negatively impacted by severe or clinical cases, and in the long-term, increased TAG concentrations are associated with decreased reproductive success and milk production in dairy cows (Bobe et al. 2004). Therefore, a better understanding of the pathology and etiology of fatty liver is important for greater profitability in the dairy industry.

11.3 Fatty Liver in Nonruminants

Fatty liver diseases are not unique to peripartal dairy cows, but can occur in other mammalian (e.g., human, mice, rats) and avian (e.g., goose, fowl) species. The most-studied cases are related to human medicine. In humans, two types of fatty liver disease are common: alcoholic liver disease (ALD), a form of fatty liver associated with excess alcohol consumption (~60 or more g/day) (O'Shea et al. 2010), and nonalcoholic fatty liver disease (NAFLD), which occurs when fat is deposited in the liver due to causes other than excessive alcohol use (Spengler and Loomba 2015). One major difference with the ruminant is that human liver (and that of the model species, rat and mouse) is lipogenic, meaning that it is able to synthetize palmitic acid de novo, a metabolic ability almost lacking in all ruminant livestock species. In fact, the lipogenic capacity of liver in bovine and ovine is approximately 1% that of their adipose tissue (Ingle et al. 1972), the primary site of de novo lipogenesis in non-lactating ruminants.

Omics technologies are being used extensively in human-related fields of research such as medicine, pharmacology, and nutrition, and their applications are not rare when it comes to liver diseases. NAFLD is able to impact the flow of physiologic information right from the onset of the disease. For example, in rats and mice fed a high-fat diet (HFD) to induce NAFLD, microarray data revealed substantial differences with the control counterparts (Kirpich et al. 2011; Xie et al. 2010). As expected, the largest numbers of differentially expressed genes were involved in lipid metabolism, specifically lipogenesis and mitochondrial β-oxidation albeit with different results. When the HFD was fed to mice for 2 months, lipogenesis was downregulated and β-oxidation upregulated (Kirpich et al. 2011), but when it was fed to rats for 4 months their trend was reversed (Xie et al. 2010).

Clearly, even among rodent species the transcriptome profiles reveal different hepatic responses to TAG accumulation. This inconsistency also was detected in dairy cows, as mitochondrial oxidation and ketogenesis were both positively or negatively associated with fatty liver (Bobe et al. 2004). Ignoring the changes in lipogenic genes as not pertinent to the ruminant liver, the observed changes in lipid oxidation indicate a challenge to the liver environment. For instance, increased mitochondrial β-oxidation of fatty acids is an important source of reactive oxygen species (ROS) (Sanyal et al. 2001). Kirpich et al. (2011) through transcriptome analysis observed that several genes involved in ROS detoxification were downregulated in a model of diet-induced liver steatosis, and that their downregulation was maintained at the protein level. This led to a possible reduction in antioxidant defenses during cases of NAFLD, when production of ROS might increase. Together with the upregulation of pro-inflammatory genes (Kirpich et al. 2011), these biologic changes could contribute to the progression of the disease and the partial or complete impairment of liver functions, with consequences for the entire system.

The disrupted antioxidant system reported by Kirpich et al. (2011) via transcriptomic and proteomic analysis was related to the glutathione antioxidant system. Both Svoboda and Kawaja (2012) and Thomas et al. (2013) observed that the same system was impaired when they induced NAFLD in mice and analyzed their proteome. Glutathione-S-transferase (GST) mu1 was, in fact, less abundant in liver of NAFLD-affected mice across three experiments. The GST catalyze the reactions between reduced glutathione and unsaturated aldehydes, quinines, and many other substrates, especially under conditions of oxidative stress (Raza et al. 2002). These enzymes are not only involved in the reduction of free-radical damage, but also in detoxification processes (Jakoby 1978). Their downregulation seems to be an important mechanism through which a disruption of hepatic lipid metabolism, with the following accumulation of TAG, may lead to cellular damage and apoptotic events in the liver.

11.4 A Holistic Approach for Bovine Fatty Liver Disease

11.4.1 The Adaptation to Lactation

Advancements in computational biology, genome sequencing, and high-throughput technologies in the last decade have provided the tools for approaching biological systems in an integrative fashion, i.e., allow access to the functional capabilities of an individual organism en masse (Bionaz and Loor 2012). In the week around parturition, several tissues such as adipose and mammary gland play a prominent role in allowing cows to adapt successfully to the onset of lactation; however, the liver plays a central role in regulating overall metabolism. Various groups worldwide have been studying the dynamic adaptations of those tissues (especially liver) during the transition to lactation in dairy cows using omics technologies in combination with bioinformatics tools. Examples of their application to general peripartal cow physiology can be found elsewhere (Bionaz and Loor 2012; Loor 2010; Loor et al. 2005; Steele et al. 2015; Akbar et al. 2013).

At parturition cows not only undergo complex changes in their metabolic physiology, but also abrupt changes in diet composition to better fulfill the requirements for lactation (e.g., milk synthesis). One of the main shifts facing the postpartal cow is the higher energy density of the diet, achieved mainly via an increase in nonstructural carbohydrate (or the grain fraction). Researchers often forget that the rumen is the first interface between the animal's tissues and the nutrients in the diet. Applying NMR-based metabolomics, Ametaj et al. (2010) evaluated for the first time metabolomics of the rumen fluid from dairy cows fed graded amounts of cereal grain. Results from that study revealed significant elevations of several harmful or potentially harmful compounds in the rumen fluid including methylamine, nitrosodimethylamine, and ethanol with increasing amounts of dietary grain.

In an effort to confirm and extend these studies, the authors undertook a subsequent analysis with a much more comprehensive, quantitative metabolomics technique including proton NMR spectroscopy, GC-MS, and direct flow injection tandem mass spectrometry (Saleem et al. 2012). In this second study the high-grain diets (>30%) resulted in increased rumen fluid concentrations of several toxic, inflammatory, and unnatural compounds including putrescine, methylamines, ethanolamine, and short-chain fatty acids. Furthermore, perturbations in several amino acids (phenylalanine, ornithine, lysine, leucine, arginine, valine, and phenylacetylglycine) were also evident.

Authors of both works pointed out how these compounds might play a role in the development of fatty liver disease. Endotoxins absorbed into the circulation are, in fact, removed from the blood by liver macrophages; however, the liver also activates a second line of defense involving plasma lipoproteins. Even though lipoproteins are mostly known for their function as lipid carriers, they also bind and neutralize endotoxin (Harris et al. 2002). The authors proposed that fatty liver could be exacerbated during activation of immune responses because of the rapid removal of endotoxin-lipoprotein complexes by the liver resulting in accumulation of high concentration of TAG in the liver (Ametaj 2005).

Management practices during the dry period and early lactation in dairy cows essentially determine the productivity of the animal throughout the lactation, with the peripartal period as the one with highest incidence of infectious and metabolic disease (Ingvartsen 2006). Thanks to the use of transcriptomics and bioinformatics our group was able to highlight how overfeeding energy in the close-up period (last 3 weeks of gestation prior parturition) predisposes the cow to hepatic accumulation of TAG, greatly increasing the risk for developing fatty liver (Loor et al. 2006; Shahzad et al. 2014). When comparing the liver response between overfed (ad libitum, 1.61 Mcal net energy/kg diet dry matter) or restricted (80% of calculated net energy requirements) cows, restricted energy intake prepartum resulted in more pronounced upregulation of hepatic fatty acid oxidation, gluconeogenesis, and cholesterol synthesis (Loor et al. 2006). Ad libitum feeding, instead, upregulated a number of genes associated with liver TAG synthesis. Authors further argued that energy restriction prepartum allows for upregulation of hepatic processes that modulate responses to a wide range of environmental and physiological stressors (e.g., parturition, copious milk production). These processes include, but are not limited to, reduced oxidative stress and DNA damage, enhanced tissue DNA repair, reduced *TP53* expression, which in nonruminants was linked to the pathogenesis of fatty liver (Yahagi et al. 2004), and reduced apoptosis.

In a more comprehensive bioinformatics analysis of the liver transcriptome comparing feed restriction (80% of estimated requirements) to a more pronounced overfeeding practice (150%) in the close-up period, Shahzad et al. (2014) confirmed results of Loor et al. (2006) via bioinformatics analysis. Furthermore, the analyses revealed that although energy restriction led to substantial catabolism of muscle mass prepartum, the liver adapted to the higher postpartal metabolic state well ahead of parturition. This adaptation was likely driven by molecular processes partly controlled by transcription regulators such as PPARA and NFE2L2, of importance in fatty acid oxidation and cellular stress. As a result, energy-restricted cows had signs of greater metabolic flux and utilization of amino acids and fatty acids but also more pronounced cellular inflammatory and endoplasmic reticulum stress response. Most of those cellular adaptations were confirmed by plasma and tissue biomarker analysis specifically during the prepartal period, which strengthened the notion that energy restriction helped "prime" the liver to cope with the change in physiological state at the onset of lactation. Overall, these changes provide evidence that not only feed restricting cows prior to parturition reduces the chances of developing fatty liver, but that in case of fatty liver it would be mild and unlikely to progress to moderate and severe forms, quickly resolving itself without compromising the productivity and profitability of the animal.

Another common predisposing factor to the development of fatty liver disease is the cow's extent of fat depots (e.g., adiposity) or body condition score (BCS). Obesity (BCS ≥4.0) does not necessarily cause fatty liver, especially when cows remain healthy or adapt their feed intake to their milk production (Smith et al. 1997). However, obese cows have a greater decrease in feed intake around calving and, therefore, have a more severe NEB (Bobe et al. 2004). Furthermore, in these animals compared with cows with normal BCS, lipolysis of adipose tissue is increased during metabolically and immunologically challenging situations, such as the peripartal period (Bobe et al. 2004). To better understand how the link between increased lipolysis and the impaired insulin signaling typical of fatty liver, Humer et al. (2016) examined the differences in plasma metabolome in cows with different degrees of lipomobilization postpartum. They identified 37 metabolites associated with excessive lipolysis, which may be used as biomarkers of cows at higher risk of developing fatty liver. Specifically, phosphatidylcholines (PC) carrying fatty acid moieties ranging from 28 to 36 carbons were enhanced in high lipid-mobilizing cows, whereas PC containing larger fatty acid components (>40 carbons) were decreased.

Further changes in the metabolite profiles such as an increase in sphingolipids and a reduction in free carnitine and propionylcarnitine suggested modifications in other metabolic pathways that are affected during excessive lipid mobilization in early-lactating cows. To this end, supplemental L-carnitine has shown potential to decrease or prevent liver TAG accumulation in feed-restricted dairy cows and laboratory animals probably through its effect on CPT-I, the mitochondrial enzyme for which L-carnitine serves as a cofactor, that facilitates oxidation of mobilized NEFA (Akbar et al. 2013; Carlson et al. 2007). Thus, by decreasing liver lipid accumulation and stimulating hepatic glucose output, carnitine supplementation might improve glucose status and diminish the risk of developing metabolic disorders during early lactation (Carlson et al. 2007). When applying an omics approach (e.g., transcriptomics) to identify effects beyond simple energy metabolism, its supplementation in mid-lactating cows led to no significant changes in the hepatic transcriptome, thus precluding identification of multifactorial mechanisms of action (Akbar et al. 2013). It would be beneficial to replicate the Carlson et al. (2007) study using RNAseq and robust bioinformatics, which were not available at the time the study was conducted.

11.4.2 Biomarker Discovery for Fatty Liver Detection

During the last century veterinary and animal science used a reductionist approach when dealing with periparturient diseases of dairy cows, i.e., to tackle these health issues the whole organism was split into smaller and simpler parts, each separately studied to understand their physiology, and diagnose, treat, and develop preventive interventions. Instead, in the past decade a "systems" approach has received increased attention, and involves using omics technologies including genomics, proteomics, transcriptomics, and metabolomics to tackle the etiopathogenesis of the disease process in transition dairy cows (Loor 2010). This holistic and integrative approach proposes that periparturient diseases develops from distress of certain pathways and networks as a result of genetic changes and/or environmental triggers (Bionaz and Loor 2012; Loor et al. 2013). The disease-perturbed networks trigger changes to the gene network outputs, which lead to the pathophysiology of the disease (Bionaz and Loor 2012; Loor et al. 2013). Therefore, studying the dynamics of the disease-perturbed networks and pathways enhances the understanding of the etiopathology of the disease and provides insights for its prevention (Ahn et al. 2006).

To our knowledge, no transcriptomic or proteomic studies have been conducted in cows that developed fatty liver spontaneously. Probably due to the difficulty in enrolling cows under those conditions. In fact, the certain diagnosis of the disease can only be achieved through liver biopsy, requiring the design of a retrospective study, without the security of reaching a significant number of animals in both a "control" and diseased group. Despite this, omics technologies have been used in cows that underwent controlled feed restriction to develop ketosis, and consequently hepatic accumulation of TAG. For example, two research groups recently applied MS and NMR to retrospectively study the metabolome of dairy cows spontaneously developing fatty liver disease (determined via liver TAG analysis). Because metabolite alterations in tissues or biofluids are indicators of variations in physiology or pathology, the metabolome represents the endpoint of the omics cascade and is also the closest point in the cascade to the phenotype. Therefore, metabolomics analysis can be a useful approach for finding effective diagnostic markers and examining unknown pathological conditions (Zhang et al. 2015).

In their first study, Imhasly et al. (2014) applied a multivariate test involving principal component and linear discriminant analyses to data generated through LC coupled to quadrupole time-of-flight MS in serum samples of control or clinical animals. This yielded 29 metabolites (amino acids, phosphatidylcholines, and sphingomyelins) that, in conjunction, were able to distinguish between healthy cows and those displaying different stages of fatty liver. In a follow-up experiment they applied the same methodology to assess differences in plasma metabolome, and specifically the lipidome (e.g., totality of lipids in a sample) due to the onset of fatty liver disease in dairy cows (Gerspach et al. 2017). Again they identified PC to be lower in plasma of cows with fatty liver disease. Furthermore, the study detected different bile acids that tended to be increased in the same cows, and two metabolites, resolvin E1 and palmitoyl-ethanolamine (PEA, related to inflammation) that, according to the authors, need to be further investigated in cattle.

The identification of PC tends to be a recurring finding in cows with altered hepatic lipid metabolism and TAG accumulation, most of all because of its involvement in the synthesis of VLDL, and the repacking of TAG for export out of the hepatocytes. Since this is one of the key regulatory points in the cow lipid metabolism (Drackley 1999), the study of this metabolic process and its possible modulations through nutrition or pharmacological means holds potential as a way to cure or prevent the disease. Furthermore, recent metabolomics studies have reported alterations in plasma PC concentrations during the onset of other health issues such as metritis, mastitis, laminitis, and displaced abomasum (Hailemariam et al. 2014a, b), hence underscoring the link between hepatic lipidosis and many metabolic and non-peripartal diseases.

Another research group, instead, focused more on the shift in the hepatic metabolome, rather than the one of circulating fluids. Through the use of NMR Xu et al. (2016) detected 31 metabolites to be quantitatively different between the fatty liver and control groups. Changes in few metabolites (e.g., higher β-hydroxybutyric acid and acetone, and lower glucose) perfectly resemble the typical scenario of clinical and subclinical ketosis, highlighting the strong connection between the two clinical scenarios. Other metabolites were instead linked to lipolysis, β-oxidation, TCA and urea cycle, and gluconeogenesis. The authors did not speculate on their role in the progression of the disease, but more on the physiologic consequences of these changes. They suggested further work on the use of these 13 metabolites (Ala, Asn, glucose, β-hydroxybutyric acid, creatinine, γ-aminobutyric acid, glycerol, acetone, citrulline, Gly, isobutyrate, trimethylamine-N-oxide, and Val) as possible biomarkers to detect animals affected by fatty liver.

11.4.3 The Ketosis and Feed Restriction Models

Both hepatic lipidosis and ketosis are peripartal diseases, associated with NEB, that result from intense lipid mobilization increasing blood NEFA concentration. Of the four metabolic fates of fatty acids within the liver, two are beneficial to the cow (complete oxidation and export as VLDL) and two are potentially detrimental (storage as TAG and partial oxidation to ketones) (Grummer 2010). Unfortunately, flux of fatty acids through the "good" pathways is limited because they essentially become saturated when there is excessive uptake of fatty acids; hence, greater TAG storage (i.e., fatty liver) and increased ketone formation (i.e., ketosis) are likely to occur at high and sustained NEFA concentrations. Research indicates that TAG storage probably precedes ketone formation as "overflow" routes during excessive NEFA uptake by hepatic tissue (Young et al. 1990; Cadorniga-Valino et al. 1997). Fatty liver, in fact, typically develops at calving while ketosis usually occurs a few weeks postcalving (Grummer 1993).

Due to the close relationship in the etiology of ketosis and fatty liver, and the fact that the latter precedes the onset of the first, studies related to clinical and subclinical ketosis are a suitable source to gather information about hepatic lipidosis. Furthermore, ketosis is easily induced through feed restriction, and its occurrence can be promptly determined via quick blood sampling and analysis, both in the lab and on-site using on-farm devices. When exploring the hepatic transcriptome of ketotic or feed-restricted cows, the bioinformatics functional analysis of the differentially expressed genes, compared with healthy and normal-fed control groups, revealed changes in energy generation by mitochondrial respiration (oxidative phosphorylation), lipid metabolism (uptake and transport, desaturation), and cholesterol metabolism (Akbar et al. 2013; Loor et al. 2007; McCarthy et al. 2010). Other molecular adaptations included regulation of genes and nuclear receptors associated with cytokine signaling, connective tissue development and function, cell signaling, cell cycle, and tissue repair (Loor et al. 2007; McCarthy et al. 2010). These alterations were linked with increased expression of abnormal cellular proliferation-associated pathways, probably caused by endoplasmic reticulum and oxidative stress, and the unfolded protein response. All these changes agree with data from nonruminant studies (human, mouse, rat) and reinforce the link between the two pathologies.

Due to posttranscriptional and posttranslational regulations, changes in expression are often not mirrored by changes in protein concentrations. For example, in human liver tissue, approximately 25% of the changes in the mRNA transcript expression are not accompanied by changes in the expression of the corresponding proteins (Shackel et al. 2006). Several studies have also detected only a modest transcriptome and proteome correspondence in yeast (Griffin et al. 2002), mammalian cells and assorted murine organs and organelles (Kislinger et al. 2006). Furthermore, as shown by Kirpich et al. (2011) in a mouse model, in the case of hepatic alteration produced by HFLD, even greater discordance can be expected. Despite this, studies comparing the proteome of ketotic or feed-restricted cows to that of the control counterparts (Kuhla et al. 2007, 2009; Xu and Wang 2008) uncovered changes and regulation in pathways similar, if not identical, to those reported by the transcriptomic studies discussed above (Akbar et al. 2013; Loor et al. 2007; McCarthy et al. 2010), e.g., disruption in energy and lipid metabolism, regulation of stress, and cell cycle related responses. This consistency was even maintained at the metabolome level (Li et al. 2014b; Sun et al. 2014; Zhang et al. 2013), indicating good agreement along the entire flux of the genetic information.

Overall, these results underscore how both feed restriction design and the use of ketotic animals are suitable ways to investigate the etiology and pathology of fatty liver. Furthermore, thanks to the consistent response at every level (transcriptome, proteome, and metabolome) they provide evidence regarding the interchangeability of the three main omics approaches.

11.5 The Epigenetics of Fatty Liver

A single approach is often not sufficient to provide a comprehensive picture of hepatic dysfunction occurring during fatty liver disease; hence, a multiple-approach combination of different techniques is recommended. Despite the many studies conducted to understand the complex molecular pathogenesis of fatty liver disease, this in part remains unclear. Instead of focusing on traditional aspects of the disease such as gene expression, protein composition, or metabolic biomarkers, research is shifting to the analysis of indirect controls of cellular processes, e.g., epigenetics.

11.5.1 DNA Methylation

The epigenetic modulation of gene expression, which can induce phenotypic changes, may occur in response to a modification of the DNA nucleotides such as DNA methylation. DNA methylation is a biochemical modification of cytosine in DNA with a methyl group. Its reaction is catalyzed by DNA methyltransferases (DNMT), which entails the addition of a methyl group to cytosine with guanine as the next nucleotide, known as CpG sites. The clustering of CpG dinucleotides (usually referred as CpG islands and CpG island shores) is commonly present at higher frequency in the promoter region of genes than at other DNA sites (Choi and Friso 2010). Hypermethylation of CpG islands is generally associated with gene silencing.

DNA methylation mechanisms have been correlated with the onset or the progression of NAFLD. Murphy et al. (2013) demonstrated how in liver tissue from patients with advanced NAFLD, many tissue repair genes were hypomethylated and overexpressed, and genes in certain metabolic pathways (including 1-carbon metabolism), were hypermethylated and underexpressed. When aligning methylation results to those of transcriptome profiling, methylation correlated with gene transcript levels for 7% of differentially methylated CpG sites, indicating that differential methylation contributes to differences in mRNA expression (Murphy et al. 2013). They further proved that functionally relevant differences in methylation can distinguish patients with advanced versus mild NAFLD.

Results from a mouse model further linked epigenetic alterations to the pathogenesis of hepatic steatosis and strongly suggested that differences in the cellular epigenetic status may be a pre-determining factor to individual susceptibility to hepatic steatosis, which was associated with changes in *DNMT1* and *DNMT3A* mRNA levels in the liver (Pogribny et al. 2009). Specific CpG in anti-fibrogenic and lipid metabolism regulator genes such as peroxisome proliferator-activated receptor α (PPARα) had significantly higher DNA methylation in the severe NAFLD patients compared with those in the mild group (Zeybel et al. 2015). Furthermore, hepatic DNA promoter methylation in PPARγ coactivator 1 alpha (PGC1-α), a key regulator of mitochondrial biogenesis, was significantly higher in NAFLD subjects compared with controls, and correlated with peripheral insulin resistance (Sookoian et al. 2010).

Animal studies have demonstrated that hepatic steatosis can be induced by dysfunction of one-carbon metabolism (da Silva et al. 2014). This metabolic pathway produces S-adenosylmethionine (SAM), the primary source of methyl groups for DNMT1 (e.g., DNA methylation). The most relevant methyl donor nutrients are folate, methionine, serine, betaine, and choline, which are ultimately utilized for synthesis of SAM in the methionine cycle. A proteomic analysis in mice fed a NAFLD-inducing diet detected a disruption of hepatic methionine metabolism (Thomas et al. 2013), while the study of liver metabolome highlighted a disruption of choline metabolism (Dumas et al. 2006). This impairment in NAFLD is likely linked to the reduction of phosphatidylcholine production, which also is associated with NAFLD development by reducing VLDL secretion (Jacobs et al. 2008). Methionine has been fed to cows in rumen-protected form to tackle the low methyl donor availability around parturition and avoid disruption of the one-carbon metabolism pathway. However, this approach did not reduce liver TAG accumulation (Osorio et al. 2013; Zhou et al. 2016), despite the potential involvement of methionine in VLDL formation through the synthesis of phosphatidylcholine, and the significantly higher methylation of the PPARα promoter regions (Osorio et al. 2016). The fact that these cows were able to reach optimal performance (even better than unsupplemented controls) suggests that a mild liver TAG accumulation does not prevent a normal adaptation to the new physiological state; hence, when properly managed TAG infiltration should not affect cow profitability.

Earlier evidence indicated that whereas insulin resistance increased fatty acid β-oxidation (and hepatic oxidative stress is present in both mild and severe cases of NAFLD), only severe cases that reach the level of steatohepatitis are associated with mitochondrial structural and molecular defects (Sanyal et al. 2001). These defects significantly decreased activity of mitochondria respiratory chain complexes (Perez-Carreras et al. 2003). Therefore, the evidence strongly suggests that mitochondrial dysfunction is involved in the pathogenesis of NAFLD progression. However, the molecular mechanisms leading to liver mitochondrial dysfunction in humans with severe NAFLD are still unknown.

Defects of the mitochondrial genome (mtDNA) are widely recognized as responsible for mitochondrial dysfunction (Scarpulla 2008). In the past, mutations and deletions of mtDNA were the only mechanisms to explain changes in the transcriptional profile of mitochondria. However, recent evidence showed that an isoform of DNMT1 appears to be responsible for mtDNA methylation of cytosine in CpG dinucleotides, which in turn regulates mitochondrial function and gene transcription (Shock et al. 2011). Following these findings, Pirola et al. (2013) observed that mitochondrial-encoded NADH dehydrogenase 6 (MT-ND6) was highly methylated in the liver of patients with severe, but not mild, NAFLD. The degree of methylation significantly impacted MT-ND6 transcriptional regulation. This effect was probably caused by the enhanced expression of the mitochondrial-targeted isoform of the DNMT1. This process may play an important role in the pathogenesis of fatty liver progression, and its relevance for bovine physiology is worthy of study, because methylation can be controlled by environmental stimuli such as dietary management.

11.5.2 miRNA

Despite the lack (to our knowledge) of application to bovine fatty liver disease, miRNome profiling has provided important insights into molecular mechanisms involved in the development of NAFLD. For instance, Alisi et al. (2011) demonstrated that a dysregulated miRNA expression pattern in rats is associated with histological damage and metabolic derangements during NAFLD. In their experiment, the levels of miRNA potentially regulating the transcripts of factors involved in fatty acid metabolism, inflammatory cytokines, cell growth and apoptosis, signal transduction and fibrogenesis were significantly altered in the group fed a HFD to induce liver TAG accumulation.

A similar miRNome analysis was conducted in human subjects affected by NAFLD, yielding similar results (Cheung et al. 2008). Despite the fact that the list of differentially expressed miRNA due to NAFLD differede between the two species (rat vs. human) miRNA-122 was common to both (Alisi et al. 2011; Cheung et al. 2008). Silencing miR-122 in vitro reproduced a pattern of key hepatic lipogenic gene mRNA and protein expression profiles similar to that seen in humans with NASH (Cheung et al. 2008). The mechanisms by which miR-122 affects lipid homeostasis need to be defined further; however, two of its target genes are the sterol receptor element binding protein (SREBP) 1c and 2. This result is of particular importance in regard to the role of miR-122 in bovine liver physiology. At least in neonatal calf hepatocytes, there is evidence that TAG accumulation might be mediated by the membrane-bound transcription factor SREBP-1c, which upregulates, when overexpressed, the expression and activity of fatty acid uptake, activation, and synthesis enzymes (Li et al. 2014a). It would not be unusual that in the neonatal bovine liver this lipogenic transcription regulator can play a biologic role because at this stage of life the animal still has not acquired the full characteristics of the mature ruminant, i.e., hepatic gluconeogenesis from rumen-derived propionate with little if any lipogenic capacity in the liver.

The in vitro overexpression of SREBP-1 induced mitochondrial lipid oxidation and hepatic VLDL synthesis in calf hepatocytes, but reduced their disposal and export thereby increasing TAG accumulation (Li et al. 2014a). Hepatic SREBP-1 was recently associated with the fat cow syndrome, as cows calving at a high BCS and subsequently developing fatty liver disease had elevated SREBP-1 protein concentration in liver (Prodanovic et al. 2016). Follow-up results from Li et al. (2015) indicated that in cow hepatocytes SREBP-1c can enhance the NEFA-induced overactivation of the NF-κB inflammatory pathway by increasing ROS, thereby further increasing hepatic inflammatory injury in cows with fatty liver. miRNA normally suppress the expression of their target genes; hence, because miR-122 was downregulated with diet-induced NAFLD (Alisi et al. 2011), the study of miR-122 in cows affected by fatty liver seems to hold value.

These results on SREBP-1c, however, need to be interpreted with caution as authors did not include any details regarding stage of lactation of the animals used, level of milk production, or clinical history. More importantly, it has long been established that bovine ruminant liver does not utilize acetate or glucose for lipogenesis (Hanson and Ballard 1967). The classical lipogenic enzymes and acetyl-CoA synthase are present in cow liver but their activities are 2–5-fold lower compared with rats. These characteristics of the ruminant liver argue against a role of SREBP-1c in lipogenesis. Whether this transcription factor is regulated through different mechanisms in bovine liver compared with monogastrics remains to be determined. What seems evident is that overfeeding of energy in the form of rapidly fermentable carbohydrate during the dry period not only increases systemic insulin concentration up to the point of calving (Loor et al. 2006), but it also leads to greater expression of *SREBF1* in the liver (Khan et al. 2014). Whether that reflects a priming of the liver for excessive accumulation of fat postcalving remains to be determined.

To better understand the possible role of miRNA during the onset of fatty liver in dairy cows around parturition, we applied an in silico approach to identify a tentative signature of miRNA activation from transcriptomics data (Fig. 11.1). The

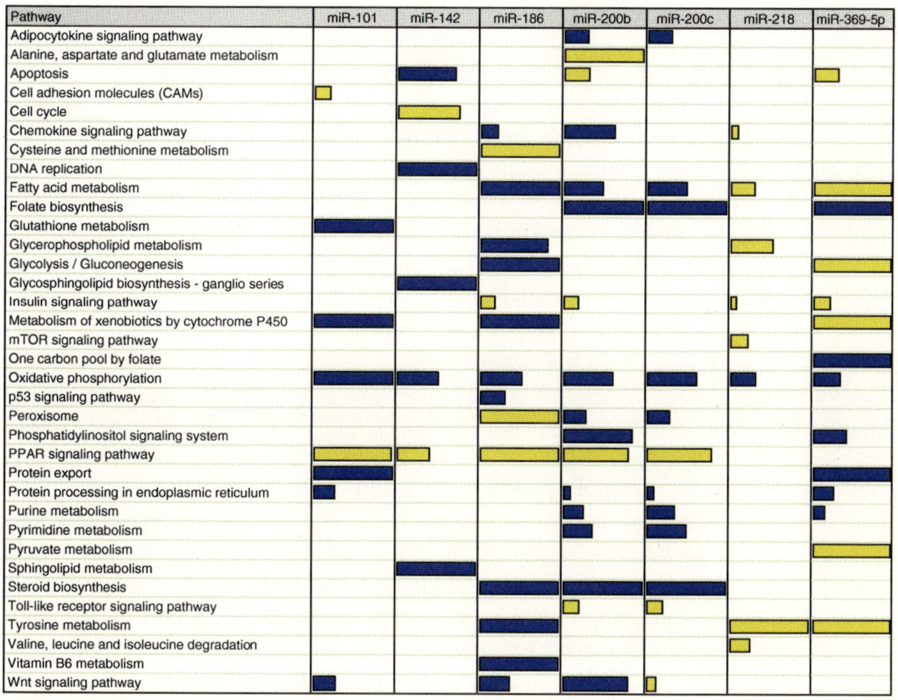

Fig. 11.1 Biological importance (impact) of pathways involved in the etiology of fatty liver regulated by the activity of miRNAs predicted (in silico) to be activated during the onset of ketosis through feed restriction in liver of peripartal dairy cows (data from Loor et al. 2007). Length of the bar represents the entity impact, while *yellow* denotes upregulation and *blue* downregulation of the pathway

transcriptome database used was generated by Loor et al. (2007). In their experiment the authors induced ketosis and fatty liver through a 50% feed restriction from day 5 up to day 14 postpartum, or until sign of clinical ketosis (anorexia, ataxia, or abnormal behavior). Liver tissue for gene expression profiling was sampled via puncture biopsy under local anesthesia at the onset of clinical ketosis (9–14 days postpartum) or day 14 postpartum for control cows.

The TAG analysis indicated that cows in the control group remained close to normal liver conditions (~1.5–2.0% w/w), while ketotic animals reached moderate levels of fatty liver (~6.0% w/w). This again underscores the link between ketosis and fatty liver in dairy cows, and how feed restriction is a suitable strategy to induce and study both pathologies. To detect miRNA activity from the transcriptome profiles we used three approaches described by Arora and Simpson (2008): Wilcoxon rank test, Ranked Ratio, and Mean absolute expression. The analysis was performed on a list of miRNA families and their predicted target genes for *Bos taurus* downloaded from the Microcosm targets website (v. 5.0). The results of the three approaches were then overlapped to find common miRNA predicted to be activated. It is noteworthy that miR-122 was not one of the predicted miRNA; however, a total of 7 miRNA were predicted as possible players in the transcriptomic response to fatty liver: miR-101, miR-142, miR-186, miR-200b, miR-200c, miR-218, and miR-369-3p.

To identify the miRNA functions, their target differentially expressed genes were used for pathway analysis via the dynamic impact approach (Bionaz et al. 2012). Among the most-impacted pathways were the classical cell functions involved in the etiology of fatty liver, "fatty acid metabolism" (miR-186, 200b and c, 218, 369), "oxidative phosphorylation" (all miRNAs) and "peroxisome" (miR-186, 200b and c), "gluconeogenesis" (miR-186, 369), "PPAR signaling pathway" (miR-101, 142, 186, 200b and c) and "insulin signaling pathway" (miR-186, 200b, 218, 369), and "apoptosis" (miR-142, 200b, 369). The most recurring miRNA, miR-186, appears to be a feasible target for further studies, as it was detected differentially expressed in both mouse (Hoekstra et al. 2012) and human (Leti et al. 2015) with NAFLD. Furthermore, pathway analysis of miR-186 target differentially expressed genes revealed its impact on "p53 signaling pathway," which is involved in the pathogenesis of fatty liver (Yahagi et al. 2004).

Of great importance also is the possible involvement of miRNA in the regulation of the one-carbon metabolism. In fact, "folate biosynthesis" (miR-200b and c, 369), "one carbon pool by folate" (miR-369), "vitamin B6 metabolism" (miR-186), "cysteine and methionine metabolism" (miR-186), and "glutathione metabolism" (miR-101) were all impacted pathways. These metabolic pathways are themselves responsible for epigenetic modifications through methylation, as previously described, a mechanism that appears to be multifactorially controlled (nutrients availability, gene expression, miRNA). It also serves as an important source of antioxidants (glutathione and taurine) to counteract the oxidative stress that arises from TAG accumulation and ROS generation during β-oxidation.

Conclusions

The application of transcriptome-, miRNome-, proteome-, and metabolome-enabled experimental techniques and bioinformatics tools for data interpretation has enhanced understanding of the pathophysiology of fatty liver disease in peripartal dairy cows. The reviewed omics studies in feed-restricted, ketotic, and liver TAG-infiltrated cows confirmed that the mechanisms of this disease are complex and interrelated, involving the accumulation and traffic of various lipids in the liver and trigger inflammatory responses. Transcriptomics, proteomics, and metabolomics data converge to provide evidence that the earliest detectable pathogenic mechanisms are mitochondrial energetic and structural dysfunction, and activation of inflammation via multiple targets. Furthermore, miRNome and methylation analyses support the involvement of epigenetic mechanism in the susceptibility, onset, and progression of the disease. These mechanisms hold the greatest potential for future research, in an attempt to translate nonruminant control networks to the dairy cow peripartal physiology.

References

Ahn AC, Tewari M, Poon CS, Phillips RS (2006) The limits of reductionism in medicine: could systems biology offer an alternative? PLos Med 3(6):709–713. ARTN e208. doi:10.1371/journal.pmed.0030208

Akbar H, Bionaz M, Carlson DB, Rodriguez-Zas SL, Everts RE, Lewin HA, Drackley JK, Loor JJ (2013) Feed restriction, but not l-carnitine infusion, alters the liver transcriptome by inhibiting sterol synthesis and mitochondrial oxidative phosphorylation and increasing gluconeogenesis in mid-lactation dairy cows. J Dairy Sci 96(4):2201–2213. doi:10.3168/jds.2012-6036

Alisi A, Da Sacco L, Bruscalupi G, Piemonte F, Panera N, De Vito R, Leoni S, Bottazzo GF, Masotti A, Nobili V (2011) Mirnome analysis reveals novel molecular determinants in the pathogenesis of diet-induced nonalcoholic fatty liver disease. Lab Invest 91(2):283–293. doi:10.1038/labinvest.2010.166

Ametaj BN (2005) A new understanding of the causes of fatty liver in dairy cows. Adv Dairy Technol 17:97–112

Ametaj BN, Zebeli Q, Saleem F, Psychogios N, Lewis MJ, Dunn SM, Xia JG, Wishart DS (2010) Metabolomics reveals unhealthy alterations in rumen metabolism with increased proportion of cereal grain in the diet of dairy cows. Metabolomics 6(4):583–594. doi:10.1007/s11306-010-0227-6

Arora A, Simpson DA (2008) Individual mRNA expression profiles reveal the effects of specific microRNAs. Genome Biol 9(5):R82. doi:10.1186/gb-2008-9-5-r82

Bell AW (1995) Regulation of organic nutrient metabolism during transition from late pregnancy to early lactation. J Anim Sci 73(9):2804–2819

Bernabucci U, Ronchi B, Basirico L, Pirazzi D, Rueca F, Lacetera N, Nardone A (2004) Abundance of mRNA of apolipoprotein b100, apolipoprotein e, and microsomal triglyceride transfer protein in liver from periparturient dairy cows. J Dairy Sci 87(9):2881–2888. doi:10.3168/jds.S0022-0302(04)73418-9

Bionaz M, Loor JJ (2012) Ruminant metabolic systems biology: reconstruction and integration of transcriptome dynamics underlying functional responses of tissues to nutrition and physiological state. Gene Regul Syst Biol 6:109–125. doi:10.4137/GRSB.S9852

Bionaz M, Periasamy K, Rodriguez-Zas SL, Hurley WL, Loor JJ (2012) A novel dynamic impact approach (DIA) for functional analysis of time-course omics studies: validation using the bovine mammary transcriptome. PLoS One 7(3):e32455. doi:10.1371/journal.pone.0032455

Bobe G, Young JW, Beitz DC (2004) Invited review: pathology, etiology, prevention, and treatment of fatty liver in dairy cows. J Dairy Sci 87(10):3105–3124

Cadorniga-Valino C, Grummer RR, Armentano LE, Donkin SS, Bertics SJ (1997) Effects of fatty acids and hormones on fatty acid metabolism and gluconeogenesis in bovine hepatocytes. J Dairy Sci 80(4):646–656. doi:10.3168/jds.S0022-0302(97)75983-6

Carlson DB, McFadden JW, D'Angelo A, Woodworth JC, Drackley JK (2007) Dietary L-carnitine affects periparturient nutrient metabolism and lactation in multiparous cows. J Dairy Sci 90(7):3422–3441. doi:10.3168/jds.2006-811

Cheung O, Puri P, Eicken C, Contos MJ, Mirshahi F, Maher JW, Kellum JM, Min H, Luketic VA, Sanyal AJ (2008) Nonalcoholic steatohepatitis is associated with altered hepatic micro rna expression. Hepatology 48(6):1810–1820. doi:10.1002/hep.22569

Choi SW, Friso S (2010) Epigenetics: a new bridge between nutrition and health. Adv Nutr 1(1):8–16. doi:10.3945/an.110.1004

Drackley JK (1999) ADSA foundation scholar award. Biology of dairy cows during the transition period: the final frontier? J Dairy Sci 82(11):2259–2273

Dumas ME, Barton RH, Toye A, Cloarec O, Blancher C, Rothwell A, Fearnside J, Tatoud R, Blanc V, Lindon JC, Mitchell SC, Holmes E, McCarthy MI, Scott J, Gauguier D, Nicholson JK (2006) Metabolic profiling reveals a contribution of gut microbiota to fatty liver phenotype in insulin-resistant mice. Proc Natl Acad Sci U S A 103(33):12511–12516. doi:10.1073/pnas.0601056103

Gerspach C, Imhasly S, Gubler M, Naegeli H, Ruetten M, Laczko E (2017) Altered plasma lipidome profile of dairy cows with fatty liver disease. Res Vet Sci 110:47–59. doi:10.1016/j.rvsc.2016.10.001

Griffin TJ, Gygi SP, Ideker T, Rist B, Eng J, Hood L, Aebersold R (2002) Complementary profiling of gene expression at the transcriptome and proteome levels in Saccharomyces cerevisiae. Mol Cell Proteomics 1(4):323–333. doi:10.1074/mcp.M200001-MCP200

Gruffat D, Durand D, Chilliard Y, Williams P, Bauchart D (1997) Hepatic gene expression of apolipoprotein B100 during early lactation in underfed, high producing dairy cows. J Dairy Sci 80(4):657–666. doi:10.3168/jds.S0022-0302(97)75984-8

Grummer RR (1993) Etiology of lipid-related metabolic disorders in periparturient dairy cows. J Dairy Sci 76(12):3882–3896. doi:10.3168/jds.S0022-0302(93)77729-2

Grummer RR (1995) Impact of changes in organic nutrient metabolism on feeding the transition dairy cow. J Anim Sci 73(9):2820–2833

Grummer RR (2010) Managing the transition cow-emphasis on ketosis and fatty liver syndrome. In: 2010 Western nutrition conference, challenging conventional nutrition dogma, Saskatoon, Saskatchewan, Canada, 21–23 September 2010, pp 191–204

Hailemariam D, Mandal R, Saleem F, Dunn SM, Wishart DS, Ametaj BN (2014a) Identification of predictive biomarkers of disease state in transition dairy cows. J Dairy Sci 97(5):2680–2693. doi:10.3168/jds.2013-6803

Hailemariam D, Mandal R, Saleem F, Dunn SM, Wishart DS, Ametaj BN (2014b) Metabolomics approach reveals altered plasma amino acid and sphingolipid profiles associated with pathological state in transition dairy cows. Curr Metabolomics 2(3):184–195

Hanson RW, Ballard FJ (1967) The relative significance of acetate and glucose as precursors for lipid synthesis in liver and adipose tissue from ruminants. Biochem J 105(2):529–536

Harris HW, Brady SE, Rapp JH (2002) Hepatic endosomal trafficking of lipoprotein-bound endotoxin in rats. J Surg Res 106(1):188–195

Herdt TH (1991) Relationship of fat metabolism to health and performance in dairy cattle. Bovine Practitioner 26:92–95

Hoekstra M, van der Sluis RJ, Kuiper J, Van Berkel TJC (2012) Nonalcoholic fatty liver disease is associated with an altered hepatocyte microRNA profile in LDL receptor knockout mice. J Nutr Biochem 23(6):622–628. doi:10.1016/j.jnutbio.2011.03.005

Humer E, Khol-Parisini A, Metzler-Zebeli BU, Gruber L, Zebeli Q (2016) Alterations of the lipid metabolome in dairy cows experiencing excessive lipolysis early postpartum. PLoS One 11(7):e0158633. doi:10.1371/journal.pone.0158633

Imhasly S, Naegeli H, Baumann S, von Bergen M, Luch A, Jungnickel H, Potratz S, Gerspach C (2014) Metabolomic biomarkers correlating with hepatic lipidosis in dairy cows. BMC Vet Res 10:122. doi:10.1186/1746-6148-10-122

Ingle DL, Bauman DE, Garrigus US (1972) Lipogenesis in the ruminant: in vitro study of tissue sites, carbon source and reducing equivalent generation for fatty acid synthesis. J Nutr 102(5):609–616

Ingvartsen KL (2006) Feeding- and management-related diseases in the transition cow - physiological adaptations around calving and strategies to reduce feeding-related diseases. Anim Feed Sci Technol 126(3–4):175–213. doi:10.1016/j.anifeedsci.2005.08.003

Jacobs RL, Lingrell S, Zhao Y, Francis GA, Vance DE (2008) Hepatic CTP: phosphocholine cytidylyltransferase-alpha is a critical predictor of plasma high density lipoprotein and very low density lipoprotein. J Biol Chem 283(4):2147–2155. doi:10.1074/jbc.M706628200

Jakoby WB (1978) The glutathione S-transferases: a group of multifunctional detoxification proteins. Adv Enzymol Relat Areas Mol Biol 46:383–414

Jorritsma R, Jorritsma H, Schukken YH, Wentink GH (2000) Relationships between fatty liver and fertility and some periparturient diseases in commercial Dutch dairy herds. Theriogenology 54(7):1065–1074

Jorritsma R, Jorritsma H, Schukken YH, Bartlett PC, Wensing T, Wentink GH (2001) Prevalence and indicators of post partum fatty infiltration of the liver in nine commercial dairy herds in The Netherlands. Livest Prod Sci 68(1):53–60. doi:10.1016/s0301-6226(00)00208-6

Khan MJ, Jacometo CB, Graugnard DE, Correa MN, Schmitt E, Cardoso F, Loor JJ (2014) Overfeeding dairy cattle during late-pregnancy alters hepatic PPARalpha-regulated pathways including Hepatokines: impact on metabolism and peripheral insulin sensitivity. Gene Regul Syst Biol 8:97–111. doi:10.4137/GRSB.S14116

Kirpich IA, Gobejishvili LN, Bon Homme M, Waigel S, Cave M, Arteel G, Barve SS, McClain CJ, Deaciuc IV (2011) Integrated hepatic transcriptome and proteome analysis of mice with high-fat diet-induced nonalcoholic fatty liver disease. J Nutr Biochem 22(1):38–45. doi:10.1016/j.jnutbio.2009.11.009

Kislinger T, Cox B, Kannan A, Chung C, Hu PZ, Ignatchenko A, Scott MS, Gramolini AO, Morris Q, Hallett MT, Rossant J, Hughes TR, Frey B, Emili A (2006) Global survey of organ and organelle protein expression in mouse: combined proteomic and transcriptomic profiling. Cell 125(1):173–186. doi:10.1016/j.cell.2006.01.044

Kuhla B, Kuhla S, Rudolph PE, Albrecht D, Metges CC (2007) Proteomics analysis of hypothalamic response to energy restriction in dairy cows. Proteomics 7(19):3602–3617. doi:10.1002/pmic.200700248

Kuhla B, Albrecht D, Kuhla S, Metges CC (2009) Proteome analysis of fatty liver in feed-deprived dairy cows reveals interaction of fuel sensing, calcium, fatty acid, and glycogen metabolism. Physiol Genomics 37(2):88–98. doi:10.1152/physiolgenomics.90381.2008

Leti F, Malenica I, Doshi M, Courtright A, Van Keuren-Jensen K, Legendre C, Still CD, Gerhard GS, DiStefano JK (2015) High-throughput sequencing reveals altered expression of hepatic microRNAs in nonalcoholic fatty liver disease-related fibrosis. Transl Res 166(3):304–314. doi:10.1016/j.trsl.2015.04.014

Li X, Li Y, Yang W, Xiao C, Fu S, Deng Q, Ding H, Wang Z, Liu G, Li X (2014a) SREBP-1c overexpression induces triglycerides accumulation through increasing lipid synthesis and decreasing lipid oxidation and VLDL assembly in bovine hepatocytes. J Steroid Biochem Mol Biol 143:174–182. doi:10.1016/j.jsbmb.2014.02.009

Li Y, Xu C, Xia C, Zhang H, Sun L, Gao Y (2014b) Plasma metabolic profiling of dairy cows affected with clinical ketosis using LC/MS technology. Vet Q 34(3):152–158. doi:10.1080/01652176.2014.962116

Li X, Huang W, Gu J, Du X, Lei L, Yuan X, Sun G, Wang Z, Li X, Liu G (2015) SREBP-1c overactivates ROS-mediated hepatic NF-kappaB inflammatory pathway in dairy cows with fatty liver. Cell Signal 27(10):2099–2109. doi:10.1016/j.cellsig.2015.07.011

Loor JJ (2010) Genomics of metabolic adaptations in the peripartal cow. Animal 4(7):1110–1139. doi:10.1017/S1751731110000960

Loor JJ, Dann HM, Everts RE, Oliveira R, Green CA, Guretzky NAJ, Rodriguez-Zas SL, Lewin HA, Drackley JK (2005) Temporal gene expression profiling of liver from periparturient dairy cows reveals complex adaptive mechanisms in hepatic function. Physiol Genomics 23(2):217–226. doi:10.1152/physiolgenomics.00132.2005

Loor JJ, Dann HM, Guretzky NA, Everts RE, Oliveira R, Green CA, Litherland NB, Rodriguez-Zas SL, Lewin HA, Drackley JK (2006) Plane of nutrition prepartum alters hepatic gene expression and function in dairy cows as assessed by longitudinal transcript and metabolic profiling. Physiol Genomics 27(1):29–41. doi:10.1152/physiolgenomics.00036.2006

Loor JJ, Everts RE, Bionaz M, Dann HM, Morin DE, Oliveira R, Rodriguez-Zas SL, Drackley JK, Lewin HA (2007) Nutrition-induced ketosis alters metabolic and signaling gene networks in liver of periparturient dairy cows. Physiol Genomics 32(1):105–116. doi:10.1152/physiolgenomics.00188.2007

Loor JJ, Bionaz M, Drackley JK (2013) Systems physiology in dairy cattle: nutritional genomics and beyond. Annu Rev Anim Biosci 1:365–392. doi:10.1146/annurev-animal-031412-103728

McCarthy SD, Waters SM, Kenny DA, Diskin MG, Fitzpatrick R, Patton J, Wathes DC, Morris DG (2010) Negative energy balance and hepatic gene expression patterns in high-yielding dairy cows during the early postpartum period: a global approach. Physiol Genomics 42A(3):188–199. doi:10.1152/physiolgenomics.00118.2010

Murphy SK, Yang HN, Moylan CA, Pang H, Dellinger A, Abdelmalek MF, Garrett ME, Ashley-Koch A, Suzuki A, Tillmann HL, Hauser MA, Diehl AM (2013) Relationship between Methylome and transcriptome in patients with nonalcoholic fatty liver disease. Gastroenterology 145(5):1076–1087. doi:10.1053/j.gastro.2013.07.047

National Animal Health Monitoring System (2008) Dairy 2007 Part I: reference of dairy cattle health and management practices in the United States, 2007. USDA—APHIS

O'Shea RS, Dasarathy S, AJ MC, Practice Guideline Committee of the American Association for the Study of Liver D, Practice Parameters Committee of the American College of G (2010) Alcoholic liver disease. Hepatology 51(1):307–328. doi:10.1002/hep.23258

Osorio JS, Ji P, Drackley JK, Luchini D, Loor JJ (2013) Supplemental Smartamine M or MetaSmart during the transition period benefits postpartal cow performance and blood neutrophil function. J Dairy Sci 96(10):6248–6263. doi:10.3168/jds.2012-5790

Osorio JS, Jacometo CB, Zhou Z, Luchini D, Cardoso FC, Loor JJ (2016) Hepatic global DNA and peroxisome proliferator-activated receptor alpha promoter methylation are altered in peripartal dairy cows fed rumen-protected methionineJ Dairy Sci 99 (1):234-244. Doi:http://dx.doi.org/10.3168/jds.2015-10157

Perez-Carreras M, Del Hoyo P, Martin MA, Rubio JC, Martin A, Castellano G, Colina F, Arenas J, Solis-Herruzo JA (2003) Defective hepatic mitochondrial respiratory chain in patients with nonalcoholic steatohepatitis. Hepatology 38(4):999–1007. doi:10.1053/jhep.2003.50398

Pirola CJ, Gianotti TF, Burgueno AL, Rey-Funes M, Loidl CF, Mallardi P, Martino JS, Castano GO, Sookoian S (2013) Epigenetic modification of liver mitochondrial DNA is associated with histological severity of nonalcoholic fatty liver disease. Gut 62(9):1356–1363. doi:10.1136/gutjnl-2012-302962

Pogribny IP, Tryndyak VP, Bagnyukova TV, Melnyk S, Montgomery B, Ross SA, Latendresse JR, Rusyn I, Beland FA (2009) Hepatic epigenetic phenotype predetermines individual susceptibility to hepatic steatosis in mice fed a lipogenic methyl-deficient diet. J Hepatol 51(1):176–186. doi:10.1016/j.jhep.2009.03.021

Prodanovic R, Koricanac G, Vujanac I, Djordjevic A, Pantelic M, Romic S, Stanimirovic Z, Kirovski D (2016) Obesity-driven prepartal hepatic lipid accumulation in dairy cows is associated with increased CD36 and SREBP-1 expression. Res Vet Sci 107:16–19. doi:10.1016/j.rvsc.2016.04.007

Raza H, Robin MA, Fang JK, Avadhani NG (2002) Multiple isoforms of mitochondrial glutathione S-transferases and their differential induction under oxidative stress. Biochem J 366(Pt 1):45–55. doi:10.1042/BJ20020533

Saleem F, Ametaj BN, Bouatra S, Mandal R, Zebeli Q, Dunn SM, Wishart DS (2012) A metabolomics approach to uncover the effects of grain diets on rumen health in dairy cows. J Dairy Sci 95(11):6606–6623. doi:10.3168/jds.2012-5403

Sanyal AJ, Campbell-Sargent C, Mirshahi F, Rizzo WB, Contos MJ, Sterling RK, Luketic VA, Shiffman ML, Clore JN (2001) Nonalcoholic steatohepatitis: association of insulin resistance and mitochondrial abnormalities. Gastroenterology 120(5):1183–1192. doi:10.1053/gast.2001.23256

Scarpulla RC (2008) Transcriptional paradigms in mammalian mitochondrial biogenesis and function. Physiol Rev 88(2):611–638. doi:10.1152/physrev.00025.2007

Shackel NA, Seth D, Haber PS, Gorrell MD, McCaughan GW (2006) The hepatic transcriptome in human liver disease. Comp Hepatol 5:6. doi:10.1186/1476-5926-5-6

Shahzad K, Bionaz M, Trevisi E, Bertoni G, Rodriguez-Zas SL, Loor JJ (2014) Integrative analyses of hepatic differentially expressed genes and blood biomarkers during the peripartal period between dairy cows overfed or restricted-fed energy prepartum. PLoS One 9(6):e99757. doi:10.1371/journal.pone.0099757

Shock LS, Thakkar PV, Peterson EJ, Moran RG, Taylor SM (2011) DNA methyltransferase 1, cytosine methylation, and cytosine hydroxymethylation in mammalian mitochondria. Proc Natl Acad Sci U S A 108(9):3630–3635. doi:10.1073/pnas.1012311108

da Silva RP, Kelly KB, Al Rajabi A, Jacobs RL (2014) Novel insights on interactions between folate and lipid metabolism. Biofactors 40(3):277–283. doi:10.1002/biof.1154

Smith TR, Hippen AR, Beitz DC, Young JW (1997) Metabolic characteristics of induced ketosis in normal and obese dairy cows. J Dairy Sci 80(8):1569–1581. doi:10.3168/jds.S0022-0302(97)76087-9

Sookoian S, Rosselli MS, Gemma C, Burgueno AL, Fernandez Gianotti T, Castano GO, Pirola CJ (2010) Epigenetic regulation of insulin resistance in nonalcoholic fatty liver disease: impact of liver methylation of the peroxisome proliferator-activated receptor gamma coactivator 1alpha promoter. Hepatology 52(6):1992–2000. doi:10.1002/hep.23927

Spengler EK, Loomba R (2015) Recommendations for diagnosis, referral for liver biopsy, and treatment of nonalcoholic fatty liver disease and nonalcoholic steatohepatitis. Mayo Clin Proc 90(9):1233–1246. doi:10.1016/j.mayocp.2015.06.013

Steele MA, Schiestel C, AlZahal O, Dionissopoulos L, Laarman AH, Matthews JC, McBride BW (2015) The periparturient period is associated with structural and transcriptomic adaptations of rumen papillae in dairy cattle. J Dairy Sci 98(4):2583–2595. doi:10.3168/jds.2014-8640

Sun LW, Zhang HY, Wu L, Shu S, Xia C, Xu C, Zheng JS (2014) (1)H-nuclear magnetic resonance-based plasma metabolic profiling of dairy cows with clinical and subclinical ketosis. J Dairy Sci 97(3):1552–1562. doi:10.3168/jds.2013-6757

Svoboda DS, Kawaja MD (2012) Changes in hepatic protein expression in spontaneously hypertensive rats suggest early stages of non-alcoholic fatty liver disease. J Proteome 75(6):1752–1763. doi:10.1016/j.jprot.2011.12.011

Thomas A, Klein MS, Stevens AP, Reinders Y, Hellerbrand C, Dettmer K, Gronwald W, Oefner PJ, Reinders J (2013) Changes in the hepatic mitochondrial and membrane proteome in mice fed a non-alcoholic steatohepatitis inducing diet. J Proteome 80:107–122. doi:10.1016/j.jprot.2012.12.027

Veenhuizen JJ, Drackley JK, Richard MJ, Sanderson TP, Miller LD, Young JW (1991) Metabolic changes in blood and liver during development and early treatment of experimental fatty liver and ketosis in cows. J Dairy Sci 74(12):4238–4253. doi:10.3168/jds.S0022-0302(91)78619-0

Woitow G, Staufenbiel R, Langhans J (1991) Comparison between histologically and biochemically determined liver fat levels and resulting conclusions. Monatsh Veterinarmed 46(16):576–582

Xie ZQ, Li HK, Wang K, Lin JC, Wang Q, Zhao GP, Jia W, Zhang QH (2010) Analysis of transcriptome and metabolome profiles alterations in fatty liver induced by high-fat diet in rat. Metabolism 59(4):554–560. doi:10.1016/j.metabol.2009.08.022

Xu C, Wang Z (2008) Comparative proteomic analysis of livers from ketotic cows. Vet Res Commun 32(3):263–273. doi:10.1007/s11259-007-9028-4

Xu C, Sun LW, Xia C, Zhang HY, Zheng JS, Wang JS (2016) (1)H-nuclear magnetic resonance-based plasma metabolic profiling of dairy cows with fatty liver. Asian Australas J Anim Sci 29(2):219–229. doi:10.5713/ajas.15.0439

Yahagi N, Shimano H, Matsuzaka T, Sekiya M, Najima Y, Okazaki S, Okazaki H, Tamura Y, Iizuka Y, Inoue N, Nakagawa Y, Takeuchi Y, Ohashi K, Harada K, Gotoda T, Nagai R, Kadowaki T, Ishibashi S, Osuga J, Yamada N (2004) p53 involvement in the pathogenesis of fatty liver disease. J Biol Chem 279(20):20571–20575. doi:10.1074/jbc.M400884200

Young JW, Veenhuizen JJ, Drackley JK, Smith TR (1990) New insights into lactation ketosis and fatty liver. In: Proceedings 1990: Cornell Nutrition Conference for Feed Manufacturers, pp 60–67

Zeybel M, Hardy T, Robinson SM, Fox C, Anstee QM, Ness T, Masson S, Mathers JC, French J, White S, Mann J (2015) Differential DNA methylation of genes involved in fibrosis progression in non-alcoholic fatty liver disease and alcoholic liver disease. Clin Epigenetics 7:25. doi:10.1186/s13148-015-0056-6

Zhang H, Wu L, Xu C, Xia C, Sun L, Shu S (2013) Plasma metabolomic profiling of dairy cows affected with ketosis using gas chromatography/mass spectrometry. BMC Vet Res 9:186. doi:10.1186/1746-6148-9-186

Zhang A, Sun H, Yan G, Wang P, Wang X (2015) Metabolomics for biomarker discovery: moving to the clinic. Biomed Res Int 2015:354671. doi:10.1155/2015/354671

Zhou Z, Vailati-Riboni M, Trevisi E, Drackley JK, Luchini DN, Loor JJ (2016) Better postpartal performance in dairy cows supplemented with rumen-protected methionine compared with choline during the peripartal period. J Dairy Sci 99(11):8716–8732. doi:10.3168/jds.2015-10525

Milk Fever: Reductionist Versus Systems Veterinary Approach

12

Elda Dervishi and Burim N. Ametaj

Abstract

Milk fever affects mostly multiparous and high-performance dairy cows that are close to calving and at the beginning of lactation. While much is known about milk fever, the real cause of failure in the Ca homeostatic mechanism remains unanswered, and, despite the considerable number of hypotheses presented over the years, milk fever still remains a complex and not very well understood disease of dairy cattle. From the recent "omics" studies, it is obvious that Ca is not the only perturbed variable in the organism of a cow affected by milk fever. Therefore, the complexity of the pathobiology of this disease makes it necessary to consider it from a systems veterinary perspective instead of the reductionist view only. A new approach is necessary in order to clarify and better understand the pathobiology of milk fever. Systems biology sciences have modestly contributed to the identification of multiple unknown alterations in cows with milk fever including unidentified QTL, a number of genes and proteins with altered expressions as well as a number of metabolic changes. A combination of the omics sciences will offer great possibilities for animal health scientists to incorporate all the information generated and study component-to-component interactions and the dynamic communications that result from them.

E. Dervishi, Ph.D. (✉) • B.N. Ametaj, D.V.M., Ph.D., Ph.D.
Department of Agricultural, Food and Nutritional Science, University of Alberta, Edmonton, AB, Canada, T6G 2P5
e-mail: dervishi@ualberta.ca; burim.ametaj@ualberta.ca

B.N. Ametaj (ed.), *Periparturient Diseases of Dairy Cows*,
DOI 10.1007/978-3-319-43033-1_12

12.1 Introduction: Historical Perspective

Periparturient hypocalcemia or parturient paresis is known in the scientific and industry community as milk fever. The origin of the name comes from the belief of the time that milk instead of flowing through the udder was coming out of vagina and causing fever in the affected cows. The disease affects mostly multiparous and high-performance dairy cows that are close to calving and at the beginning of lactation. The incidence of milk fever varies among farms and countries as reviewed by Pryce et al. (2016). As an average, around 5% of the cows in a herd exhibit clinical signs of milk fever (DeGaris and Lean 2008). Today the dominant thinking is that milk fever is a metabolic disease related to Ca deficiency. Indeed, milk fever is characterized by hypocalcemia, typically lower than 1.5 mmol/L (Goff 2008; Reinhardt et al. 2011) as well as progressive neuromuscular dysfunction with flaccid paralysis, recumbency, circulatory collapse, cold extremities, and depressed consciousness (Goff 2008). If the cow is not treated, it might end up in coma and death (Horst et al. 1997).

The first scientific report on milk fever is documented in Germany in 1793 by Eberhardt (Hutyra et al. 1938). It is believed that the disease became known when increasing amounts of feedstuff were fed to the cows in order to increase milk production (Hutyra and Marek 1926).

One of the earliest and extended reviews on milk fever in the twentieth century was written by Hibbs (1950). By that time milk fever had gained the interest of researchers and during the 1950s a total of 30 hypotheses on milk fever had been proposed. These hypotheses included general inflammation, derangement of the nervous system, general circulatory disturbances, cerebral anemia, cerebral congestion, apoplexy, thrombosis, fat embolism, spinal traumatism, general infection, bacterial infection of uterine origin, infection of mammary origin, anaphylaxis, mammary-neurasthenia, anhydremia, auto-intoxication, defective oxidation in the tissues, excess oxytocic principle in blood after parturition, faulty protein metabolism, ovarian dysfunction, hyper functioning of the anterior pituitary, disturbed cholesterin metabolism, hyperadrenalinemia, magnesium narcosis, alkalosis, acidosis and auto-asphyxiation and hypoglycemia (Hibbs 1950). Out of more than 30 different hypotheses on milk fever, only three of them have survived time: (1) parathyroid deficiency and hypocalcemia, (2) dietary alkalosis, and (3) bacterial toxic compounds. We will discuss further about these three hypotheses in this chapter.

Currently there are more than 1000 scientific publications on milk fever, which shows both the interest of scientific community on this disease and its economic importance especially to dairy industry. Milk fever has been found to be associated with other diseases such as dystocia, uterine prolapse, retained placenta, metritis, displaced abomasum, and mastitis (Chamberlain 1987; Pryce et al. 1997; Zwald et al. 2004; DeGaris and Lean 2008; Hossein-Zadeh and Ardalan 2011). Economic losses caused by milk fever include the cost of medication, additional labor, reduction in milk yield, the cost of the dead animal, and early culling as well as increased susceptibility of cows to other metabolic and infectious diseases (Fikadu et al. 2016).

12.2 Reductionist Understanding

The disease was mentioned for the first time in 1793 by Eberhardt. In 1806, Price, based on external signs of the disease (cool extremities), suggested treatment of cows with hot packs and blanketing. In 1814, Clater suggested that the disease originates from an imbalance of body humors (based on the four body humors theory: blood, phlegm, yellow bile, and black bile), recommending bloodletting or phlebotomy (4–5 L/day) for 8–10 days. More than 80 years later (1897), Schmidt suggested that milk fever was caused by a viral infection of the mammary gland, and to fight the infection he recommended injection of potassium iodide into the udder of sick cows (as revised by Hibbs (1950)). This treatment reduced mortality rate by 60–70%. Later, Marshak (1956) interpreted the preventive effects of injected potassium iodide into the mammary as a reduction of milk and calcium secretion by the pressure exerted on the alveoli. Based on that assumption, he recommended the technique of insufflation of the udder for treatment of disease. This technique was abandoned because of increased incidence of mastitis and diminished milk production after the treatment.

The year 1925 marked a turning point in the history of milk fever when Little and Wright made the observation that concentrations of plasma calcium (Ca) were greatly reduced in cows suffering from the disease. Based on this observation, Dryerre and Greig (1925) proposed the use of Ca solutions (especially calcium borogluconate) as a treatment for milk fever. Even though this is the most widespread method used for treatment of milk fever today, high plasma Ca resulting from infusion of 23% calcium borogluconate solutions can cause cardiac arrest and about 25% of the cows relapse and require additional treatment.

The discovery of both vitamin D and its active metabolite 1,25-dihydroxyvitamin D_3 generated considerable interest in using them as means to prevent milk fever. In studies summarized by Littledike and Horst (1982), many of the cows treated prepartum with the vitamin D_3 or 1α-hydroxylated vitamin D metabolites showed hypocalcemia and clinical signs of milk fever at 10–14 days postpartum. Littledike and Horst (1982) demonstrated that cows treated with vitamin D compounds were unable to produce endogenous 1,25-dihydroxyvitamin D_3 and were, therefore, unable to recover from hypocalcemic episodes. This inhibition was thought to be the result of inhibition of the kidney 1α-hydroxylase by hypercalcemia that was associated with the use of these compounds as well as direct feedback inhibition by the high circulating concentrations of vitamin D compounds resulting from the pharmacological doses administered. This research fostered the investigation into the next generation of vitamin D analogs like 24-F-1,25-dihydroxyvitamin D_3 and 1α-hydroxyvitamin D_3. The use of these two analogs, however, shared the same disadvantages of earlier compounds: hypercalcemia and inhibition of the endogenous synthesis of 1,25-dihydroxyvitamin D_3.

Oral administration of large amounts of calcium salts, usually $CaCl_2$, have been used effectively to increase concentration of Ca in the blood during the peripartal period and to prevent milk fever, but $CaCl_2$ solutions have disadvantages of being very caustic and ulcerative in the mouth and mucosa of the gastrointestinal tract (Goff and Horst 1994). Also, excessive oral $CaCl_2$ can induce metabolic acidosis, which can cause anorexia at a time (i.e., transition period) when feed intake is already compromised (Goff and Horst 1993).

In 1925, Dryerre and Greig suggested that hypocalcemia of milk fever might be the result of insufficient parathyroid hormone (PTH) secretion. However, subsequent reports concluded that concentrations of PTH in plasma of animals suffering from milk fever are equal or higher than in normal cows. Injection of cows with PTH resulted in less responsiveness from older cows and no prevention of milk fever, which led investigators to hypothesize that responsiveness of the target tissue to PTH stimulation, may be deficient or delayed in the peripartal cow. Goff et al. (1986, 1989) revisited the PTH hypothesis and showed that injection or infusion with PTH could prevent milk fever, but the manufacturing of a subcutaneous slow-release product delayed the benefits of this treatment. Recently Goff et al. (2014) showed that metabolic alkalosis induced by high-DCAD (dietary cation-anion difference) diet rich in K triggers a state of pseudohypoparathyroidism, at the beginning of lactation, resulting in hypocalcemia and milk fever. Pseudohypoparathyroidism is characterized by end-organ resistance to the effects of PTH.

Another interesting, but short-lived hypothesis on the etiology of milk fever was that of hypersecretion of calcitonin in paretic cows. Capen and Young (1967) reported that thyroid glands and plasmas of cows with milk fever are depleted of calcitonin. Barlet (1967) produced hypocalcemia, hypophosphatemia, and paretic signs in young cattle by infusing calcitonin. From these observations evolved the theory that a sudden release of calcitonin at parturition was the cause of milk fever. However, Mayer et al. (1975) and Hollis et al. (1981) showed no increased concentrations of calcitonin in cows with milk fever. In fact, cows suffering from milk fever had lower calcitonin concentrations than did normal cows.

Excessive production of cortisol at parturition also has been suggested as a contributing factor in development of milk fever. Concentrations of cortisol increase during parturition and plasma cortisol is higher in milk fever cows (Littledike et al. 1970; Horst and Jorgensen 1982). In humans, glucocorticoid therapy results in decreased intestinal absorption of Ca and a profound loss of skeletal mass (Hahn et al. 1981; Gluck et al. 1981). However, Horst and Jorgensen (1982) showed that induced hypocalcemia stimulates cortisol secretion, whereas exogenous injections of cortisol cause no hypocalcemia. As can be seen, the exact role of cortisol in the etiology of milk fever remains uncertain.

One of the most significant and yet least understood findings in the history of prevention of milk fever is the observation made by Ender et al. (1971) that feeding cows before parturition with inorganic acids (a mixture of sulfuric and hydrochloric acids) lowers the incidence of milk fever. Ender et al. (1971) proposed that the ratio of cations (Na^+ and K^+) relative to anions (Cl^- and SO_4^-) plays a significant role in the incidence of milk fever. This concept is referred generally to as the dietary cation–anion difference. Block (1984) found that cows fed anionic salts had lower incidence of milk fever. Subsequently, a number of other publications addressed the concept of DCAD and demonstrated the utility of adding anions to prepartal diets for prevention of milk fever. This hypothesis relies on the assumption that a major underlying cause of milk fever is metabolic alkalosis, which lowers the response of tissues to PTH. On the other hand, enhancement of anions in the diet results in reduction of blood pH and therefore a lower incidence of milk fever. The exact mechanism of how dietary anions work is still unknown, but results from Gaynor et al. (1989) and Goff et al. (1991) indicated that inducing a mild metabolic acidosis by the addition of Cl^- and SO_4^- to prepartal diets increases the responsiveness of tissues to PTH.

In conclusion, while much is known about milk fever, the real cause of failure in the Ca homeostatic mechanism remains unanswered, and, despite the considerable number of hypotheses presented over the years, milk fever still remains a complex and not very well understood disease of dairy cattle.

A summary of present reductionist understanding of milk fever and current methods of prevention and treatments are shown in Fig. 12.1.

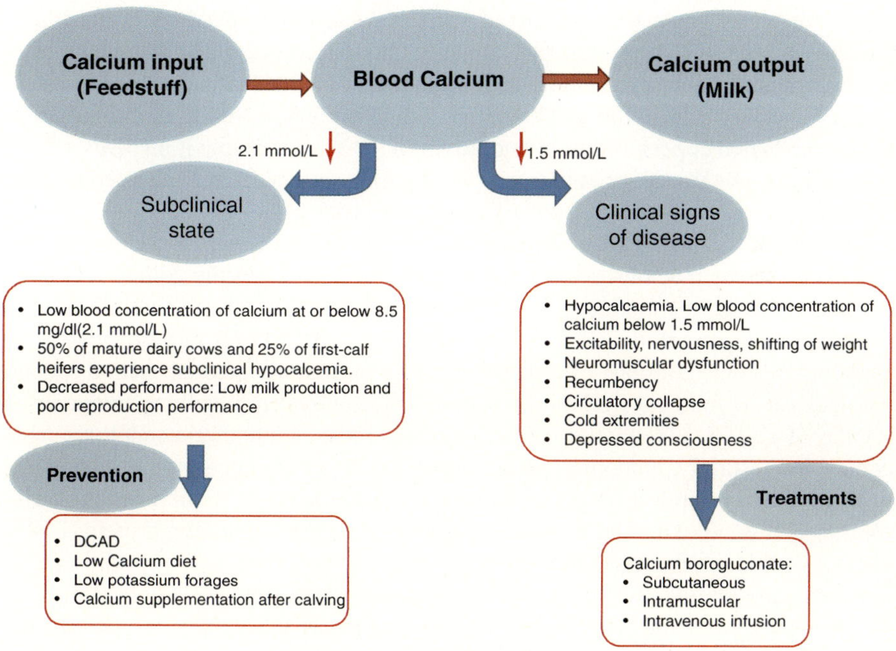

Fig. 12.1 Current reductionist understanding of milk fever

12.3 Potential Contribution of Inflammatory Conditions in the Pathobiology of the Disease

An interesting question is what causes the rapid decline of Ca around calving, and why? Many investigators have proposed that high-milk producing cows have high demand for Ca at the onset of lactation for milk production; however, the fact that not all the high producing cows are affected by hypocalcemia and milk fever suggests the existence of other triggering factors.

A new interesting line of thought regarding the etiology of milk fever was proposed by Aiumlamai et al. (1992) where they speculated about a possible role of endotoxin in the pathology of the disease. However, they did not elaborate about what would be the mechanism(s) of how endotoxin contributes to the pathobiology of milk fever. In a later study, Ametaj et al. (2003) reported that cows with milk fever had greater plasma serum amyloid A (SAA) and lower concentrations of calcitonin gene-related peptide (CGRP) in the plasma compared with clinically normal cows. Moreover, Zebeli et al. (2013) reported downregulation of CGRP secretion in dairy cows approaching the day of calving and affected by milk fever. The greater levels of SAA suggested presence of an inflammatory state related to presence of endotoxin in the blood circulation, whereas low CGRP indicated inhibition of its secretion in sick cows. Another interesting report was that intravenous (iv) infusion of LPS in dairy cows was associated with hypocalcemia (Waldron et al. 2003). Ametaj et al. (2010) proposed a mechanism of how endotoxemia and hypocalcemia are interrelated. More specifically they suggested two pathways involved in clearance of endotoxin from bloodstream, depending on concentration of endotoxin in the blood. When concentration of endotoxin in the plasma is low it involves activation of macrophages. Since endotoxin (i.e., LPS) is very negatively charged it binds Ca easily to form large aggregates (Rosen et al. 1958; Munford et al. 1981). Those aggregates are removed from circulation by macrophages, which become highly reactive releasing a whole array of proinflammatory cytokines including TNF, IL-1, and IL-6. The latter cytokines are proinflammatory and in high concentrations trigger overall sickness in the organism (Ametaj et al. 2010; Eckel and Ametaj 2016). However, if concentrations of endotoxin in the plasma are greater, then, a second pathway for its removal from blood circulation is activated, involving lipoproteins (Gallay et al. 1994). The latter pathway is triggered by monomeric endotoxin molecules. Interestingly, monomerization of endotoxin is facilitated by low concentrations of Ca (withdrawal of plasma Ca helps the process of monomerization and clearance of endotoxin) and presence of lipopolysaccharide binding protein (LBP), mCD-14, and lipoproteins (Munford et al. 1981). Ametaj et al. (2010) proposed that hypocalcemia of milk fever might be a combination of Ca-impaired mobilization by metabolic alkalosis and Ca withdrawal from plasma, as a protective response of the host for safely removing endotoxin from blood circulation during conditions of endotoxemia. Recent studies conducted by our lab demonstrated that cows affected by milk fever have alterations of innate immunity reactants and metabolites related to carbohydrate metabolism several weeks prior to clinical appearance of milk fever. Most interestingly, serum TNF, SAA, LBP, Hp, and lactate in cows with milk fever

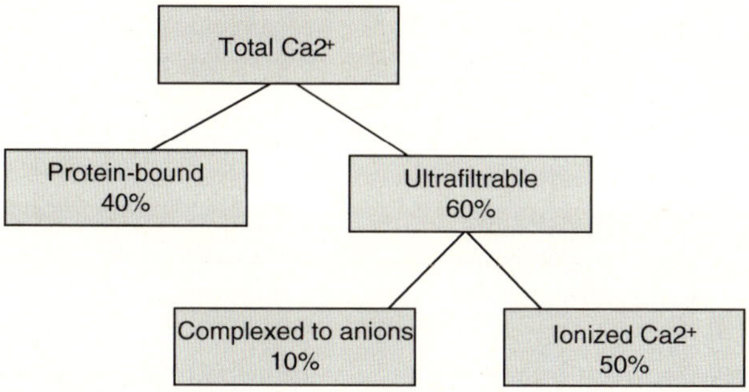

Fig. 12.2 Forms of calcium in the blood

were elevated in pre-milk fever cows starting at around 9–10 weeks before diagnosis of disease (Zhang et al. 2014).

Another finding in support of a role of inflammatory state in hypocalcemia of transition period is fluctuations of plasma albumin during acute phase response (APR). It is known that during inflammatory states concentrations of albumin in the blood decrease as part of the APR. Indeed albumin is one of the negative acute phase proteins. In a review by Aldred and Schreiber (1993) they demonstrated that mRNA expression for albumin in the liver was downregulated during induced inflammation, decreasing concentrations of albumin in the plasma. Albumin is one of the most significant carriers of Ca in the plasma. The total blood Ca includes a fraction that is bound to plasma albumin (i.e., 40%) and a fraction that is unbound, free, or ultrafiltrable (i.e., 60%; Fig. 12.2). The ultrafiltrable component of Ca is further divided between a fraction that is bound to anions (10%), such as phosphate and sulfate, and the remainder that is free ionized Ca (50%). The biologically active form of Ca is only the free ionized Ca, whereas any Ca bound to albumin or complexed to anions is inactive. Regrettably, most researchers, studying milk fever, have reported only the total concentration of Ca in the plasma but not the ionized Ca. In fact, the real hypocalcemia is related to the decrease of ionized Ca and not to the overall hypocalcemia.

There are several conditions that are associated with changes in the concentration of the free ionized Ca in the blood. For example, an increase in the concentration of anions in the plasma is associated with a greater fraction of Ca complexed with anions and therefore decreased concentration of ionized Ca in the plasma. Additionally, acid-base changes also affect concentration of ionized Ca (Fig. 12.3). Moreover, plasma albumin has multiple negatively charged sites that can bind either H^+ or Ca^{2+}. During acidemia, for example, when there is excess H^+ in the blood, more H^+ is bound to plasma albumin; thus, the free ionized Ca increases. On the other hand, during alkalemia (as during milk fever), when there is a deficit of H^+ in the blood, H^+ is released and Ca^{2+} is bound to albumin; thus, the ionized Ca^{2+} decreases.

Another condition that is associated with changes in the concentration of Ca in the plasma is endotoxemia. For instance, in a study conducted by Waldron et al.

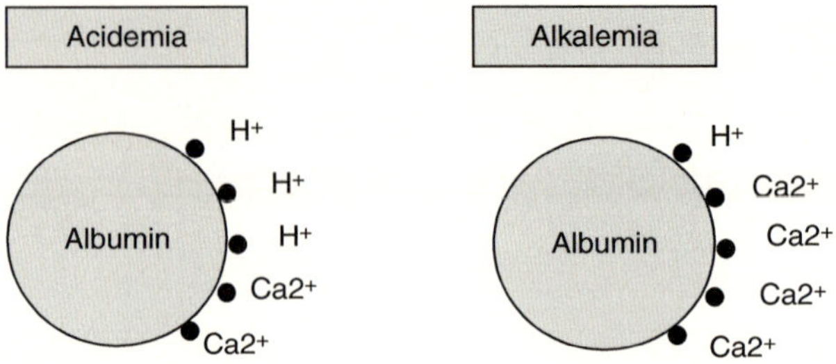

Fig. 12.3 Effects of acid-base disturbances on calcium binding to plasma albumin

(2003) iv administration of lipopolysaccharide (LPS) was associated with decreased plasma Ca. What is the mechanism by which LPS decreases Ca concentration in the plasma? One mechanism might be that LPS disturbs the acid-base balance. This speculation is supported by data reported by Ohtsuka et al. (2013) who demonstrated that iv administration of 0.025 mg/kg of LPS in Holstein cows was associated with a significant increase in arterial blood pH, or alkalosis, during 24–48 h after its administration.

One of the major practical applications for prevention of milk fever has been utilization of DCAD. No theoretical explanation has been given of how the DCAD works. Therefore, an attempt to explain the mechanism by which DCAD works is necessary. It should be noted that minerals including Ca are used by the host to neutralize acids that enter into the blood circulation. Therefore, blood mineral fluctuations during dietary induced metabolic acidosis or alkalosis are expected to occur. Elevated or decreased plasma Ca during metabolic acidosis (low DCAD) or alkalosis (high DCAD), respectively, are none but host responses to those metabolic states and not Ca deficiency as commonly interpreted for transition dairy cows. For example, Arnett (2003) in his review article indicates that it has been known since the early twentieth century that systemic acidosis causes depletion of the skeleton (i.e., loss of minerals) and that resorption pit formation depends on external acidification. Metabolic acidosis can be caused by acidogenic diets (i.e., low DCAD diets for cows), excessive protein intake, and aging in human subjects. Indeed, parity is known to be associated with lower plasma Ca and greater incidence of subclinical hypocalcemia and milk fever in dairy cows (Reinhardt et al. 2011). According to Arnett (2003) the unusual stimulatory effect of acidosis on osteoclast represents an evolutionary response of vertebrates to correct systemic acidosis by release of alkaline bone minerals (including Ca^{2+}) when the lungs and kidneys are unable to remove sufficient H^+ equivalents. On the other hand, high-DCAD diets (i.e., alkaline diets) trigger the opposite effects of acidogenic diets. They stimulate Ca deposition in the bones, lower all PTH activities including mobilization of Ca from bones, Ca excreted through urine, and the amount of 25-hydroxyvitamin D_3 conversion into the active form (i.e., 1,25-dihydroxyvitamin D_3) in the kidneys and, as a consequence, lowering absorption of Ca from intestines and the concentration of Ca in the plasma.

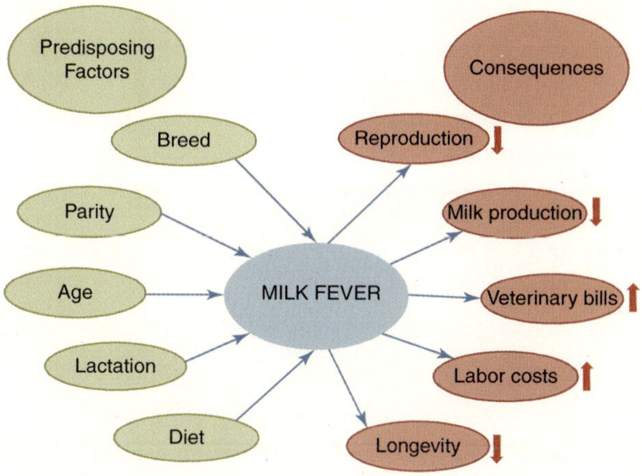

Fig. 12.4 Milk fever predisposing factors and impact on cow's performance

12.4 Predisposing Factors

A summary of predisposing factors and economic consequences of milk fever are summarized in Fig. 12.4.

12.4.1 Parity

Milk fever is a complex metabolic disorder and reasons for greater susceptibility to hypocalcemia vary and may be multifactorial (Shire and Beede 2013). Arnett (2003) indicated that age is an important factor in mineral metabolism. It is very well documented that multiparous cows have greater susceptibility to milk fever based on differences in cows' abilities to mobilize Ca from bone (van Mosel et al. 1993). The cow is not able to extract enough Ca from their bones and diet to replace the Ca lost to milk (Goff 2008). In addition, those cows that have previously been affected by milk fever will develop hypocalcemia at subsequent parturitions. This is presumably due to a decreased ability of these particular cows to respond immediately to biological signals and increase vitamin D receptor (VDR) numbers in a timely manner (Goff et al. 1995).

12.4.2 Diet

Diet is another factor that has been associated with the incidence of milk fever. In order to control the incidence of milk fever most attention has focused on manipulating the levels of dietary Ca (Horst et al. 1997). Many studies have shown that feeding low Ca diets lowered the incidence of milk fever (Goings et al. 1974;

Kichura et al. 1982). Low Ca diets have been used successfully to reduce the incidence of hypocalcemia (Goings et al. 1974; Thilsing-Hansen et al. 2002). The mechanism behind lowering the amount of Ca in the diet is not clear. Some authors have argued that it needs some time for the Ca to be mobilized from the bone tissue and lowering the amount of Ca in the diet, before calving, might activate those mechanisms. In addition, in the recent years the use of dietary anions (i.e., Cl^- and SO_4^{2-}) has been used in controlling milk fever. An effective means of offsetting the detrimental effects of K^+ is to increase the anionic (Cl^- and SO_4^{2-}) content of diets (Horst et al. 1997).

12.4.3 Genetic Factors

For decades the main focus in dairy cattle breeding programs has been on improving/increasing milk production. However, no efforts were made to select cows for resistance to metabolic diseases. The incidence of milk fever has been mostly unchanged for decades. A study conducted by USDA, National Animal Health Monitoring System (2002) estimated a 5% incidence rate of milk fever in the United States (Reinhardt et al. 2011). Involvement of genetic factors in the incidence of milk fever has been previously studied by various researchers. Milk fever heritability has been estimated to be low, however, across lactations it varies between 0.08 and 0.47 (Lyons et al. 1991; Uribe et al. 1995; Pryce et al. 1997). In addition, Thompson (1984) estimated that the heritability of milk fever is 0.13 in Holstein cows; however, in older cows the heritability of the disease was reported to be at a moderate level 0.42 (Lin et al. 1989). Genetic correlation between milk fever and milk yield and/or milk production traits is very limited, and therefore more research is warranted in this area. For example, Uribe et al. (1995) reported a strong negative genetic correlation between milk fever and milk yield. In addition, it has been shown that there is a genetic correlation between milk fever and mastitis, and milk fever and metritis (Pryce et al. 1997; Zwald et al. 2004; Hossein-Zadeh and Ardalan 2011). Even though these metabolic diseases are different, the existence of genetic correlations indicates the existence of some disease resistance factor with a genetic component, indicating a potential for improvement by estimating breeding values for metabolic disease traits and considering these traits in the selection index. This would require collaboration among producers, research institutions, animal scientists, and governments in implementing a reliable system of data collection and processing.

Among various factors, breed has been reported to be an important factor in the incidence of milk fever. For example, Channel Island, Swedish Red and White, and Jersey cows are known to be more susceptible to incidence of milk fever than the other breeds of dairy cattle (Henderson 1938; Hibbs et al. 1946; Metzger 1936; Kusumanti et al. 1993; Lean et al. 2006). This susceptibility has been related to the fact that Jersey cows have fewer intestinal receptors for $1,25(OH)_2D_3$ when compared to Holsteins (Goff et al. 1995). The regulation of vitamin D receptor (VDR) could be a plausible explanation for differences among age, breed, and body condition, length of dry period, milk yield, parity, and dietary factors.

Hypocalcemia can be considered as one of the most costly diseases in dairy cows and many factors have been discussed as possible causes. For example in humans a mutation and a rare loss of function in the PTH gene or in the transcriptions factors such as GCM2 and GATA3 is the key to development of hypoparathyroidism. In cows there are no reports about the presence of potential mutations, or single nucleotide polymorphisms (SNPs); therefore it would be interesting to explore the existence of SNPs in the PTH gene in dairy cows.

Autosomal-dominant hypocalcemia (ADH) is a common inherited form of hypoparathyroidism, which in humans is caused by a mutation in the Ca-sensing receptor. Moreover, mutation in the gene encoding the guanine-binding protein G11 has recently been identified as a cause of hypoparathyroidism (Nesbit et al. 2013; Mannstadt et al. 2013). In addition, polymorphism in the VDR (vitamin D receptor) has shown to influence Ca homeostasis in humans, and for milk fever prevention, strategies including vitamin D supplementation have been tested successfully. In an attempt to explain why some dairy cows are unable to maintain Ca homeostasis after parturition, Deiner et al. (2012) investigated polymorphism in the VDR gene. They found eight single-base aberrations in the VDR gene of 26 dairy cows, four of which were located on exons and cause a potential change in the amino acid sequence on one allele. However no significant correlation with the incidence of hypocalcemia was detected. The absence of the correlation was attributed to lower number of animals involved in the study. Therefore, it would be interesting that these four alterations found in the coding regions of VDR gene be tested in a greater number of cows and different breeds.

12.5 A Systems Veterinary Approach to the Pathobiology of the Disease

Reductionist approach has left us with 30 hypotheses on milk fever and counting. It is time to make a paradigm shift in the scientific philosophy and the instrumentation we use to better understand the causes as well as pathways and networks involved in the pathobiology of milk fever.

Systems veterinary approach involves using "omics" technologies including genomics, transcriptomics, proteomics, and metabolomics to tackle the pathobiology of the disease process in animals (Ametaj 2015). Despite fast advancement in the instrumentation, few studies have been carried out to address the pathobiology of milk fever. The present modest contribution of "omics" sciences to understanding of pathobiology of milk fever is shown in Fig. 12.5.

Genomic studies have helped to identify the most useful markers associated with milk fever, for QTL detection and, eventually, for marker-assisted selection for improvement of economically important traits such as health and fertility. Elo et al. (1999) described a QTL mapped to bovine chromosome 23 for veterinary treatment,

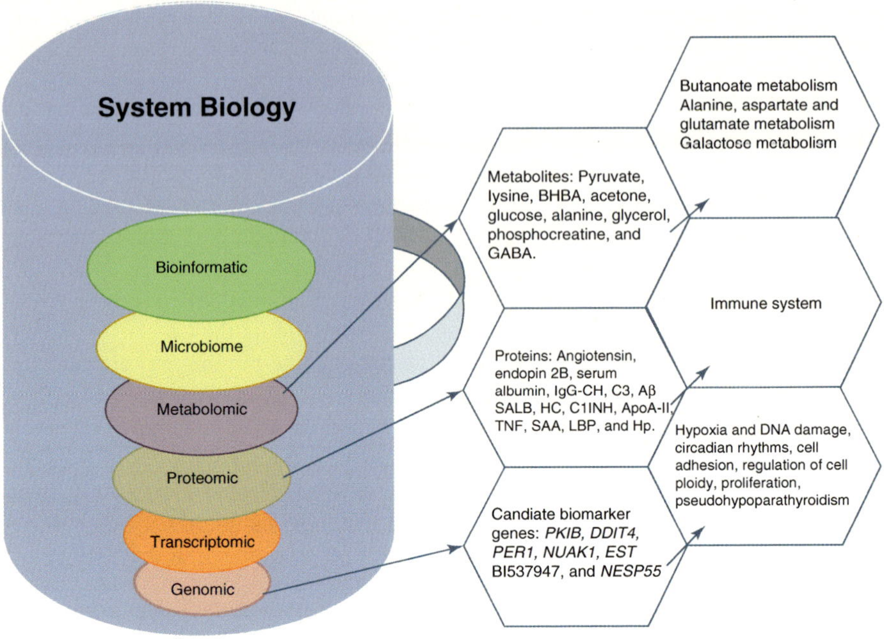

Fig. 12.5 Systems veterinary contribution to new understanding of pathobiology of milk fever

and they reported that the putative QTL probably affects susceptibility to milk fever or ketosis. In addition, Genome Wide Association Studies (GWAS) identified CYP2J2 as a gene controlling serum vitamin D status in beef cattle (Casas et al. 2014), which makes it an interesting target for dairy cows as well. In cattle, genomic approach has been utilized to study the pathophysiology of many diseases (Almeida et al. 2007; Loor et al. 2007; Lutzow et al. 2008; Ishida et al. 2013). However, little information exists on microarray-based gene expression profiling for milk fever. A search within Gene Expression Omnibus (GEO) and the ArrayExpress archives revealed only one article published involving milk fever in dairy cows (Sasaki et al. 2014). These authors studied gene expression in peripheral blood mononuclear cells of spontaneous milk fever cows. Milk fever cows exhibited significantly different expression of 98 genes when compared to healthy cows, among which, 61 genes were upregulated. We did an enrichment analysis of those genes and found out that the upregulated genes are involved in regulation of cellular protein metabolic processes, regulation of phosphorylation, regulation of protein amino acid phosphorylation, intracellular signaling cascade, apoptosis, and cell death. The 37 downregulated genes identified are involved in metal-ion binding, cation-binding, DNA-binding, and transcription regulation. In addition, the authors reported specific candidate biomarker genes for induced hypocalcemia and milk fever including protein kinase (cAMP-dependent, catalytic) inhibitor β (*PKIB*), DNA-damage-inducible transcript 4 (*DDIT4*), period homolog 1 (*PER1*), NUAK family, SNF1-like kinase 1 (*NUAK1*), and expressed sequence tag EST (BI537947) that were strongly related to both experimental hypocalcemia and milk fever. These genes were significantly upregulated in cows with milk fever compared with clinically healthy parturient cows, whereas the neuroendocrine secretory protein 55 (NESP55) was downregulated in cows with milk fever. The authors suggested that the altered mRNA expression of the genes may be related to immune suppression. In the future, genomic biomarkers have the potential to provide new insights into the pathophysiological processes of milk fever and microarray technology offers the possibility to study expression of thousands of other genes.

The same prospect applies to the utilization of proteomics analysis to milk fever cows. Unfortunately, there is a scarce number of published articles that have explored alterations of proteome profile in cows with milk fever. For example, Xia et al. (2012) reported novel pathophysiological changes in the plasma proteome of cows affected by milk fever including increased serpin peptidase inhibitor (angiotensin), which regulates blood pressure and maintains fluid and electrolyte homeostasis, and endopin 2B which is involved in neural regulation. Serpins are a group of proteins able to inhibit proteases and control a wide array of biological processes, including inflammation and immune response. Bacterial membrane-associated proteases are critical for bacterial growth, division, and survival (Dalbeya et al. 2012) and help in colonization, evasion of host defenses, facilitation of dissemination, and host tissue damage during infection. Hwang et al. (2005) suggested that endopin 2B inhibits endogenous secretory vesicle cathepsin L that participates in the production of biologically active enkephalin peptide neurotransmitter (Yasothornsrikul et al. 2003). The mature, processed enkephalin peptide is stored within these vesicles and undergoes stimulated secretion to mediate neurotransmission and cell–cell communication in the regulation of analgesia, behavior, and immune-cell functions (Yasothornsrikul et al. 2013). It is possible that the increase in serpin peptidase inhibitor and endopin 2B is an attempt of the host to overcome bacterial infection and control inflammation in cows with milk fever. The downregulated proteins, in this study, were serum albumin, which is considered a negative acute phase protein and acts as a transport protein for Ca, steroid and thyroid hormones, and fatty acids in the plasma. Fibrinogen beta chain, which is involved in blood coagulation and IgG heavy-chain C-region (IgG-CH) were also downregulated. At the onset of lactation cow's colostrum contains high levels of immunoglobulins in order to protect the newborn calf from potential pathogenic bacteria in the GI tract. The calf receives no passive transfer of immunity via the placenta prior to birth; therefore, the necessary antibodies need to be orally taken immediately after birth. Cows with milk fever experience a tremendous decrease in milk production; therefore, downregulation of IgG-CH in the serum of cows with milk fever might be a consequence of the decrease in milk production. This suggests a strategy of the organism to prioritize between production of milk and synthesis of IgG-CH for the offspring or to mount an appropriate immune response to defend itself. Presumably, the energy is directed to mount a successful immune response instead of milk and protein production.

In another study, Shu et al. (2014) investigated alterations in the plasma proteomic profiling of cows with milk fever and reported that complement C frag (C3), amyloid beta a4 protein (Aβ), serum albumin frag (SALB), and hepcidin (HC) were upregulated in cows with milk fever, whereas plasma protease c1 inhibitor frag (C1INH) and apolipoprotein a-2 (ApoA-II) were downregulated. The complement component C3 is directly involved in the activation of the complement cascade (Sahu and Lambris 2001), which is part of the innate immune response to infection caused by different pathogenic bacteria. A similar increase in the complement component C3 was observed in the plasma of cows with mastitis (Turk et al. 2012). The latter is activated by proteolysis and triggers large conformational changes in the proteins which are very important in mediating the opsonization process by covalent attachment to targets such as immune complexes and carbohydrates on the surface of invading microorganisms (Fredslund et al. 2006). It is possible that the increase of complement C3 in the plasma of cows with milk fever might be a sign of the host's response to invading bacteria. Apolipoprotein A-II is the second major protein of HDL with antioxidant properties (Garner et al. 1998). During inflammatory states apolipoprotein A-II is replaced by serum amyloid A, which helps in expedited removal of acute phase HDL from systemic circulation. High-density lipoprotein contributes to binding and neutralization of LPS and its expedited removal from blood circulation is a must for protecting the host from harmful effects of endotoxin.

Overall, proteomics studies reveal an activation of the immune response during milk fever; however, it would be of great interest to apply a proteomics approach in order to find out if the activation of the immune response precedes milk fever. In a recent study conducted by our lab we demonstrated that cows affected by milk fever had alterations of innate immunity reactants and metabolites related to carbohydrate metabolism that preceded clinical appearance of milk fever by more than 8 weeks (Zhang et al. 2014). In addition, serum TNF, SAA, LBP, Hp, and lactate, in cows with milk fever, were elevated compared to healthy animals starting at around 9–10 weeks prior to diagnosis of disease.

An increasing number of research articles have been published recently with regard to application of metabolomics technologies in the area of periparturient diseases of dairy cows including ketosis and mastitis. However, such studies on milk fever cows are very scarce. In one of these studies, Sun et al. (2014) applied metabolomics analysis in the plasma of cows affected by milk fever and reported that glucose, alanine, glycerol, phosphocreatine, and gamma-aminobutyrate (GABA) were decreased, whereas concentrations of β-hydroxybutyrate (BHB), acetone, pyruvate, and lysine were increased in cows diagnosed with milk fever. Most of these perturbations were related to carbohydrate, fat, and protein metabolism involved in the energy metabolism. Metabolomics approaches despite being relatively new in the animal health sciences have a great potential to identify biomarkers for the risk of milk fever.

Taken together the systems biology sciences have modestly contributed to the identification of multiple previously unknown alterations in cows with milk fever including unidentified QTL, a number of genes and proteins with altered expressions as well as a number of metabolic changes. Previously most metabolic diseases have been related to perturbation of one specific metabolite. For example, milk fever had been strongly linked and explained by perturbation of Ca homeostasis

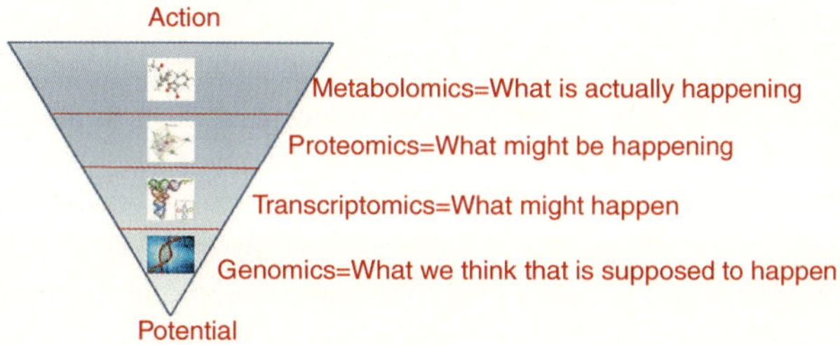

Fig. 12.6 Information generated from the omics

only as well as a number of hormones related to its metabolism. Despite the tremendous miniscule knowledge and information revolving around Ca hypothesis, the root cause of milk fever remains unknown. From the recent omics studies, it is obvious that Ca is not the only perturbed variable in the organism of a cow affected by milk fever. Therefore, the complexity of the pathobiology of milk fever makes it necessary to consider it from a systems veterinary perspective instead of the reductionist view only. Moreover, genomic selection is a powerful tool for generating breeding values and can be utilized to identify animals with better metabolic health. However, the levels of accuracy of genomic selection remain to be improved. When we select from the genomic level, the information that is generated doesn't take into account the physiology of the animal and/or interaction of the genotype with the environment. Therefore, information coming from the genes indicates what is supposed to happen based on the genetic potential of the animal, but not what is actually happening within the animal body. On the other hand, the information generated from metabolomics (i.e., metabotyping) is closer to the phenotype and reflects better the physiology of the animal (Fig. 12.6), which if used in combination with genomics have the potential to improve the accuracy of genomic selection. Another advantage of metabolomics is that it offers the possibility to study and understand the epigenetics of milk fever and how the cellular environment contributes to the disease.

Conclusions

A new approach to milk fever is necessary in order to clarify and better understand the pathobiology of milk fever. The systems veterinary approach is in its early stages of development and it has just started to use "omics" sciences including genomics, proteomics, transcriptomics, and metabolomics in tackling the many issues related to various periparturient diseases of transition dairy cows. A combination of the "omics" sciences can offer great opportunities for animal health scientists to incorporate all the information generated and study component-to-component interactions and the dynamic communications that result from them.

Acknowledgements We recognize and thank Alberta Livestock and Meat Agency Ltd. (ALMA, Edmonton, AB, Canada), Genome Alberta (Calgary, AB, Canada), and Natural Sciences and Engineering Research Council of Canada (Ottawa, ON, Canada) for their financial support of our research efforts in relation with metabolomics.

Glossary

Disease	Any impairment of normal physiological function affecting all or part of an organism, a specific pathological change caused by infection, stress, etc., producing characteristic symptoms; illness or sickness in general.
Genomics	A branch of biotechnology concerned with applying the techniques of genetics and molecular biology to the genetic mapping and DNA sequencing of sets of genes or the complete genomes of selected organisms, with organizing the results in databases, and with applications of the data (as in medicine or biology).
Metabolomics	Metabolomics is the scientific study of chemical processes involving metabolites. Specifically, metabolomics is the "systematic study of the unique chemical fingerprints that specific cellular processes leave behind", the study of their small-molecule metabolite profiles. It is the study of all the metabolites present in cells, tissues, and organs.
Milk fever	A disorder following parturition. A disease of fresh cows, sheep, or goats that is caused by excessive drain on the body mineral reserves during the establishment of the milk flow.
Proteomics	A branch of biotechnology concerned with applying the techniques of molecular biology, biochemistry, and genetics to analyzing the structure, function, and interactions of the proteins produced by the genes of a particular cell, tissue, or organism, with organizing the information in databases, and with applications of the data.
Reductionist approach	A procedure or theory that reduces complex data and phenomena to simple terms.
System biology	It is the computational and mathematical modeling of complex biological systems. An emerging engineering approach applied to biological scientific research, systems biology is a biology-based inter-disciplinary field of study that focuses on complex interactions within biological systems, using a holistic approach (holism instead of the more traditional reductionism) to biological research.
Transcriptome	It is the set of all messenger RNA molecules in one cell or a population of cells.

References

Almeida PE, Weber PSD, Burtona JL et al (2007) Gene expression profiling of peripheral mononuclear cells in lame dairy cows with foot lesions. Vet Immunol Immunopathol 120:234–245

Ametaj BN, Goff JP, Horst RL et al (2003) Presence of acute phase response in normal and milk fever dairy cows around parturition. Acta Vet Scand 44:66

Ametaj BN, Zebeli Q, Iqbal S (2010) Nutrition, microbiota, and endotoxin-related diseases in dairy cows. Bras Zootec 39:433–444

Ametaj BN (2015) A systems veterinary approach in understanding transition cow diseases: metabolomics. Proceedings of the 4th international symposium on dairy cow nutrition and milk quality. Advances in fundamental research, Beijing, China, 2015

Aiumlamai S, Fredriksson G, Kindahl H et al (1992) A possible role of endotoxins in spontaneous paretic cows around parturition. Zentralbl Veterinarmed A 39(1):57–68

Arnett T (2003) Regulation of bone cell function by acid-base balance. Proc Nutr Soc 62:511–520

Aldred AR, Schreiber G (1993) The negative phase proteins. In: Acute phase proteins, molecular biology, biochemstry, and clinical applications. CRC Press, London, pp 21–37

Bartlet JP (1967) Induction experimentale d'um syndrome analogue a'la fieure vitulaire par administration de thyrocalcitonin a'des vaches en cours de lactation. CR Acad Sci 265:1075–1078

Block E (1984) Manipulating dietary anions and cations for prepartum dairy cows to reduce incidence of milk fever. J Dairy Sci 67:2939–2948

Capen CC, Young DM (1967) The ultrastructure of the parathyroid glands and thyroid parafollicular cells of cows with parturient paresis and hypocalcemia. Lab Invest 17:717–737

Casas E, Leach RJ, Reinhardt TA et al (2014) A genomewide association study identified CYP2J2 as a gene controlling serum vitamin D status in beef cattle. J Anim Sci 91:3549–3556

Clater F (1814) Every man his own cattle doctor. E and H Hods, London, pp 143–150

Chamberlain AT (1987) The management and prevention of the downer cow syndrome. In: Proceedings of the British Cattle Veterinarians Association, Nottingham, England, pp 20–30

Dalbeya RE, Wanga P, van Dijl JM (2012) Membrane proteases in the bacterial protein secretion and quality control pathway. Microbiol Mol Biol Rev 76:311–330

DeGaris PJ, Lean IJ (2008) Milk fever in dairy cows: a review of pathophysiology and control principles. Vet J 176:58–69

Deiner C, Reiche M, Lassner D et al (2012) Allelic variations in coding regions of the vitamin D receptor gene in dairy cows and potential susceptibility to periparturient hypocalcaemia. J Dairy Res 79:423–428

Dryerre H, Greig JR (1925) Milk fever: its possible association with derangements in the internal secretions. Vet Rec 5:225–231

Elo KT, Vilkki J, de Koning DJ et al (1999) A quantitative trait locus for live weight maps to bovine chromosome 23. Mamm Genome 10:831–835

Eckel EF, Ametaj BN (2016) Role of bacterial endotoxins in the etiopathogenesis of periparturient diseases of transition dairy cows. J Dairy Sci 99:5967–5990

Ender F, Dishington IW, Helgebostad A (1971) Calcium balance studies in dairy cows under experimental induction and prevention of hypocalcaemic paresis puerperalis. The solution of the aetiology and the prevention of milk fever by dietary means. Z. Z Tierphysiol Tierernahr Futtermittelkd 28:233–256

Fikadu W, Tegegne D, Abdela N et al (2016) Milk fever and its economic consequences in dairy cows: a review. Global Vet 16:441–452

Fredslund F, Jenner L, Husted LB et al (2006) The structure of bovine complement component 3 reveals the basis for thioester function. J Mol Biol 361:115–127

Gallay P, Heumann D, Le Roy D et al (1994) Mode of action of anti-lipopolysaccharide-binding protein antibodies for prevention of endotoxemic shock in mice. Proc Natl Acad Sci U S A 91:7922–7926

Gaynor PJ, Mueller FJ, Miller JK et al (1989) Parturient hypocalcemia in Jersey cows fed alfalfa haylage-based diets with different cation to anion ratios. J Dairy Sci 72:2525–2531

Gluck OS, Murphy WA, Hahn TJ et al (1981) Bone loss in adults receiving alternate day glucocorticoid therapy. A comparison with daily therapy. Arthritis Rheum 24:892–898

Goff JP, Littledike TE, Horst RL (1986) Effect of synthetic bovine parathyroid hormone in dairy cows: prevention of hypocalcemic parturient paresis. J Dairy Sci 69:2278–2289

Goff JP, Reinhardt TA, Horst RL (1989) Recurring hypocalcemia of bovine parturient paresis is associated with failure to produce 1,25-dihydroxyvitamin D. Endocrinology 125:49–53

Goff JP, Horst RL, Mueller FJ et al (1991) Addition of chloride to a prepartal diet high in cations increases 1,25-dihydroxyvitamin D response to hypocalcemia preventing milk fever. J Dairy Sci 74:3863–3871

Goff JP, Horst RL (1993) Oral administration of calcium salts for treatment of hypocalcemia in cattle. J Dairy Sci 76:101–108

Goff JP, Horst RL (1994) Calcium salts for treating hypocalcemia: carrier effects, acid-base balance, and oral versus rectal administration. J Dairy Sci 77:1451–1456

Goff JP, Reinhardt TA, Beitz DB et al (1995) Breed affects tissue vitamin D receptor concentration in periparturient dairy cows: a milk fever risk factor? J Dairy Sci 78:184

Goff JP (2008) The monitoring, prevention, and treatment of milk fever and subclinical hypocalcemia in dairy cows. Vet J 176:50–57

Goff JP, Liesegang A, Horst RL (2014) Diet-induced pseudohypoparathyroidism: a hypocalcemia and milk fever risk factor. J Dairy Sci 97:1520–1528

Garner B, Waldeck AR, Witting PK et al (1998) Oxidation of high density lipoproteins. II. Evidence for direct reduction of lipid hydroperoxides by methionine residues of apolipoproteins AI and AII. J Biol Chem 273:6088–6095

Goings RL, Jacobson NL, Beitz DC et al (1974) Prevention of parturient paresis by a prepartum, calcium-deficient diet. J Dairy Sci 57:1184

Hahn TJ, Halstead LR, Baran DT (1981) Effects of short term glucocorticoid administration on intestinal calcium absorption and circulating vitamin d metabolite concentrations in man. J Clin Endocrinol Metab 52:111–115

Henderson JA (1938) Dairy cattle reproduction statistics. Cornell Vet 28:173–195

Hibbs JW, Krauss WE, Pounden WD et al (1946) Studies on milk fever in dairy cows. II. The effect of vitamin D on some of the blood changes in normal and milk fever cows at parturition. J Dairy Sci 29:767–782

Hiibbs JW (1950) Milk fever (parturient paresis) in dairy cows—a review. J Dairy Sci 33:758–789

Hollis BW, Draper HH, Burton JH et al (1981) A hormonal assessment of bovine parturient paresis: Evidence for a role of oestrogen. J Endocrinol 88:161–171

Horst RL, Goff JP, Reinhardt TA et al (1997) Strategies for preventing milk fever in dairy cattle. J Dairy Sci 80:1269–1280

Horst RL, Jorgensen NA (1982) Elevated plasma cortisol during induced and spontaneous hypocalcemia in ruminants. J Dairy Sci 65:2332–2337

Hossein-Zadeh NG, Ardalan M (2011) Bayesian estimates of genetic parameters for metritis, retained placenta, milk fever, and clinical mastitis in Holstein dairy cows via Gibbs sampling. Res Vet Sci 90:146–149

Hutyra F, Marek J (1926) In: Eger A (ed) Special pathology and therapeutics of diseases of domestic animals, 3rd edn, Chicago, p 514

Hutyra F, Marek J, Manniger R (1938) In: Eger A (ed) Special pathology and therapeutics of the diseases of domestic animals. IJR Greig, Chicago, p 217

Hwang SR, Garza CZ, Wegrzyn JL et al (2005) Demonstration of GTG as an alternative initiation codon for the serpin endopin 2B-2. Biochem Biophys Res Commun 327:837–844

Ishida S, Yonezawa T, Eirai S et al (2013) Hormonal differences in peripheral blood and gene profiling in the liver and lymphocytes in Japanese black cattle with growth retardation. J Vet Med Sci 75:17–25

Kusumanti E, Agger JF, Jensen K (1993) Association between incidence risk of milk fever and lactation number, breed and season. Acta Vet Scand 89:141

Kichura TS, Horst RL, Beitz DC et al (1982) Relationships between prepartal dietary calcium and phosphorus, vitamin D metabolism, and parturient paresis in dairy cows. J Nutr 112:480

Lean IJ, Degaris PJ, Mcneil DM et al (2006) Hypocalcemia in dairy cows: meta-analysis and dietary cation anion difference theory revisited. J Dairy Sci 89:669–684

Littledike ET, Whipp SC, Witzel DA et al (1970) Insulin, corticoids and parturient paresis. In: JJB A (ed) Parturient hypocalcemia. Academic, New York

Littledike ET, Horst RL (1982) Inappropriate plasma 1,25(OH)2D response to parturient hypocalcemia in cows treated with vitamin D_3, 1,25$(OH)_2D_3$, or 1,25,26-$(OH)_3D_3$ prepartum. In: Vitamin D, chemical, biochemical and clinical endocrinology of calcium metabolism, p 475

Lin HK, Oltenacu PA, Van Vleck LD et al (1989) Heritabilities of and genetic correlations among six health problems in holstein cows. J Dairy Sci 72:180–186

Loor JJ, Everts RE, Bionaz M et al (2007) Nutrition-induced ketosis alters metabolic and signaling gene networks in liver of periparturient dairy cows. Physiol Genomics 32:105–116

Lutzow YCS, Donaldson L, Gray CP et al (2008) Identification of immune genes and proteins involved in the response of bovine mammary tissue to Staphylococcus aureus infection. BMC Vet Res 4:18

Lyons DT, Freeman AE, Kuck AL (1991) Genetics of health traits in Holstein cattle. J Dairy Sci 74:1092–1100

Mannstadt M, Harris M, Bravenboer B (2013) Germline mutations affecting Galpha11 in hypoparathyroidism. N Engl J Med 368:2532–2534

Marshak RR (1956) Studies on parturient paresis in dairy cows with particular reference to udder insufflation. JAVMA 128:423

Mayer GP, Blum JW, Deftos LJ (1975) Diminished prepartal plasma calcitonin concentration in cows developing parturient hypocalcemia. Endocrinology 96:1478–1485

Metzger HJ (1936) A case of tetany with hypomagnesemia in a dairy cow. Cornell Vet 26:353–357

Munford RS, Hall CL, Dietschy JM (1981) Binding of Salmonella typhimurium lipopolysaccharides to rat high-density lipoproteins. Infect Immun 34:835–843

Nesbit MA, Hannan FM, Howles SA et al (2013) Mutations affecting G-protein subunit alpha11 in hypercalcemia and hypocalcemia. N Engl J Med 368:2476–2486

Ohtsuka H, Fukuda S, Kudo K et al (2013) Changes in leukocyte populations of cows with milk fever or displaced abomasum after calving. Can J Vet Res 77:226–230

Pryce JE, Veerkamp RF, Thopson R et al (1997) Genetic aspects of common health disorders and measures of fertility in Holstein Friesian dairy cattle. Anim Sci 65:353–360

Pryce JE, Parker Gaddis KL, Koeck A et al (2016) Invited review: opportunities for genetic improvement of metabolic diseases. J Dairy Sci 99:6855–6873

Reinhardt TA, Lippolis JD, McCluskey BJ et al (2011) Prevalence of subclinical hypocalcemia in dairy herds. Vet J 188:122–124

Rosen FS, Skarnes RC, Landy M et al (1958) Inactivation of endotoxin by a humoral component. III. Role of divalent cations and a dialyzable component. J Exp Med 108:701–711

Sasaki K, Yamagishi N, Kizaki K et al (2014) Microarray-based gene expression profiling of peripheral blood mononuclear cells in dairy cows with experimental hypocalcemia and milk fever. J Dairy Sci 97:247–258

Sahu A, Lambris JD (2001) Structure and biology of complement protein C3, a connecting link between innate and acquired immunity. Immunol Rev 180:35–48

Schmidt JJ (1897) Studier og forsog over kalvnlngsfeberens, aarsag og behandling. Maanedsskrift fiir Dyrlaeger 9:228

Shire JA, Beede DK (2013) DCAD revisited: Prepartum use to optimize health and lactational performance. In: Proceedings of the 28th annual Southwest Nutrition and Management, vol 86, pp 1–11

Shu S, Xia C, Zhang H et al (2014) Plasma proteomics analysis of dairy cows with milk fever using SELDI-TOF-MS. Asian J Anim Vet Adv 9:1–12

Sun Y, Xu C, Li C et al (2014) Characterization of the serum metabolic profile of dairy cows with milk fever using 1H-NMR spectroscopy. Vet Q 34:159–163

Thilsing-Hansen T, Jørgensen RJ, Enemark JMD et al (2002) The effect of zeolite a supplementation in the dry period on periparturient calcium, phosphorus, and magnesium homeostasis. J Dairy Sci 85:1855–1862

Thompson JR (1984) Genetic interrelationships of parturition problems and production. J Dairy Sci 67:628–635

Turk R, Piras C, Kovačić M et al (2012) Proteomics of inflammatory and oxidative stress response in cows with subclinical and clinical mastitis. J Proteomics 75:4412–4428

Uribe HA, Kennedy BW, Martin SW et al (1995) Genetic parameters for common health disorders of Holsteins. J Dairy Sci 78:421–430

Van Mosel M, Van't Klooster AT, Van Mosel F et al (1993) Effects of reducing dietary [(Na+ + K+)−(Cl- +SO4=)] on the rate of calcium mobilisation by dairy cows at parturition. Res Vet Sci 54:1–9

Xia C, Zhang HY, Wu L et al (2012) Proteomic analysis of plasma from cows affected with milk fever using two-dimensional differential in-gel electrophoresis and mass spectrometry. Res Vet Sci 93:857–861

Waldron MR, Nonnecke BJ, Nishida T et al (2003) Effect of lipopolysaccharide infusion on serum macromineral and vitamin D concentrations in dairy cows. J Dairy Sci 86:3440–3446

Yasothornsrikul S, Greenbaum D, Medzihradszky KF et al (2003) Cathepsin L in secretory vesicles functions as a prohormone-processing enzyme for production of the enkephalin peptide neurotransmitter. Proc Natl Acad Sci USA 100:9590–9595

Zebeli Q, Beitz DC, Bradford BJ et al (2013) Peripartal alterations of calcitonin gene-related peptide and minerals in dairy cows affected by milk fever. Vet Clin Pathol 42:70–77

Zhang G, Hailemariam DW, Dervishi E et al (2014) Activation of innate immunity in transition dairy cows before clinical appearance of milk fever. In: Abstract of the joint annual meeting (JAM), Kansas City Missouri, USA, 20–24 July 2014

Zwald NR, Weigel KA, Chang YM et al (2004) Genetic selection for health traits using producer-recorded data. II. Genetic correlations, disease probabilities, and relationships with existing traits. J Dairy Sci 87:4295–4302

Index

B.N. Ametaj (ed.), *Periparturient Diseases of Dairy Cows*,
https://doi.org/10.1007/978-3-319-43033-1

Vet. & Animal

Printed by Printforce, the Netherlands